A Century of
Chemical Engineering

A Century of
Chemical Engineering

Edited by

WILLIAM F. FURTER
Royal Military College of Canada
Kingston, Ontario, Canada

PLENUM PRESS • NEW YORK AND LONDON

Library of Congress Cataloging in Publication Data

Main entry under title:

A Century of chemical engineering.

Papers selected primarily from an international symposium held at the American Chemical Society national meeting in Las Vegas in 1980.
Bibliography: p.
Includes index.
1. Chemical engineering—History—Congresses. I. Furter, William F. II. American Chemical Society.
TP15.C46   660'.09   81-23444
ISBN 0-306-40895-3   AACR2

Based on the proceedings of an international symposium on the history of chemical engineering, held August 24 – 29, 1980, at the 108th Meeting of the American Chemical Society, in Las Vegas, Nevada

© 1982 Plenum Press, New York
A Division of Plenum Publishing Corporation
233 Spring Street, New York, N.Y. 10013

All rights reserved

No part of this book may be reproduced, stored in a retrieval system, or transmitted, in any form or by any means, electronic, mechanical, photocopying, microfilming, recording, or otherwise, without written permission from the Publisher

Printed in the United States of America

PREFACE

Chemical engineering has reached the approximate Centennial of its emergence as a profession in its own right, that is, separate from the profession of chemistry and from all other forms of engineering. The date must be considered approximate, because the exact point of origin cannot be established definitively. The reason is that chemical engineering resulted from the fusion of two major components, which developed along differing time scales. One was industrial, or applied, chemistry; the other was the physical, unit operations. While some historians would place the origin of the profession as recently as 1908 with the founding of the American Institute of Chemical Engineers, others would locate it much earlier, and perhaps even back into antiquity. However, a reasonably general consensus seems to exist for placing it in conjunction with events taking place in England about a century ago. Frank Morton, writing in this volume, states that "The year 1880 is regarded by many as the year in which Chemical Engineering was first recognized." Among other reasons, this was the year in which the first attempts were made to found a 'Society of Chemical Engineers' in London. Others regard the famous lecture series on chemical engineering given by George E. Davis in 1887 at the Manchester Technical School, which formed the basis for his later *Handbook of Chemical Engineering* (1901), as the point of emergence. Miall (1931) reports that the first firm to advertise its services as 'chemical engineers' in the *Journal of the Society of Chemical Industry* was Messrs. Kirkham and Co. of Runcorn, England, in 1884.

In order to honor the profession of chemical engineering at its approximate Centennial, the American Chemical Society commissioned a two-part, international symposium on the History of Chemical Engineering, which was subsequently held at the ACS National Meetings in Honolulu in 1979, and Las Vegas in 1980. The chapters of the present book have been selected primarily from the papers of the second, or Las Vegas, part of the symposium, while an earlier book (Furter, 1980) presented papers from the first, or Honolulu, part. Hence, the two books can be considered companion volumes in presenting the American Chemical Society's

v

centennial tribute to chemical engineering, and in recording the accomplishments of its first century of existence as a recognized and distinctive profession.

The two main roots of chemical engineering, industrial chemistry and the unit operations, are examined in various chapters. The more recent root, the physical unit operations, and claims to its mainly American development (although not necessarily its founding; Lewis, 1959), are addressed in the first chapter. This chapter, by Martha Trescott, was the keynote paper of the Las Vegas portion of the symposium. Frank Morton outlines early British contributions, and Max Appl and André Thépot some of the early German and French contributions, respectively. The remaining chapters have been selected to range widely over chemical engineering's first century of independent existence, both in topic and geographically. Of particular note is the chapter by Marcinkowsky and Keller outlining developments which took place in the Kanawha Valley region of West Virginia around and before 1930 which led to the establishment of the North American petrochemical industry. Other chapters address the early development of chemical engineering education, and still others deal with significant developments in the chemical engineering fundamentals and their applications.

This book is dedicated to my wife Pamela, my daughters Lesley, Jane, and Pamela, and to three of my former (and late) teachers: Edwin R. Gilliland, Albin I. Johnson, and W. Reginald Sawyer.

William F. Furter

July 1981

REFERENCES

Davis, G.E., 1901, "A Handbook of Chemical Engineering", Davis Bros., Manchester, England.
Lewis, W.K., 1959, Evolution of the unit operations, Chem. Eng. Prog. Symp. Ser., 55:1.
Miall, S., 1931, "A History of the British Chemical Industry", Ernest Benn Ltd., London, England.
Furter, W.F., ed., 1980, "History of Chemical Engineering", Advan. Chem. Ser. 190, American Chemical Society, Washington, D.C.

CONTENTS

Unit Operations in the Chemical Industry:
    An American Innovation in Modern
    Chemical Engineering 1
  M.M. Trescott

A Short History of Chemical Engineering in
    the North-West of England 19
  F. Morton

The Haber-Bosch Process and the Development
    of Chemical Engineering 29
  M. Appl

The Evolution of Chemical Engineering in the
    Heavy Inorganic Chemical Industry in France
    During the Nineteenth Century 55
  A. Thépot

Early Chemical Engineering Education in London
    and Scotland 65
  S.R. Tailby

Four Score and Seven Years of Chemical Engineering
    at the University of Pennsylvania 127
  A.N. Hixson and A.L. Myers

History of Chemical Engineering at the University
    of Michigan 139
  D.L. Katz, J.O. Wilkes, and E.H. Young

Chemical Engineering at the University of
    Wisconsin: The Early Years 159
  E.E. Daub

Evolution of Chemical Engineering from Industrial
    Chemistry at Kyoto University 195
  F. Yoshida

Doctoral Thesis Work in Chemical Engineering
   in the United States from the Beginning
   to 1960 211
   J.O. Maloney

Historical Highlights in the Chemical
   Engineering Literature 225
   H. Skolnik

Pioneers in the Field of Diffusion 243
   F. Mohtadi

Distillation - Some Steps in its Development 259
   D.F. Othmer

Chemical Engineering and the Chemical
   Industry in South Africa 275
   O.B. Volckman and F. Hawke

Ethylene and its Derivatives: Their Chemical
   Engineering Genesis and Evolution at
   Union Carbide Corporation 293
   A.E. Marcinkowsky and G.E. Keller, II

Some Historical Notes on the Use of Mathematics in
   Chemical Engineering 353
   A. Varma

The Development of the Notions of Multiplicity
   and Stability in the Understanding of
   Chemical Reactor Behavior 389
   R. Aris

Historical Development of Polymer Blends and
   Interpenetrating Polymer Networks 405
   L.H. Sperling

The Origins and Growth of Naptha Reforming 419
   H.B. Kendall

Chemical Engineering and Food Processing 437
   C.J. King

A History of Ultrafiltration Separations 449
   A.R. Cooper

List of Contributors 459

Index 461

UNIT OPERATIONS IN THE CHEMICAL INDUSTRY:

AN AMERICAN INNOVATION IN MODERN CHEMICAL ENGINEERING

Martha M. Trescott

University of Illinois at Urbana-Champaign
College of Engineering
1308 W. Green St.
Urbana, IL 61801

I

My work on the history of chemical engineering has stemmed from my dissertation research on the history of the American electrochemicals industry and, therefore, especially draws upon that sector and the manufacture of heavy chemicals. However, the particular viability of American electrochemicals production before the turn of the century points to forces in the American technical and business climate in general which were strengthening this industry and also other chemical sectors before World War I. After many years of research and a view of developments from the vantage point of electrochemicals, which combine chemical, mechanical, metallurgical and electrical technologies in an impressive blend, and drawing upon my own work in chemistry and in the histories of various mechanical and other technologies, I became very convinced about this paper's subtitle, "An American Innovation in Modern Chemical Engineering", and about approaching a book on the history of the American electrochemicals industry as a study in the cultural context of technological and industrial change.

Just as by 1900 distinctively American forms of art, music and literature were becoming more and more visible, so in the realm of technology, the same kind of phenomenon could be seen. In the days between the emergence of the "American System" of manufacture and Henry Ford's moving assembly line, we had become a nation of machine builders, and also scientific management had begun to be articulated and implemented by some of our foremost mechanical engineers in the metallurgical and mechanical sectors.[1] Indeed, foreign visitors remarked repeatedly on our distinctive mechanical and managerial expertise, both before and after 1900. Even the German chemist, Fritz Haber, discussed such American expertise in relation to our electrochemicals industry in 1902.[2]

I believe that in the rise of chemical engineering, the U. S. contributed the widespread adoption of the "unit operations" concept, as it emerged from our particular mechanical process industries (particularly metal-working firms) and from our metal-making (particularly steel) sectors, both of which focused on mass production with coordination of units of production. I like to emphasize the historic foundation of American chemical engineering, especially in mechanical engineering and also the historic linkages between developments in two closely related sectors--metallurgical production and chemical manufacture.

I feel that historians of technology have too long considered developments in the chemical, mechanical, and metallurgical areas separately, and I think that the electrochemicals study has forced me to think about likanges and overlaps among them.[3]

---

[1] Paul J. Uselding, Studies in the Technological Development of the American Economy During the First Half of the Nineteenth Century, doctoral idssertation, Northwestern University, 1970, published in the series Dissertations in American Economic History by the Arno Press, 1975, especially cf. pp. 142-149. Professor Uselding's comments and advice have been most indispensable throughout this work on unit operations. Also, cf. Martha M. Trescott, "Lillian Moller Gilbreth and the Founding of Modern Industrial Engineering," Berkshire Conference, June, 1981, mimeo.

[2] It is interesting to note, as does H. J. Habbakkuk in American and British Technology in the Nineteenth Century, 1962, that English visitors were remarking on the uniqueness of certain American machines as early as the 1830s, as he noted on p. 4. See also Nathan Rosenberg, The American System of Manufactures (Chicago, 1969) and John E. Sawyer, "The Social Basis of the American System of Manufacturing," Journal of Economic History, XIV (1954), 361-379. In addition, Paul Uselding's "Studies in Technology in Economic History," Research in Economic History, Supplement 1 (Greenwich, Connecticut, 1977), pp. 160-180 is useful. "Prof. Haber on Electrochemistry in the United States," Electrochemical Industry, I (1903), 350.

[3] Generally, in histories of the chemical industries such as those by Williams Haynes, American Chemical Industry,

II

Elevating various American contributions to modern chemical engineering in no way undermines the popular notion that Germany was the seedbed of many aspects of modern chemical engineering. It is a fact that one of Germany's most notable contributions lies in the early use of chemical science, especially physical chemistry and thermodynamics, in chemical production. We know that, as Alfred H. White and other important chemical engineers of this century have said, physical chemistry is the very foundation of modern chemical engineering.[4]

William J. Reader, Imperial Chemical Industries, A History, 1970, and L. F. Haber's two, The Chemical Industry During the Nineteenth Century, 1958, and The Chemical Industry, 1900-1930 (1971), it is most difficult to trace what happened to chemical equipment. If one searches long enough, one might find some clues as in Haynes, Vol. I, p. 361. But this information is typically hard to come by, and once one has, it is difficult to determine the context (that is, the historical evolution). Similarly, in essays and books on mechanical engineering, one often does not see a discussion of the chemical industry's equipment per se, as in Aubrey F. Burstall, A History of Mechanical Engineering, 1965, and others.

The metallurgical and chemical industries are typically separated explicitly, as in Williams Haynes, American Chemical Industry, which does not explicitly include the metallurgical sector but in actuality does include some metallurgical production, which is inseparable from chemical production in some cases (as with certain electrochemicals). Also, Census classifications have separated metallurgical from chemical industries.

[4]Alfred H. White, "Chemical Engineering Education", ed. American Institute of Chemical Engineers, Twenty-Five Years of Progress in Chemical Engineering, 1933, pp. 353-4 and 359-60. Particularly in the synthetic dyestuffs industry, physical chemistry found early industrial importance, as Paul M. Hohenberg indicates in Chemicals in Western Europe, 1850-1914, An Economic Study of Technical Change, 1967, p. 112, e.g. Also, cf. Aaron J. Ihde, The Development of Modern Chemistry, 1964, p. 417, e.g. Of course, numerous works such as John J. Beer's The Emergence of the German Dye Industry, 1959; William J. Reader's Imperial Chemical Industries, A History, 1970; and the report by William E. Wickenden, A Comparative Study of Engineering Education in the United States and Europe, 1929, conducted for the Society for the Promotion of Engineering Education, along with the excellent monograph by Edward H. Beardsley, The Rise of the American Chemistry Profession, 1850-1900 (1964), all note the

Yet, physical chemistry, as important as it is here, is <u>not</u> sufficient for successful chemical engineering and production. A statement from Olaf Hougen is useful to understand what is really involved. In his popular textbook, <u>Chemical Process Principles</u>, 1943, he and his co-author, Kenneth M. Watson, assert:

> The design of a chemical process involves three types of problems, which although closely interrelated, depend on quite different technical principles . . .
>
> These three types may be designated as process, unit-operation, and plant design, respectively . . . Process problems are primarily chemical and physico-chemical in nature; unit-operation problems are for the most part purely physical; the plant-design problems are to a large extent mechanical.[5]

This statement is useful to historians trying to sort out the different strands as they evolved and came together in modern chemical engineering. In the case of the rise of process technology with emphasis on physical chemistry in chemical production, Germany can largely be credited as the wellspring. But in the rise of the <u>concept</u> of unit operations, credit must primarily go to the British for its general introduction and to Americans for its widespread popularization, both in the educational and industrial sectors, and for its concomitant integration with physical chemistry.

The Englishman, George E. Davis, who had served as a factory inspector under the Alkali Works Regulation Act in England, has been credited with the "first written description of these unit physical changes" in the first edition of his book, <u>Handbook of Chemical Engineering</u>, 1901, in two volumes.[6] Earlier, in 1887, he had first presented his ideas on unit operations in a series of lectures at the Manchester Technical School but did not publish these ideas until 1901.

---

incorporation of chemical science into the German chemical industries before 1900, but we are mainly interested here in stressing physical chemistry.

[5]Hougen and Watson, <u>Chemical Process Principles</u>, p. v.

[6]R. Norris Shreve, "Unit Operations and Unit Processes", <u>Encyclopedia of Chemical Technology</u>, XIV, 422. Also, see History of Chemical Engineering Symposium and Symposium volume by the American Chemical Society, ed. William F. Furter, Advances in Chemistry series, 1980, especially D. C. Freshwater, "George E. Davis, Norman Swindin and the Empirical Tradition in Chemical

Almost all of Davis's two volumes deal with various equipment and apparatus used in chemical production, such as evaporators, distillation equipment, crushers, grinders and so on. Since unit operations in chemical engineering are operations which deal with physical changes in materials, such as boiling, freezing or grinding operations, one can sense that considerations of machines and equipment are integral to unit operations and can really determine the definition of a "unit" and how the units for carrying out a given physical change will be related to one another.

The rapid improvement in chemical equipment in industry, especially after the 1870s and 1880s, as Williams Haynes has noted,[7] undoubtedly helped foster the rise of unit operations for organizing chemical production. Yet the literature from the late nineteenth century on chemical equipment in both European and American industry is scarce.

The triangulation of opinions from secondary sources such as Haynes, Paul Hohenberg and Aubrey Burstall, who has discussed mechanical engineering in the rise of process engineering in various industries,[8] tends to be corroborated by contemporaries in the late nineteenth and early twentieth century, such as the Briton George Davis, the American Magnus Swenson, the German Parnicke, and the German-American Oskar Nagel. Each of these sources discusses the importance of machines and/or mechanical engineering in chemical production and engineering for the period between 1880 and 1910.[9]

Engineering," now published as History of Chemical Engineering, pp. 97-111.

[7]Williams Haynes, American Chemical Industry, I, 361: this discussion does not highlight very well changes in chemical equipment.

[8]Burstall, A History of Mechanical Engineering, especially p. 358.

[9]A. Parnicke, a civil engineer, wrote Die Maschinellen Hilfsmittel der Chemischen Technik in 1898, and Oskar Nagel published in the U.S. The Mechanical Appliances of the Chemical and Metallurgical Industries in 1909. Parnicke's book is cited by Nagel as the only other book besides his "on the market along similar lines." (Preface to Nagel's 1909 work.) In comparing the Nagel and Parnicke works, one cannot help but be struck by certain differences. One of these is that the German work focuses a good bit on pumps and other machines which deal with gases (as measuring devices, e.g.) and thereby concern thermodynamics, while the American book would seem to include more milling, crushing, and grinding machines. But this is an entirely subjective impression

And, of course, chemical engineering curricula in American universities have historically been heavily oriented to mechanical engineering.[10]

Davis and Swenson both felt that much more than chemical science is needed for successful chemical engineering, and both stressed the importance of mechanical engineering here. Davis commented in his 1904 edition that "though the Chemical Engineer must possess an almost perfect knowledge of applied chemistry, it must not be forgotten that he deals principally with the suitability of plant and apparatus to perform certain operations, of which the construction and maintenance are the most important features."[11] Consequently, his two volumes deal mainly with such apparatus and equipment.

Davis apparently never employed the term "unit operation" in

---

on the author's part. It might be desirable to count the different kinds of equipment illustrated and discussed in these works. Also, and another subjective judgment, it seems that the German equipment illustrated was in some sense less substantial, smaller or more complex than many of the American machines in Nagel's volume.

Nagel, probably a German immigrant, judging from his name and his ability to translate German, was a prolific writer on the chemical process industries, judging from the appendix in his 1909 book. He translated a German work by Hanns v. Juptner on Heat Energy and Fuels, 1909, advertised as a "complete and up-to-date treatise on the chemistry and production of fuels for use in Metallurgy, Boilers, Gas Producers, Etc." (appendix, p. vi) This is just one more piece of evidence that the Germans concerned themselves with the thermodynamic considerations perhaps more so than with matters of equipment design. For an interesting discussion of the use, or non-use, of thermodynamics in machine design before 1860, cf. Lynwood Bryant, "The Role of Thermodynamics in Evolution of Heat Engines," Technology and Culture, XIV (1973), 152-165.

[10]Catalogues from the University of Pennsylvania, the University of Wisconsin, Lehigh University and other schools show that programs in chemical engineering in the late nineties and/or early 1900s, contain a goodly amount of mechanical engineering courses, not infrequently the first two years of an M.E. program. The author's survey of college and university engineering curricula can be made available to interested scholars.

[11]Davis, Handbook of Chemical Engineering, I, second edition, p. 3.

his discussions. Instead, Arthur D. Little evidently coined the phrase in his 1915 report to the President of Massachusetts Institute of Technology, as F. J. Van Antwerpen and others have noted.[12] And it is useful to note that Little's statement was very much in the vein of scientific management and came during the heyday of its rise.

The credit for a beginning at systematizing the study of chemical engineering around unit operations as such apparently should not only go to Little but also to his partner William H. Walker. Having been a business partner with Little prior to 1904, when Walker was asked to develop the chemical engineering program at M.I.T., Walker was one of the authors of the 1923 textbook which is universally said by chemical engineers to have been first to treat comprehensively (in any language) unit operations: Principles of Chemical Engineering.[13]

In the first edition of their text, Walker, Lewis and McAdams commented:

> The unit operations of chemical engineering have in some instances been developed to such an extent in individual industries that the operation is looked upon as a special one adapted to these conditions alone, and is, therefore, not frequently used by other industries. All important unit operations have much in common, and if the underlying principles upon which the rational design and operation of basic types of engineering equipment depend are understood, their successful adaptation to manufacturing processes becomes a matter of good management rather than of good fortune.[14]

---

[12] F. J. Van Antwerpen, "The Origins of Chemical Engineering," read at the American Chemical Society, Honolulu, April 1-6, 1979 (also see the symposium volume, edited by William F. Furter, History of Chemical Engineering, pp. 1-14).

[13] White, "Chemical Engineering Education," p. 358; the Encyclopedia of Chemical Technology, XIV, p. 423; A. Eucken and M. Jakob, Der Chemie-Ingenieur, Ein Handbuch der Physikalischen Arbeitsmethoden in Chemischen und Verwandten Industriebetrieben (1937), with an introduction by Fritz Haber, p. vii. Also, Professor Richard C. Alkire of the University of Illinois, Department of Chemical Engineering, corroborates the idea that the Walker, Lewis and McAdams text set a precedent in discussion of unit operations as such. In addition, see White, "Chemical Engineering Education," p. 356; Shreve, "Unit Operations and Unit Processes," pp. 422-23; and Van Antwerpen, "The Origins of Chemical Engineering," p. 5, of typescript sent to the author.

Two things in this statement are especially interesting: 1) the <u>equipment</u> associated with unit operations and 2) the <u>prior evolution of industrial practice</u>. Concerning the latter, the industrial case studies from U.S. and German industry, noted later on, suggest that American industrial practice in the chemical process industries differed significantly from German counterparts prior to World War I (as well as afterward) and that certain aspects of this difference may well suggest reasons for the rapid popularization of the unit operations concept among American chemical engineers.

In the statements made by Little in 1915 and Walker and his co-authors eight years later, one not only notices the explicit phrase "unit operation," but, related to this terminology and highly significant for the comparative view, are such phrases as "coordinate series," "rational design and operation," and "good management." It is important that these terms appeared in sources by American, and not European, engineers. In the 1930s, the American Institute of Chemical Engineers adopted the following definition of modern process chemical engineering:

> Chemical engineering is that branch of engineering concerned with the development and application of manufacturing processes in which chemical or certain physical changes of materials are involved. These processes may usually be resolved into a coordinated series of physical operations and unit chemical processes. The work of the chemical engineer is concerned primarily with the design, construction, and operation of equipment and plants in which these unit operations and processes are applied. Chemistry, physics, and mathematics are the underlying sciences of chemical engineering, and economics its guide in practice.[15]

Texts by American chemical engineers such as R. Norris Shreve and Olaf Hougen, subsequently followed the basic themes expressed in the official A.I.Ch.E. statement of 1938, stressing mainly the bases of modern chemical engineering as 1) unit operations (and its related concept unit processes), 2) equipment and plant design and the importance of mechanical engineering here, 3) physical chemistry, and 4) economics.[16]

---

[14]William H. Walker, Warren K. Lewis, and William H. McAdams, <u>Principles of Chemical Engineering</u>, second edition (1927), p. ix.

[15]Shreve, <u>Chemical Process Industries</u>, p. 1, citing a definition written for the A. I. Ch. E. in 1938.

It would appear that unit operations, as explicitly stated and as related to coordinated series of such operations, in chemical engineering came to be a part of engineering thought and curricula, in a widespread way first in the U.S. In fact, the important and comprehensive two-volume text on modern chemical engineering in German by Eucken and Jakob in 1937 (with an introduction by Fritz Haber), cited the precedence of the American textbook by Walker, Lewis and McAdams in treating unit operations. According to Jakob and Eucken, the orientation to modern process chemical production as based on coordination of unit operations first took hold in the U.S. And they saw it as having occurred mainly in the 1920s.[17]

Like these German authors, others too have seen this transition and change in the chemical industries, with significant streamlining of operations, as having taken place no earlier than after the First World War.[18] However, our case studies show that, to some extent, the concept--both intuitive and explicit--of unit operations (and perhaps by implication unit processes, such as electrolysis) was operative in the U.S. chemical industry around 1900.

---

[16]Cf., e.g., Shreve, Chemical Process Industries, p. 9 and Hougen and Watson, Chemical Process Principles, p. v.

[17]Eucken and Jakob, Der Chemie-Ingenieur, p. vii, with title cited in full in fnt. 13. Similarly, neither Van Antwerpen, "The Origins of Chemical Engineering," nor Jean-Claude Guedon, "Conceptual and Institutional Obstacles to the Emergence of Unit Operations in Europe," note the likelihood that unit operations was already a part of industrial chemical practice in the U.S. before World War I.

[18]For example, Eduard Farber in "Man Makes His Materials," Technology in Western Civilization, ed. Melvin Kransberg and Carroll W. Pursell, Jr., II (1967) provides a brief mention of Arthur Little's 1915 statement about unit operations, saying that these ideas simplified chemical instruction and planning thereafter but fails to point out that these ideas emanated from the existing chemical sector and the evolution of thought and practice in chemical instruction and industry. Moreover, it is not clear which "new field of chemical industry," to Farber's mind, really spearheaded the development of "unit operations." (p. 194) If he means synthesis, then the question becomes synthesis of which products? Our data here show that the chemical industry in the U.S. was undergoing revolutionary changes in heavy chemicals around 1900, changes which are suggestive of a unit operations approach in industry.

Therefore, in viewing the rise of modern chemical production and engineering, it is not enough to stress the German scientific contributions. As has been seen above, chemistry--including physical chemistry and the industrial research laboratory--constitute only one major avenue through which modern chemical production arose. The other major components have practically nothing to do with chemistry directly: 1) coordination of unit operations, 2) mechanical engineering in equipment and plant design, and 3) economics and management, as managerial practices and theory evolved from American leadership in the scientific management movement. I feel that the American technical and industrial climate by the early 1900s possessed a comparative advantage in these three areas.[19]

III

Some of the most prominent aspects of distinctive American manufacturing practice in general, within which our modern chemical firms grew, I delineate as: 1) a mass market mindset, which entails rapid, volume production of an inexpensive good, 2) building of cheap machines, as opposed to mere machine use; or, stated another way, mechanical know-how distinctive from the European, 3) volume production of inexpensive metals, especially steel, and other inorganic materials, 4) experience with large-scale plant design (as seen particularly in the steel sectors), and 5) unit operations and their

---

Also, by concentrating too heavily on the lack of newness of products at the turn of the century, Farber has failed to capture the revolution in production of these "same old products." True, the alkalies were not new, but electrochemical production of some of them constituted new methods of production. (cf. p. 184, e.g.) Further, one cannot necessarily judge accurately the timing of the introduction of revolutionary ideas by quantity of production, as amount produced will only begin to reflect a change in production technique some time after the innovation has been adopted and implemented. Of course, much of our thought about innovations in the chemical sector has been conditioned by our knowledge of the growth of petroleum refining and innovations in that area, along with the rise of petrochemicals, mostly after World War I. Cf. John T. Enos, Petroleum, Progress and Profits, A History of Process Innovation, 1962. Also the fact that the concept of unit operations did not become explicitly stated in chemical engineering texts until the 1920s has had a lot to do with our having been misled into dating the beginnings of industrial changes in the 1920s.

[19]David F. Noble, America by Design, 1977, notes the origins of Taylorism in the shop culture, p. 40.

integration into coordinated flow (which was a focus of American scientific management).

There is much evidence of historical patterns of American mass production, marketing techniques, and organization of units of work in flow patterns which differed significantly from European practice, whether in Britain, Germany or elsewhere.[20] By the 1920s, Henry Ford's Highland Park and River Rouge plants had constituted one manifestation of unique American production techniques. But earlier, so did Standard Oil and Carnegie Steel. Eastman Kodak, ALCOA, and Dow Chemical, among many others, also sprang from the American environment, as did Union Carbide and General Chemical. Precisely how the diffusion of ideas concerning systems of work occurred in the U.S., will hopefully be addressed more fully by case studies of other scholars. At this time, I can only suggest some avenues of influence which include, in the period before World War I, in part, the transfer of personnel between--for example--the steel and other electrochemical sectors, the transfer of ideas and information among chemical, metallurgical and machinery makers who supplied inputs and products to each other, and the transfer of knowledge about--for instance--unit operations among various engineering disciplines, via professional papers, publications and university curricula. In my book, I note specific examples of each of these things.[21]

IV

With the five points about our unique technological environment in mind, we may view case studies from the turn-of-the-century chemical industry. Even though the articulation of many aspects of modern chemical engineering has been attributed to academics such as William Walker, it is well to bear in mind that Walker himself and other teachers had previously served in industry. Also, as Walker, Lewis and McAdams pointed out in their chemical engineering text of 1923, "unit operations of chemical engineering have in some instances been developed in individual industries" already quite extensively.[22] Further, while chemical engineering courses helped

---

[20]See Chapter 6, Martha M. Trescott, The Rise of the American Electrochemicals Industry, 1880-1910: Studies in the American Technological Environment (forthcoming, Greenwood Press, series in economic history, 1981).

[21]Ibid., especially Chapter 3.

[22]Walker, Lewis and McAdams, Principles of Chemical Engineering, 2nd ed., p. ix.

transform the chemical industry here, it nevertheless should be realized that these courses themselves evolved out of the general American industrial and technical climate, drawing upon industry for professors in many cases.

Although aspects of the organization of work and plant layout are notoriously difficult for historians to discern, in the case studies which follow, various clues about industrial practice are pinpointed such that one can infer characteristics of the American industrial climate, particularly as pertain to the rise of the modern chemicals sectors.

The first case study concerns early Union Carbide activities. Writing in 1900, William R. Kenan, then Chemical Superintendent of Union Carbide at Niagara Falls, noted the electric furnace as a "unit" and went on to describe the furnace innovations of William S. Horry of Union Carbide for the production of calcium carbide. Union Carbide had been able to discard the use of the pot type of furnace, which only produced in batches, for the semicontinuous Horry furnace during 1898-9.[23] Having just delineated the technology of the batch process, Kenan, who had been personally involved in these early innovations, said:

> The same method is employed wherever Carbide is being produced commercially, and no radical change is to be looked for. The apparatus, however, has been greatly improved and along this line will come the better results and therefore a decrease in the cost of the product.
>
> As far as I have been able to judge from my personal experience, the American Manufacturers [sic] are far ahead of all others, and their success has been due entirely to the apparatus employed. Much money and labor has been expended on experimental work of a strictly scientific nature. It is of interest to note that in Germany, England and Canada the apparatus and method of procedure which I have described above is still in vogue. In America, we have passed through the era of brick furnaces and steel pots; having discarded them for a continuous Rotary Furnace,[24] (italics added).

---

[23]Anonymous, Historical Notes on the Origin of the Union Carbide Company and Electrometallurgical Company at Niagara Falls, pp. 20-23, pamphlet of 47 pp., courtesy of Union Carbide Company and Mrs. Mildred H. Phillips, librarian there.

[24]William R. Kenan, Jr., "Some Interesting Facts Relative to Calcium Carbide and the Process of Manufacture," p. 17, pamphlet

UNIT OPERATIONS: AN AMERICAN INNOVATION

This 1900 account may seem more impressive when one realizes that it was written at the request of the German Acetylene Company of Berlin and used by them at that time in Germany. It was not published in the U.S. until 1939.[25]

One sees here the importance of the <u>machinery</u>, emphasizing point number (2) in the list. Next, the decrease in cost of the product and hence selling price is underscored. This supports point (1). After that, a kind of research and development process, or experimental work, is noted, supporting the notion that American chemical R & D was greatly enhanced by American mechanical engineering expertise. Then one sees the advent of continuous production (in which—I should note—the electrochemical sector was one pioneer),[26] and this again focuses attention on mechanical ingenuity and also on cutting the time of the production process (points 1 and 2).[27] Implicit in this example is also the fourth point, since this developmental work did entail materials handling on a large scale and large-scale plant design.[28]

Another instance from Union Carbide history will illustrate some of these five points. Union Carbide was a leader in establishing corporate R & D in the U.S., especially after 1906, when a precursor company of Union Carbide and Carbon Corporation, Electro Metallurgical Company, acquired the expertise of Frederick M. Becket and his Niagara Research Laboratories. Along with Becket came the patent applications for his silicon reduction processes for the production of ferroalloys. Becket had been educated as an electrical engineer, physical chemist, and metallurgist and consequently employed physical chemistry in his work. He was one

---

[25] <u>Ibid.</u>, p. 11.

[26] This is exemplified in the Hall process for production of cheap aluminum prior to 1900, in the Acker caustic-chlorine process, and in the process used by Electrical Lead Reduction Company (with whom Becket also consulted), as noted in <u>The Niagara Falls Electrical Handbook</u>, pp. 114-5, 1904, published by the American Institute of Electrical Engineers. This handbook is an excellent resource for historians. Of course, the Horry furnace was continuously operated. These are a few of numerous examples of continuous electrochemical production at this time.

[27] Cf. Uselding, Studies in the <u>Technological Development of the American Economy</u>, section on time in production, pp. 130-142.

[28] This scale of operations can be seen throughout the document from the Union Carbide Archives, <u>Historical Notes on the Origin of the Union Carbide Company</u> (fnt 23).

of our leading innovators in corporate R & D, and one of his greatest contributions in this area was the implementation of an <u>intermediate-scale</u> plant in which production processes which had been worked out in the lab could be tested without interrupting regular plant operations. This was embodied in the Niagara Research Laboratories, to which I devote a chapter of my book.[29] At these labs, Becket and his co-workers had brought to the point of commercial success reduction of many ferroalloys, including ferrosilicon, the production of which Electro Met was able to establish very <u>quickly</u> after 1907. At this time, as Becket himself said about ferrosilicon, the company had to meet "almost disastrous competition from abroad."[30] The <u>rapid</u> development of these ferroalloy processes to produce a product of high quality continued to characterize the efforts of Becket and his staff, and during World War I, they were able to double production of ferrosilicon within a few weeks.[31]

Here again, as in the case of the Horry furnace, we see the element of <u>time</u> in the production process. Becket and his co-workers were able to gear up for the market more <u>quickly</u> than European competitors. They were able to produce in <u>volume</u> a product of high quality <u>quickly</u>. Second, in doing so, they were very successful in translating laboratory procedures to the <u>engineering scale</u>.

To be sure, physical chemistry was important in Becket's work, but this case helps to show how physical chemistry merged with a solid base of mechanical, metallurgical and managerial expertise to help transform chemical production.

I could continue to illustrate aspects of points 1-5 with case studies of other electrochemical firms such as Dow and Hooker, but

---

[29] Martha Moore Trescott, "Frederick M. Becket and the Niagara Research Laboratories: A Case Study in the Evolution of Modern Corporate R & D," paper read before the American Chemical Society, Division of the History of Chemistry, Philadelphia, April 8, 1975; a version of this paper has been published as "Frederick M. Becket and Electrothermics," <u>Selected Topics in the History of Electrochemistry</u>, pp. 456-477. Also, see Trescott, <u>Rise of the American Electrochemicals Industry</u>, Chapter 7.

[30] "Origin and Development of Electro Metallurgical Company, 1906-1937," pamphlet probably authored by Frederick M. Becket as dictated to H. R. Lee, 23 pp., p. 3 (1937), courtesy of Mrs. Phillips and Union Carbide Archives.

[31] Frederick M. Becket, "Perkin Medal Address," <u>Industrial and Engineering Chemistry</u>, XVI (1924), 200, a reprint of which was supplied the author by Union Carbide.

UNIT OPERATIONS: AN AMERICAN INNOVATION

I want to focus on a case from another sector to show that these developments were not just specific to electrochemicals. The following case comes from the sulfuric acid industry. Yet it similarly serves to illustrate fairly impressively the relative success of other American chemical producers at large-scale manufacture and also our comparative advantage in equipment and plant design.

One of the most impressive American achievements in chemical manufacture in the early 1900s, aside from electrochemicals, was the large-scale production of concentrated sulfuric acid. It is worth quoting at length about these developments, which had begun in Europe many years before, to show another example of American chemical engineering oriented to volume manufacture of high-quality, low-cost materials in a significantly different way than abroad. As with Kenan and Becket, an account by a participant in these developments is important to draw upon.

General Chemical Company, which innovated a contact process in the U.S. when they could not afford the rights to the Badische Anilin und Soda Fabrik contact process, was able to outproduce Badische in a very short time. J. M. Goetchius of General Chemical Company commented about the aftermath of a patent infringement suit settled out of court shortly after the turn of the century by exchange of patents and other agreements:

> General was meticulously careful to submit their monthly reports, but Badische practically ignored their obligation in this respect. So the years rolled on without comment or observation from Badische, though we were out-manufacturing them, as to costs, yields, and output, by a wide margin. In 1911, General received a cable that Herr So and So--and I must use that name--and his staff were coming over to inspect our plants . . . We took them to several different prize plants of which we were proud. That trip took quite a time, and as they watched all this grand apparatus at work, their faces remained stolid and they passed no comments . . . They returned to New York and said in effect that they wished us to translate all our designs from inches to meters; to build 4 units for installation at Ludwigshaven; to build all the apparatus here; to design all buildings to cover these plants; to send our engineers over to install the operation and even to provide the contact mass of platinized asbestos. Each of these 4 units was for 20,000 tons per year; each quadruple the size of their biggest units . . . We later received a similar order for the Bayer plant at Eberfeld . . . To our chagrin it was those 8 units

that supplied the German war machine with most of this essential munition during the First World War[32] (italics added).

The above account illustrates that American chemical engineering by the early 1900s was different in practice than German, for these two companies--both giants in the chemical industry in their respective countries at this time--had exchanged patents and had access to full information about each other's processes under the out-of-court negotiation. But the Americans were more successful at producing on a large scale. Since this particular manufacture was not electrochemical, one cannot argue that its success was due <u>primarily</u> to superior electrical engineering or to the availability of Niagara power (as has often been argued with the rise of the U.S. electrochemicals sector). Rather, it is fairly clear that the design of the large mechanical apparatus and plant constituted a central feature. Also, again the element of <u>time</u> in the production process looms large, since the Americans were able to outproduce Badische <u>quickly</u>--to institute volume production of a lower cost good more quickly, as with calcium carbide and the ferroalloys mentioned earlier. Also, as mentioned with the Horry furnace, "unit" operations were being at least intuitively dealt with. In the case of Union Carbide the unit concept may be somewhat easier to see since the electric furnace is easily seen as a single unit (as indeed it was by George Davis in his <u>Handbook</u>). Further, the Union Carbide electric furnaces were utilized, even before World War I, in the true spirit of the concept of unit operations, <u>i.e.</u>, the same furnace was used to make different products, in this case calcium carbide and ferroalloys. Such a use captures the real economies implied in the unit operations notion--that one piece of equipment, or a system of equipment can be used to make different products when the processes are similar.

V

Germany certainly excelled in the nineteenth century in industrial chemistry and in higher education of chemists and chemical technologists. However, since modern process chemical engineering is a complex mix of engineering, chemistry, management, and economics, it is misleading to focus only on the explicitly "chemical" contributions in the rise of chemical engineering. Indeed, this paper has argued that the U.S. contributed significantly

---

[32] J. M. Goetchius told this to Williams Haynes on January 17, 1942, as reported in Hayes, <u>American Chemical Industry</u>, I, pp. 265-266 in a lengthy footnote.

to modern chemical process engineering, since by the 1900s, America had developed an expertise in large-scale, volume production of certain chemicals, with units of operation coordinated in a flow pattern, in various industries, including certain metallurgical and chemical sectors. In fact, as we know, Arthur D. Little and his colleagues and followers stressed "coordinated series of unit operations" as being central for modern chemical engineering in 1915 during the heyday of scientific management, with its linear-flow process charts and concepts. Scientific management was, of course, a decidedly American development. The roots of scientific management, however, lie embedded in our unique technical culture, a culture which manifested itself in the "American System" of manufacture <u>before</u> the War, in the mass production of all manner of goods <u>after</u> the War, and in the rise of unit operations in chemical engineering and the modern chemical industry even <u>before</u> World War I.

The Germans chose to study and imitate American chemical production a good many years before the War began. And American chemical producers, grounded in U.S. traditions of manufacturing and mass production of metals and other goods, were in some cases outperforming European competitors significantly before the outbreak of the War.

The U.S. chemical industry could not have outperformed European counterparts in general during World War I, had there not been a solid base of pre-existing, manufacturing practice.[33] And, further,

---

[33]Van Antwerpen, "The Origins of Chemical Engineering," notes that "the chemical process industries in the United States are popularly assumed to have developed <u>after</u> the First World War. Up to that time Germany popularly is credited with being the preeminent chemical power. This is not so. Many developments that we assume are modern, were firmly established by 1908, the year AIChE was founded." (p. 9) Then he goes on to note a number of chemical processes and firms, of which many were electrochemical. However, he adds that while the U.S. did have a "flourishing inorganic chemical industry," the war greatly stimulated our own production of organic chemicals, cut off, as we were, from Germany. (p. 10) And he says that against the background of our heavy chemicals industry, "chemical engineers came on stage." (p. 10)

David Noble says that "although a healthy heavy-chemical and powder industry existed throughout the second half of the nineteenth century in the United States, it was not until World War I, with the seizure of the German-owned patents and the establishment of a protective tariff, that a domestic dyestuff industry, and thus a chemical industry proper, was established on a par with the

this base must necessarily have evolved from the American engineering and industrial climate, which can be seen to have contained a variety of unique aspects by 1900 by examination in a comparative way.

---

electrical industry." (America by Design, pp. 5-6.) However, chemical independence of a nation is crucially dependent upon production of the heavy chemicals, particularly sulfuric acid and alkalies, which are also, of course, essential to the establishment of an organic chemicals industry. Also, the electrochemicals industry, among other U.S. sectors, was producing organic chemicals prior to World War I. Dow produced synthetic indigo, an important dye, commercially in 1916. And acetylene was being commercially produced by electrothermics in the nineties, such that we could make our own acetic acid and other organic chemicals from this compound. Third, it is likely that the U.S. was developing sufficient chemical expertise and industries to be able to synthesize organic chemicals on a large scale, even if we had not obtained access to the German dyestuffs patents, but our organic chemicals industry might have evolved somewhat differently than the German. Surely, in terms of plastics, we were making progress before 1918, and our petroleum sector was impressive. Fourth, it is doubtful that "a chemical industry proper" should be defined so much in terms of capacity to make dyestuffs. And, fifth, from this study, one can see that in various ways, particularly in pioneering corporate research and development, our chemical industry could be considered, at least in some areas, as on a "par" with the electrical industry before 1918. Before the war's end, we had established ALCOA, Union Carbide and Carbon Corporation, DuPont, General Chemical, and many other large firms. Cf. also Chandler, The Visible Hand, pp. 354-356, who also says that the "modern chemical industries did not come into their own in the United States until the 1920s" but that, "even so," by 1917, "the basic structure of the American chemical industry was becoming clear." (p. 354) He indicates that technological change was rapid in chemicals and was producing obsolescence in many processes from 1917. However, the electrochemicals industry here had basically evolved its major technologies in electrolytic and electrothermic production, excluding various improvements and refinements, before World War I, though the structure of the industry continued to be in real flux for some time.

A SHORT HISTORY OF CHEMICAL ENGINEERING IN THE NORTH-WEST OF ENGLAND

Frank Morton

Emeritus Professor of Chemical Engineering
The University of Manchester
Institute of Science and Technology
Manchester, England

The year 1880 is regarded by many as the year in which Chemical Engineering was first recognised. Certainly it is a good year with which to begin a brief history of Chemical Engineering in the North-West of England, an area comprising South Lancashire and North Cheshire from the Pennines to the Irish Sea. The Chemical Industry in the North-West had achieved considerable success by the mid-nineteenth century. From Liverpool, through Widnes, Warrington and St. Helens to Manchester, factories produced the soda, acids and bleaching agents required by the textile industry, provided raw materials for an expanding glass industry, for soap manufacture, etc. At this time the industry was largely inorganic, the various processes employed being almost all batch operations very heavily dependent upon cheap labour.

By the year 1880, however, the industry was changing. The successful operation of the Solvay (ammonia-soda) process by the Brunner Mond Company at Northwich illustrated the advantages of continuous flow processes, and threatened to make the LeBlanc process uneconomic. Great changes were also beginning in other sections of the Alkali industry with the introduction of electrochemical processes made possible by the development of large reliable dynamoes. The small but highly profitable organic dyestuffs industry, which had developed in Manchester following W.H. Perkins success with the commerical manufacture of mauve and other synthetic dyestuffs, was seriously threatened by the rapidly expanding German chemical industry.

Although it was recognised that the technical manpower requirements of British industry could only be met by expansion of the Universities and Technical Colleges it was the problems of the organic chemical industry which focussed attention on the German educational advances. In Germany the Universities provided a steadily increasing flow of well-trained chemists enthusiastic for research, whilst the Technisch Hochschulen provided the manager-technologists for industry. The combination of engineer-technologists with research-motivated industrial based chemists contributed materially to the development of the German chemical industry. In contrast, the majority of men who made the industrial revolution in Britain had no systematic education in science or technology, whilst the chemical industry had a very poor reputation in the Universities. In part, this was due to a lack of understanding arising from the limited spread of chemical knowledge, for the industry was very secretive - and in part also due to the noxious nature of contemporary manufacturing processes.

Whatever the causes, the industrial chemists began to seek for some form of society which would enable discussion of the problems within the industry, and provide a means of publication of such communciations as the company secrecy and patent requirements would permit. In the introduction to his handbook[1] George E. Davis records that "the first public recognition of the Chemical Engineer seems to have been made in 1880, in which year an attempt was made to found a "Society of Chemical Engineers" in London. It was soon found that membership of such a society would be too limited in its numbers, and the project, as it was then drafted, was abandoned". However, a meeting was held at Owens' College, Manchester, in 1880 under the chairmanship of Professor H.E. Roscoe F.R.S. and attended by many of the senior personnel of the chemical industry at which it was proposed to form a "Society of Chemical Industry". Surprisingly quickly the Society of Chemical Industry was established - in 1881 - with Professor Roscoe as its first President, and George E. Davis as the first Secretary. From the beginning the Society was envisaged as a means of ensuring the exchange of ideas and information among those employed in the chemical industry, and not in any sense a Professional Institution, since the original constitution specifically excluded the Society from acting as a qualifying body. It is probable that most of the chemists supporting the formation of the new society were members of the Institute of Chemistry (now the Royal Institute of Chemistry) which had been formed in 1871, and which provided qualification in chemistry (by examination) equivalent in standard to a University degree, conferring a status similar to the Associateship of the Institution of Mechanical Engineers. Naturally the chemists did not wish the new society to compete in any way with the Institute of Chemistry.

At the first Annual General Meeting of the Society, Davis reported that 297 members had joined, and listed the profession by which the applicants for membership described themselves. Fourteen described themselves as Chemical Engineers. The first attempt to provide a course of instruction for chemical engineers was made by H.E. Armstrong at the Central College, London, in 1885[2]. He proposed a three-year course leading to a Diploma in Chemical Engineering. Armstrong proposed a balanced course in a range of subjects, rather than the closer integration of conventional engineering, with chemistry and physics, but he did not succeed in establishing the course.

In the North-West Ivan Levinstein gave the first analysis of Chemical Engineering which he believed the chemical industry required. In a lecture to the Society of Chemical Industry[3] in 1886 he commented upon the development of the German dyestuff industry which by that time was producing three quarters of the world's requirements of synthetic dyestuffs, and he considered how the British chemical industry could compete. He defined chemical engineering as "the conversion of laboratory processes into industrial ones", and he argued that it was not only necessary for chemical companies to have chemical engineers within their own organisations, but that "it might be desirable to attach to our chemical laboratories an engineering laboratory where all sorts of appliances can be devised and constructed, and the conversion of laboratory processes into practical industrial ones can be effected". Levinstein suggested the need for specialist chemical engineering establishments which could design and manufacture for the process industries.

In the same lecture, Levinstein outlined his own belief that much "technological investigation" was necessary, comparing the use in the German dyestuffs industry of many organic chemists. Both Levinstein and Roscoe were impressed by the scientific manpower used in the German industry, Roscoe believing that the success of that industry was due to the recognition by the German businessmen of the value in industry of scientific training, and their willingness to pay quite high salaries to research chemists in their works.

This contrasted starkly with the practice in the British industry where it has been stated that the chemists were paid only a little above the manual workers. If this was so, it is not surprising that by the year 1900 Germany had four thousand chemists working in industry, whilst the total number of chemists in England was about five hundred and fifty, of whom the majority were employed in teaching.

About this same time, and probably encouraged by Roscoe, who was interested in developing relationships between the newly-created Victoria University of Manchester and surrounding chemical industries, George E. Davis gave a series of lectures on chemical engineering. The lectures were given in 1887 at the Manchester Technical School with which Levinstein and Roscoe were connected. In these lectures Davis analysed the processes of the contemporary chemical industry, presenting them as a series of basic operations. He recognised that "the chemical processes could be regarded as combinations or sequences of a comparatively small number of procedures".

Several of the lectures were published in the Chemical Trade Journal during the year 1887, but the entire series of lectures was not published until 1901 when Davis produced the course in the first edition of his Handbook of Chemical Engineering, and later in the second and enlarged edition in 1904. Although it is now generally accepted that "the Handbook presented the essential concept of unit operations, and, particularly, furnished an understanding of its value for Education"[4], the subject matter was neglected in Britain - and in the North-West of England - for many years. Not until the end of the first world war was there any serious attempt to revive the idea of chemical engineering as a profession or as an educational discipline. It has been suggested that the British Chemical Industry and, indeed, British industry as a whole were stupid, and "neglectful of the nation's future and security"[5] by not ensuring that the British dyestuffs industry - and related organic chemical developments - were protected against the German competition. In fact, the British chemical industry was exporting inorganic chemicals of much greater value than the import of dyestuffs. Moreover, the imported dyestuffs were used in the textile industry, and constituted only about 1% of the finished textiles. Whilst Baeyer boasted of the German dyestuffs sales in Britain, North-West textile manufacturers pocketed the profits of exporting finished textiles all over the world. The engineers of the period were engaged in the heavy engineering industry - and in the new heavy electrical industry which provided the North-West with increasing business. A few consultant engineers thrived by supplying design and fabrication expertise to the gas industry - to the expanding "bleaching, dyeing, and finishing" section of the textile industry - and to specialist operations such as Brewing, etc. One such consultant was Henry Simon, a German who, like Ivan Levinstein, arrived in Manchester in 1865 and established a small consultancy in offices in St. Peter's Square. After a brief two year period in Russia working on railway development he returned to Britain and by 1875, he had built the first Roller Flour Mill in Britain, using rollers imported from France, and by 1881 had designed and built one of the earliest coke ovens. The business he established is today one of the North-West's most important chemical engineering companies.

The contribution which Davis made to chemical engineering has been, belatedly, recognised and well documented[6], but at the time appeared to have had little effect on the development of the chemical industry, or on education. Levinstein continued to interest himself in the education of "chemical technologists", and was closely connected with developments in Manchester which led to the expansion of the Manchester Technical School into the College of Technology, with a splendid new building opened in 1902 by the then Prime Minister, Mr. Bonar Law.

The Department of Applied Chemistry - of which A.J. Pope F.R.S. was the first Professor - offered undergraduate courses in General Chemical Technology, Metallurgy, and Chemistry of Colouring Matters, all of which followed the usual pattern of a chemistry degree course with additional subjects according to the option chosen.

The emphasis was on industrial chemistry with instruction in machine drawing using chemical plant as examples, and descriptions of chemical manufacture. A special course in the technology of Brewing was equipped with a complete small-scale brewery and was the forerunner of the modern biochemical engineering subject.

The experiences of chemists and chemical engineers during the first world war led to a rapid development in the recognition of chemical engineering as a profession. The Chemical Engineering Group of the Society of Chemical Industry was formed in 1918 and four years later (1922) the Institution of Chemical Engineers was established with powers to act as a qualifying body. The foundation of the Institution was opportune, and followed the recognition by Lord Moulton, Director General of Explosives, during the war, of the importance of the design, construction and operation of chemical plant, and the necessity of training chemical engineers. Chemical Engineering Courses were established at Battersea Technical College, University College, and Imperial College. J.W. Hinchley had been made an Associate Professor at the Imperial College of Science and Technology (1917), and had handed over his lecturing responsibilities at Battersea to Hugh Griffiths. E.C. Williams became Professor of Chemical Engineering at University College, London, in 1923, and J.W. Hinchley became a full Professor of Chemical Engineering at Imperial College in 1926.

In the North-West progress was much slower. A number of early members of the Institution were from the North-West, but no attempt was made to introduce the subject in the courses in Applied Chemistry at either Manchester or Salford.

In the year 1927 Professor James Kenner, F.R.S. was appointed to the Chair of Applied Chemistry in the Manchester College of

Technology. Kenner was an organic chemist of considerable reputation, and had followed the classic traditions of study in Germany for his Doctorate in 1910. Whilst carrying out research at Heidleberg he met Raschig and several other leading German industrial chemists at Leverkusen, where Kenner spent a little time - at Raschig's invitation. On the outbreak of war, Kenner was assigned to the Ministry of Munitions, and worked with K.B. Quinan, who was considered by many to be the most distinguished chemical engineer in the United Kingdom at that time. Kenner was impressed, not only by the man but also by the methods he used to re-organise the work of the Factories Branch of the Ministry. After the war Kenner was appointed to the Chair of Organic Chemistry in Sydney, returning to England in 1927 as Professor of Technological Chemistry at Manchester. Although wishing to continue his own researches and teaching of Organic Chemistry, Kenner was also keen to introduce the teaching of Chemical Engineering into the undergraduate curriculum. He sought the help of Herbert Levinstein and other industrial chemists in the area, and of H.W. Cremer, then in charge of chemical engineering studies at Kings College, London. Kenner's proposals were approved by Senate, and the new undergraduate degree course in Chemical Engineering commenced in 1934, by coincidence the year in which Herbert Levinstein, the son of Ivan Levinstein, became Vice-President of the Institution of Chemical Engineers.

Kenner, of course, was not involved in the teaching of chemical engineering subjects, but he strengthened the teaching of physical chemistry and was fortunate in appointing W. Cowen as the first lecturer in chemical engineering. With very limited funds, and with gifts of equipment from industrial friends, Cowen succeeded in equipping a small chemical engineering laboratory. By 1939 the course was well established, some research had begun and was continued throughout the war years. In 1944 a number of members of the Institution of Chemical Engineers resident in the North-West petitioned Council to permit a Branch of the Institution to be formed in the North-West - to be the North-Western Branch of the Institution of Chemical Engineers.

The petition pointed out that "considerable support has been given over the past three years by both corporate and non-corporate members to those Institution meetings organised by the Graduates and Students Section which have been held in Manchester". The first meeting of the Branch - the first Branch of the Institution - was held in Manchester on 25th September 1944, when J. McKillop was elected Chairman, and Mr. Rees Jones the first Hon. Secretary. Thus the Branch arose - in part at least - from the enthusiasm of the young chemical engineers in the district, and this enthusiasm was to characterise the Branch throughout subsequent highly successful years.

The post-war development of the petroleum and petrochemical

industry in Britain - the result in part of the Hydrocarbons Duty Act - was responsible for an immediate increase in the activity of Chemical Engineers. In the North-West the design and erection of the petrochemical complexes of Shell (at Stanlow), and - at Partington - Petrochemicals Limited, the growth of Atomic Energy Activities at Risley, Calder Hall, and Windscale (the site of a wartime ordinance factory) attracted chemical engineers to the area and provided careers for the increasing numbers of chemical engineering graduates.

After many difficulties largely associated with finance, W. Cowen returned to industry, and the degree course in Chemical Engineering was continued under the direction of J.A. Storrow until 1956. By that time the course was well-established, and had earned a considerable reputation for the quality of both teaching and research. The early post-war years were very difficult, but with some industrial help - in the form of lecturers seconded from industry, and small but important contributions to research funds the group continued to expand. Although Storrow must have been delighted by the quality of young men attracted by the course, he must have been disappointed that the industrial encouragement which led to the foundation of Chairs in Chemical Engineering at Cambridge and Imperial College, and to the expansion of the newly-created course at Leeds, did not materially help at Manchester.

Successive Chairmen of the North-West Branch had urged the College authorities to establish a separate Department of Chemical Engineering, believing that the natural growth of the subject was sufficient to justify such a step.

In the year 1953, following the appointment of B.V. Bowden (now Lord Bowden) as Principal of the College and encouraged by Government interest in expanding technological studies, steps were taken to transfer the College from Municipal control to the central control of the University Grants Committee. A separate charter was granted in the year 1957 creating the new Institute of Science and Technology with special funds to expand teaching and research facilities. One of the first areas chosen for expansion was Chemical Engineering which, in 1956, became a separate department with Frank Morton as its first Professor. Expansion of the Department was rapid, using a converted old (1902) cotton mill to house Chemical Engineering, Metallurgy, Biochemistry, and ancillary drawing offices. Student numbers increased until, in 1959, 120 students were admitted to the first year. A second Chair was created in 1963, to which T.K. Ross, formerly a Reader in the Department, was appointed. In 1963-4 the Department was given a large grant of funds by the University Grants Committee to build and furnish a Pilot Plant Complex. This new building, in which a number a medium-sized chemical engineering operations can be conducted has provided the practical training required by the nature of the course, which had earlier been provided by vacational works experience, and has proved a success with students, research workers, and industrial colleagues who occasionally use the facilities.

Parallel developments occurred at the Royal College of Technology, Salford (now the University of Salford). Here the Chemistry Department, which for many years had prepared students by part-time study for the external degrees of the University of London, expanded the diploma course in Gas Engineering into two separate courses, one leading to a degree in chemical engineering - with Professor Holland as its first Head - the other serving the supply side of the gas industry, and leading to degrees in Gas Engineering. Both courses are of four-year duration, including one year spent in industry. The Gas Engineering Course is supported by the Gas Board, which provides most of the undergraduates from its own training schemes. Recently A.J. Nicklin has been appointed Professor in charge of this specialised aspect of Chemical Engineering.

The North-Western Branch of the Institution has benefited from the expansion of teaching facilities, and, from 1956 onwards, served by a series of young and enthusiastic members, has planned and carried out most ambitious programmes of meetings and symposia. The "Hazards Symposia" best illustrates the character of the North-West chemical engineers - perhaps the first group of professional scientists to recognise the need for public examination of the nature of chemical hazards and the possible control. Equally practical, the major symposia on "Control of Gaseous Emissions" - like the Hazards Symposia - have attracted contributors from all over the world, and provided the chemical industry with an opportunity to examine the problems of controlling gaseous emissions to within statutory limits. These and many other meetings over the past twenty years are the result of active young engineers, proud of their profession, prepared to devote much effort in advancing the interests of chemical engineers and chemical engineering.

At the conclusion of a short history it is perhaps permitted to ask "is there anything to be learned from the past?". In chemical engineering the answer must be that, unless care is taken, modern trends in chemical engineering education, training, and practice, may well lead to a repetition of the errors of the early years. Expansion of subject matter, mostly in areas of physics, professional requirements of practical experience, management studies and economics, coupled with an increasing degree of specialisation in ancillary topics such as biochemical, nuclear and corrosion engineering, have resulted in an increasing neglect of fundamental chemical understanding. It is perhaps worth recording the comments of George Davis[7] when reporting the discussions of the Manchester meeting of 1880 -

"I could give many instances in which processes have failed from want of knowledge, either of physics or of mechanical engineering, on the part of those who have essayed to carry them out, but on the other hand, the chemical knowledge required was so thorough that I am sure no mechanical or civil engineer with a mere smattering of chemistry would have been one whit more successful". Many years later, in 1965, Professor O.F. Hougen[8] was to remark that the trend of chemical engineering education in the U.S.A. "was to return to his original emphasis on chemistry combined with its engineering implications". It may well be that, in the recent advances on so broad a front, chemical engineering is losing something of the "proficiency in a combination of chemistry with engineering" which Hougen considered to be the chemical engineer's principal claim to uniqueness.

REFERENCES

1. G. E. Davis, "A Handbook of Chemical Engineering," 2nd Edition, 2 vols, Davis Bros. Ltd., Manchester (1904).
2. D. M. Newitt, Trans.Inst.Chem.Engrs., 33:400 (1953).
3. I. Leinstein, J.Soc.Chem.Ind., 5:(1886).
4. W. K. Lewis, Chemical Engineering Progress, No. 26, Vol.55, (1959).
5. D. W. F. Hardie and J. Pratt Davidson, "A History of the Modern British Chemical Industry", Pergamon Press, (1966).
6. N. Swindin, The George E. Davis Memorial Lecture, in: Trans. Inst.Chem.Engrs., Vol.31, (1953).
   See also A. J. V. Underwood, Trans.Inst.Chem.Engrs., Vol.43 (1965).
7. G. E. Davis, Chemical News, June (1880).
8. O. F. Hougen, "Chemical Engineering Education in the United States," The Chemical Engineer, September (1965).

THE HABER-BOSCH PROCESS AND THE DEVELOPMENT OF CHEMICAL

ENGINEERING

        Dr. Max Appl

        BASF Aktiengesellschaft
        D-6700 Ludwigshafen/Rhine

One of the most important achievements of industrial chemistry is the production of ammonia on a large scale. The development of this process by Carl Bosch in the "Badische Anilin and Sodafabrik", starting with the experimental laboratory work of Fritz Haber, is a landmark in the history of Chemical Engineering. The synthesis of ammonia not only solved a fundamental problem in securing our food supply, but also opened a new phase of technical chemistry by laying the foundation for high-pressure technology. Continuous production with high space velocities and space yields combined with the ammonia oxidation developed immediately thereafter enabled chemical industry for the first time to compete successfully against a cheap natural bulk product, namely sodium nitrate from Chile. The synthesis of ammonia thus became exemplary for all subsequent chemical mass production processes.

Although for the numerous processes subsequently developed in Germany, just as in the case of ammonia, the shortage of certain raw materials provided an important impetus, the decisive factor in their rapid and successful development was the progress in chemical engineering and technology.

It is thus especially instructive to review the development of the Haber-Bosch Process in connection with the evolution of the science of chemical engineering and chemical technology. For this purpose a short digression to sketch the early days of industrial chemistry may be helpful.

The Development of Industrial Chemistry

Early in the last century small chemical factories began to evolve out of small workshops and pharmacies. Perhaps the technical realization of the Le Blanc soda process can be considered as the beginning of a true chemical industry. Around the middle of the last century the work of Justus von Liebig on the application of chemistry to agriculture led to the establishment of factories for the manufacture of soda ash, sulfuric acid, potash salt, nitrogenous and phosphate fertilizers. About 1860 the manufacture of synthetic dyestuffs started, which began to spread rapidly after tar, a necessary raw material, had become available as byproduct from the manufacture of town gas.

The processes were discontinuous and the equipment rather primitive. Locally available craftsmen and laborers in mechanical workshops worked according to the direction of chemists who designed equipment as copies of the apparatus used in the laboratory. This applied also to the factory buildings. The only engineer involved in those days was the civil engineer.

With the expansion of the dyestuff industry towards the end of the century the picture changed: the growing need for steam, water and mechanical energy to drive the agitators, pumps, mills and conveyors called for the participation of mechanical engineers.

Since it was necessary to drive agitators, pumps and mills via transmission belts from centrally located steam engines, the equipment could not be arranged according to the process flow. Unnecessary, cumbersome and frequently complicated transport of process liquids and intermediate products, by means of pumps or pneumatically, was the consequence.

The introduction of electric power brought about an important change. Due to direct drive of machines and apparatus it became possible for the first time to arrange the equipment according to the requirements of the process, thus minimizing transport. The replacement of steam-driven power was signalled when the first electric generator, driven by a gas engine, was installed in BASF in 1889. This development now required the contribution of the electrical engineer.

The growing demand for raw materials was a further consequence of the expansion of dyestuff production, and this in turn led to new and, for the first time, continuous processes. The development of the chlorine-alkali electrolysis, of the liquefaction of chlorine and of the catalytic oxidation of naphthalene to phthalic anhydride were indirect consequences of the development of the technical indigo synthesis. The manufacture of alizarin required large quantities of oleum and thus promoted the development of the catalytic oxidation of SO_2. The work on these processes was already characteristic of modern chemical technology. The concept of "Chemical Engineering" to characterize the activities comprising the planning, designing, optimizing and also the operation of a chemical plant, was admittedly still unknown at that time. It was probably first used by Davis who in 1888 gave a course on this subject at the Manchester Technical School in England.

Introduction of Catalytic Processes in Industry

With the sulfuric acid process the prototype of a catalytic gas phase process was introduced to chemical industry. For the first time it was necessary to learn how to handle large volumes of gases and to purify these in order to remove catalyst poisons. A systematic study of heat removal, of reaction equilibria and kinetics became necessary. The objective was to achieve optimal economics. Rudolf Knietsch brilliantly performed this work in BASF and described it in his famous lecture to the German Chemical society in Berlin in the year 1901.

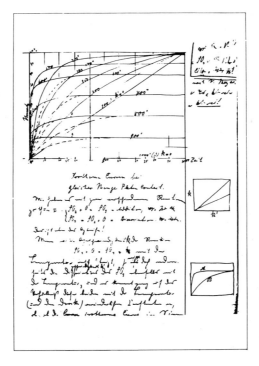

Fig. 1. Knietsch's notes on SO_2 - oxidation (1889).

Fig. 1 shows curves sketched by hand for the draft of this lecture. Knietsch was able to disprove the previously postulated stoichiometric approach and showed that dilution with inert nitrogen has no practical effect. A large part of this chemical engineering investigation may be characterized as "Applied Physical Chemistry", but his thoughts regarding the optimal temperature control of the exothermic reaction by means of a tubular reactor, which was externally cooled by fresh feed gas correspond to the most up-to-date concepts of process engineering.

Fig. 2 shows catalytic reactors designed by Knietsch in 1898. Bosch later applied the principle developed in this case to the ammonia synthesis converter and it may be significant in this connection that Carl Bosch also worked in the sulfuric acid plant around 1908.

His professional training predestined Knietsch for his work in "Chemical Engineering". After interrupting his schooling he had at first learnt the trade of mechanic and had worked as such for the railroads. After one year he continued his schooling and subsequently studied chemistry at the "Gewerbeakademie Berlin", where courses in technical chemistry were already given.

Early High-Pressure Processes

The fractionation of N_2 and O_2 is another early process which is closely connected with the Haber-Bosch process and shows typical characteristics of modern process technology. Exploiting the cooling effect caused by the adiabatic expansion after flow

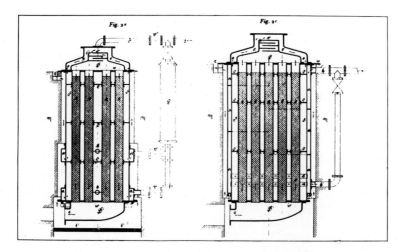

Fig. 2. SO_2 - oxidation converters designed by R. Knietsch (1898).

through a nozzle, Linde had succeeded in liquefying air in 1885. In spite of an unfavorable equilibrium curve Linde was later able in 1902 to obtain pure top and bottom product from a distillation column with more than 100 actual plates, effecting mass transfer by counter-current flow. An especially noteworthy example of Chemical Engineering was Linde's concept to achieve the separation at higher pressure, in spite of a less favorable equilibrium under this condition. The elevated pressure, however, permitted a two-column construction with a built-in condenser-evaporator. Operating the two columns at different pressure levels the components nitrogen and oxygen could be produced in a pure state.

But the demands made on the equipment were minor in comparison to the ammonia synthesis, since high temperatures were not involved, and absolutely leakproof installations to handle inflammable and expensive gases were not necessary.

State of the Art in the Construction of Chemical Apparatus prior to the Haber-Bosch Process

A good survey of the technical level of industrial chemistry early in this century, that is, when the work on the Haber-Bosch process began, is provided by a number of textbooks and handbooks. Thus, for instance, in 1903 A. Wolfrum's "Chemisches Praktikum mit einem Atlas darstellend die Apparate der Chemischen Technik" (Chemical Manual with an Atlas presenting the Apparatus Used in Chemical Technology) was published. It contains more than 700 illustrations and drawings of chemical apparatus and components.

Workmanship and construction of the equipment shown are in some cases already on a high technical level. The collection includes, besides simple agitated vessels and conveyors, also vacuum film evaporators, centrifuges and compressors. Practically all presently known basic operations are mentioned. Quite remarkably the classification corresponds in part to the present-day concept of "Unit Operations", although the latter term was coined by Arthur D. Little only in 1915. The classification into "Unit Operations" is even more pronounced in the book "Die Maschinellen Hilfsmittel der Chemischen Industrie" (Machinery in Chemical Industry) by H. Parnicke (1905). In some publications a theoretical and mathematical treatment is already found: for example in Hausbrand, "Die Wirkungsweise der Destillations- und Rektifizierapparate" (1883) (The Operating Characteristics of Distillation and Rectification Apparatus) or O. Nagel, "Der Transport von Gasen, Flüssigkeiten und Feststoffen mittels Dampf, Druckluft oder Wasserdruck" (The Transport of Gases, Liquids and Solids by Means of Steam, Compressed Air or Hydraulic Pressure).

Similar growth as in chemical industry occurred in other branches of industry. A remarkable development took place in the energy sector, where the consumption of electric power and gas increased and the gas industry made progress. The semi-continuous water gas process entered the scene in addition to coking, and continuous producer gas generators were developed to provide fuel for the gas engines, which were gaining acceptance as power packages and which were later to prove important for the ammonia process as compressor drivers.

The Call for the Fixation of Nitrogen

Growth occurred also in the population and in agricultural production. Liebig had already in 1840 pointed out the need for artificial fertilizer in his pamphlet "Die Chemie in ihrer Anwendung auf die Agricultur und Physiologie" (Chemistry and its Application to Agriculture and Physiology). His remarks fell on fertile soil, and thanks to the increasing use of fertilization the cultivation of sugar beets, for instance, increased very rapidly. In 1840 the production of 145 factories yielded 12 500 t of sugar, in 1900 this figure was already 1.6 mio tons in 399 factories.

In spite of this general progress there was increasing concern about the availability of nitrogen fertilizer. After all, in 1900 368 000 tons of Chilean nitrate and 37 000 tons of guano had to be imported into Germany in order to supplement the 118 000 tons of ammonium sulfate supplied by the gas works and the coke oven industry. The British chemist Sir William Crookes in his lecture to the British Association for the Advancement of Science in Bristol in 1889 about the problem of wheat, which was later to become famous, appealed urgently to all chemists: "The fixation of atmospheric nitrogen is one of the great discoveries awaiting the genius of chemists". Ten more years had to pass before the solution to this problem was found, and 15 years after this lecture the first plant in the world in which nitrogen from the air was converted to ammonia started production in BASF.

Early Eyperiments in Nitrogen Fixation

As late as towards the end of the last century an economical method for combining hydrogen and nitrogen was considered an impossibility. Liebig was also of the opinion that it was not feasible to combine hydrogen directly with nitrogen to form ammonia. Insufficient quantitative data on this problem were available, since these had yet to be supplied by physical chemistry, a branch of chemistry which just then started to evolve. Wilhelm Ostwald, working in Leipzig, was one of the mentors of this new discipline in Germany.

Considerable interest was, therefore, aroused in BASF when in March 1900 Ostwald offered BASF a process for the synthesis of ammonia, in which nitrogen and hydrogen were passed over metallic iron at elevated temperature and atmospheric pressure, claiming the formation of several percent of ammonia. Ironically, it was the first task for Carl Bosch, then a young chemist, to prove Ostwald wrong, in spite of the fact that the later successful ammonia synthesis was also based on an iron catalyst. The relatively large yield of ammonia resulted from the hydrogenolysis of iron nitride, which had been formed on the iron during its preliminary treatment before the experiment, when the decomposition of ammonia on the same catalyst sample had been examined. Ostwald complained to the board of BASF: "Naturally nothing will come of it if you charge an inexperienced chemist, who is new to the company and knows nothing, with such a task". But eventually he confirmed the findings of Bosch and withdrew his patent application, not knowing how important that application could have been later.

Haber's Laboratory Experiments

Fritz Haber then investigated the ammonia equilibrium in 1904 at the Technical University in Karlsruhe. He found only 0.012 % of ammonia at 1000°C and normal pressure, using a stoichiometric nitrogen-hydrogen ratio. In this case again the value was wrong, namely too high by a factor of three. After a heated quarrel with Nernst, who on the basis of theoretical reasoning postulated a lower value, Haber was able to correct his results by new measurements. He passed pure ammonia over heated iron, then removed any ammonia not decomposed, and introduced the remaining mixture of hydrogen and nitrogen into a second reactor likewise filled with finely-divided iron. The concentration of ammonia formed in the second reactor was exactly equal to that observed after the first. The low yield scarcely promised an economic future for this process. Extrapolation of the measured equilibrium constants to lower temperatures showed that somewhat higher concentrations could be expected, but no efficient catalyst was available and the application of pressure was out of reach at that time.

Alternative Methods for the Fixation of Nitrogen

Seeing no possibility of realizing the process, industry turned its interest to other methods of nitrogen fixation. By 1910, two processes had been commercially established: The cyanamid process of Frank-Caro and the electric arc process of Birkeland-Eyde. Other routes, such as that using barium cyanide produced from barytes, coke and nitrogen, or using titanium nitride, were investigated in Ludwigshafen by Bosch and Mittasch but did not look promising.

In cooperation with Norsk-Hydro the electric arc process had been considerably improved in the meantime by Schönherr at BASF.

At temperatures of 3000°C oxygen and nitrogen combined to nitrous oxide. However, energy consumption was still enormous, 700 G J per ton of ammonia. This is about twenty times as much as the energy consumed in a modern steam reforming plant. Nevertheless, BASF was at that point determined to exploit the process commercially at two locations where inexpensive hydroelectric power was available. This was to be done jointly with Norsk-Hydro, who were proprietors of the original Birkeland-Eyde process.

The High-Pressure Recycle Process

Nernst would not rest before he had been able to prove his heat theorem. In 1907 Nernst published that he was able to obtain up to 0.9 vol.% of ammonia at a pressure of 75 bar in the presence of various metals, above all iron. But in view of such low concentration Nernst did not contemplate any commercial use.
Not so, however, Fritz Haber: the equilibrium measurements showed to him that, unlike in the case of the sulfuric acid synthesis, where almost 100 % conversion is possible in a single pass, the low ammonia concentration of ammonia permitted a viable technical process only if the high-pressure gas were recycled and the ammonia formed were removed at high pressure.

Haber thus realized that production could be optimized by working under conditions that favored high reaction rates even though the thermodynamic equilibrium was unfavorable.

This should be especially true if higher pressure could be applied and better catalysts found than previously. He formulated these ideas in his famous "circulation" patent of 1908, which incorporated the main features still present in every ammonia synthesis plant to this day. Apart from the high-pressure gas recycle, the patent disclosed and claimed heat recovery by countercurrent heat exchange between the cold gases entering the catalyst zone and the hot gases leaving it. The use of ammonia evaporation as a means of cooling the product gas was also claimed in this application.

At this time Haber turned to BASF for financial support. Searching for a better catalyst he obtained excellent yields with Osmium. This was indeed surprising since platinum, for instance, belonging to the same group of the periodic system and known in other cases for its excellent catalytic properties, showed only little activity for the formation of ammonia. At a pressure of 175-200 bar and a temperature of 550-600°C yields of 6 vol. % and more were obtained with sufficient space-time yield. The question, whether it was conceivable to work on a technical scale at more than 100 atmospheres and at such high temperatures led to heated discussions within BASF. Bosch finally convinced the board with his statement: "I think it will work; I know the capability

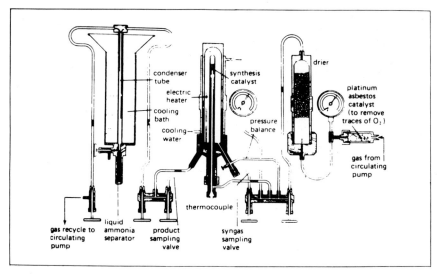

Fig. 3 Haber's laboratory ammonia synthesis apparatus

of the steel industry and we should try it." Haber was now able to continue his work. He built a laboratory apparatus which made it possible to work with recycle gas under high pressure according to his concept.

With the new laboratory-scale apparatus he was able to demonstrate ammonia synthesis successfully to BASF representatives in Karlsruhe on July 2, 1908.

Fig. 3 shows this apparatus with which 80 g of ammonia per hour were obtained operating at a pressure of 175 atmospheres. This experiment gave BASF final justification to concentrate all efforts on the direct ammonia synthesis and to abandon all other activities for nitrogen fixation including the relatively advanced electric arc project which was stopped in 1912. By extraordinary team work chemists, engineers and physicists succeeded in scaling up this process from the laboratory to a commercial plant of 30 000 to/year within the incredibly short time of 4 years. Yet until this goal was reached, the path was hard and arduous and the temptation to give up was sometimes great.

Research and Development Programm

The decision of the BASF board to support this research and development project with major financial resources was admirably progressive. The objective was a process for which no firm concept existed: Neither the catalyst nor the apparatus nor the synthesis gas supply route had been settled at that time. Carl Bosch, then 35, became project leader and was entrusted with extraordinary authority. It was significant that the management nominated a project leader in the person of Carl Bosch who, although he held a chemical degree, could be considered by training and career just as much a chemical engineer, although he would not have liked to be called this to his face during his life time. After graduation from high school Bosch had at first served as an apprentice in a metallurgical plant where he also did practical work as a mechanic. Afterwards he had studied metallurgy and mechanical engineering for two years in Berlin, before he started to study chemistry in Leipzig in 1896.

Bosch immediately organized a task force of capable collaborators and set up a research program which comprised three major objectives:

1. to find an efficient and stable catalyst based on easily available elements

2. to develop suitable equipment for the high pressure synthesis

3. to devise methods for producing inexpensive hydrogen and nitrogen in quantity.

Bosch and his chief engineer Franz Lappe devoted themselves to the apparatus problem. With the catalyst problem he entrusted his colleague Alwin Mittasch.

The ammonia catalyst

Osmium showed excellent activity, but was no real choice for a technical catalyst. It was difficult to handle, because in contact with air it was readily converted to the volatile tetroxide. Its main disadvantage, however, was that the world's total stock of this rare metal at that time was only about 100 kilograms which BASF already had acquired.

During the first tests with externally heated high-pressure tubes unexpected hydrogen embrittlement caused rupture after a few hours.

This was another reason why experiments with osmium could not be considered advisable. A second catalyst proposed by Haber was uranium - expensive too, but available in larger quantities. It was similar in activity but fairly sensitive to traces of oxygen and water.

For catalyst testing it became necessary to have available small high-pressure reactors which could be used for continuous around-the-clock tests and which should be designed for easy change of catalyst. The solution was found by Georg Stern, who worked with Mittasch.

CHEMICAL ENGINEERING IN GERMANY

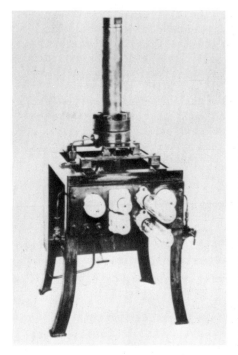

Fig. 4a

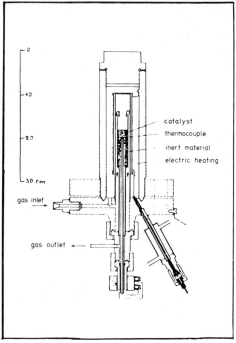

Fig. 4b. Lab-scale high-pressure apparatus for the catalyst testing (1910).

Fig. 4 shows this laboratory-scale unit. A vertical high pressure tube was equipped with internal electric heating and air cooling. With only a few operations it was possible to open and to reassemble it. A small cartridge designed to hold about two grams of catalyst could be easily removed, emptied and reloaded for the next run. 30 test reactors of this type were operated simultaneously, thus allowing an extensive series of investigations. In spite of the failure of iron in Ostwald's experiments, which had actually been confirmed by the investigation of Bosch, the latter himself and Mittasch believed in iron as a suitable base metal for a technical catalyst. But the results of Haber and Nernst with pure iron could also not provide encouragement. Nevertheless, when coming upon a newly published atlas of spectra, Bosch stated: "An element (iron) which has such a complicated emission spectrum should be of some use".

When experiments with various types of iron and iron compounds were performed, a special Swedish magnetite which happened to be at hand gave rather good results. This was surprising because magnetites from other sources were complete failures. For this reason Mittasch now systematically investigated the influence of various promotors on pure iron. Element by element was checked, first using only one additional compound, but later on multiple promoting was also included. The program also covered variation in the weight ratio of the additives and different procedures for catalyst preparation. In 1910 an iron-aluminum catalyst had finally been developed which had an activity similar to osmium and uranium. In addition to this, the systematic test series had led to some understanding of the phenomenon of catalyst poisoning: sulfur compounds, chlorine, phosphorus compounds, arsenic and oxygen compounds were found to block the catalysis. Therefore it seemed extremely important to remove all these poisons from synthesis gas.

In 1911 a relatively stable iron catalyst activated with aluminum and potassium was available. Then by using calcium as a third promotor Mittasch optimized the catalyst to essentially its presentday composition. The whole program was continued even after successful commissioning of the first commercial ammonia plant and finally ended in 1922 after 20 000 runs. With the introduction of these small and easy-to-handle testing units using only 2 g of catalyst Mittasch had introduced a method which corresponds to the screening technique used in catalyst research even today.

The Development of the High Pressure Apparatus

With such a small laboratory reactor several hundred runs could be performed without any problem. It was therefore completely unexpected that the first larger experimental units with a catalyst charge of 1 kg burst after 80 hours of service. The investigation showed that the material at the inner surface had changed, having completely lost its tensile strength. Obviously this phenomenon propagated from the inner wall to the outside. When the remaining sound material became too thin rupture started. Therefore a chemical attack seemed to be the probable cause, and nitrogen was believed to attack the steel by nitride formation. However, chemical analysis could not trace any nitrogen in the embrittled material. Metallographic investigation by thin section, a technique which was not familiar to chemists at that time, but was well known to Bosch, the trained metallurgist, provided the explanation.

The carbon steel used had a structure which showed carbon containing perlite dispersed in the ferritic phase. Under the microscope it became visible that the perlite in the lighter-colored parts at the inside of the tube had disappeared and the structure was broken up by small cracks. Decarbonization had occurred, but surprisingly the result was not soft iron but rather a hard and embrittled material. This was caused by hydrogen which by diffusion had entered into the iron forming there a brittle alloy. An additional weakening of the steel was effected by the

CHEMICAL ENGINEERING IN GERMANY 41

methane formed in the decarbonization process. This gas became entrapped under high pressure within the structure of the metal, thus contributing to its deterioration. The fact that these first small converters had to be heated externally by gas complicated the problem additionally. Systematic laboratory investigation and material tests demonstrated that all carbon steels will be attacked by hydrogen at higher temperatures and that the destruction is just a matter of time. It will be of interest that the phenomenon of hydrogen-embrittlement was not totally unknown at this time. In a 1910 handbook on material testing this subject is already treated but without any explanation of the mechanism involved. Here again it becomes obvious that Bosch was already a chemical engineer as this profession is understood today. Thanks to his education and practice he was able to grasp in addition to chemistry other scientific and technical disciplines which were necessary to push ahead with the development.

The unconventional solution to the embrittlement problem was that Bosch used a carbon steel pressure shell with a soft iron liner, which was not subject to decarbonization by hydrogen. To prevent the hydrogen which had penetrated the soft iron liner from attacking the pressure shell, measures had to be taken to release this hydrogen safely to normal pressure. This was done by providing small channels on the outer side of the liner which was in contact with the inner wall of the pressure shell and by drilling small holes, later known as "Bosch Holes", through the pressure shell, through which the hydrogen should escape to the atmosphere. The losses of hydrogen encountered thereby were negligible.

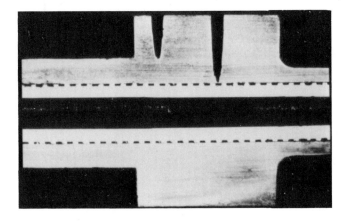

Fig. 5

Fig. 5 shows this design, using as example a small test model. A metallurgical solution of the hydrogen embrittlement problem was not possible at that time, for stainless steels were only just then about to be developed.

Characteristically for Bosch he did not content himself with an answer once it had been found: In spite of the success of the double wall reactor concept he continued to turn his attention to the question of materials.

For tests on the strength and other physical properties of materials and for research on the structure of metals he set up a special laboratory headed by an engineer. This institution grew to considerable size and was equipped with the most modern physical and technical apparatus then available. Even the steel industry did not at that time have at its disposal better equipment for testing and analysis.

The formation of ammonia is a distinctly exothermic reaction. But at low conversion and in relatively small-sized equipment heat losses were preponderant and additional heat had to be supplied.

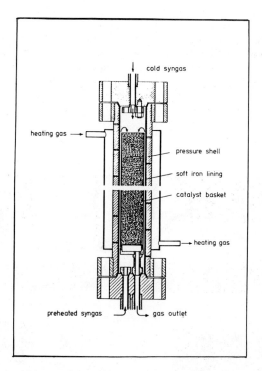

Fig. 6. First pilot plant converter with soft iron lining and external heating

Initially the reactor vessel was externally heated with gas (Fig. 6). New difficulties arose when it turned out that even without attack by hydrogen the steel shells were not adequate to withstand the high pressures at the large temperature gradients which were made necessary by the external heating. By introducing a small amount of air through a nozzle some of the synthesis gas was burnt under pressure within the reactor itself. This was called a "reverse" flame. Ignition was initiated by means of an electric wire. Some water was formed which was even then known to be detrimental to the catalyst, but this was tolerated in this early stage.

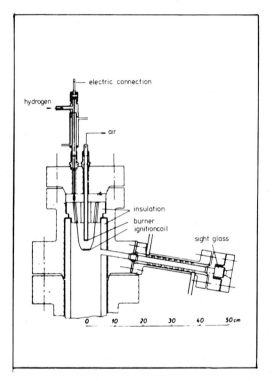

Fig. 7. Internal heating by "inversed" flame (Test apparatus) (1911).

   In an experimental apparatus which included a sight glass (Fig. 7) and which Lappe had specially designed, this technique was first put to the test with simultaneous observation of the flame. It was in technical use until 1922, and in the larger reactors which came into service after 1915 and which actually provided surplus heat, it was only used to supply heat for start-up. In this case, however, the flow direction during heating was reversed in order to keep the water from contacting the catalyst.

A close cooperation with leading steel companies developed thereafter with the objective of finding steels that were resistant to hydrogen. Many different alloy components were tested, and in the course of the following years it was shown that addition of chromium, molybdenum and tungsten imparted great strength as well as improved resistance to hydrogen.

In following reactor designs, which also incorporated internal heat exchange in the reactor, Bosch partially followed the known design of Knietsch for the sulfuric acid contact process.

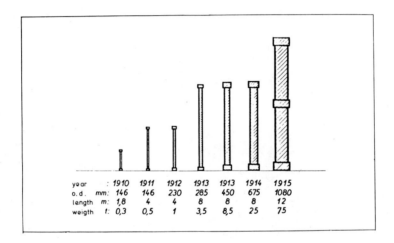

Fig. 8. Development of the ammonia converter size

Subsequent improvement consisted in providing additional protection for the ammonia converter wall by cooling with nitrogen along the space between the catalyst basket and the pressure shell.

Several years of painstaking trials with various models of high-pressure compressors were necessary to get reliable performance for this important part of the ammonia synthesis. Unlike with the small old air compressors, in the big new machines perfect sealing became important to prevent loss of hydrogen and avoid explosion hazard.

Fig. 8 shows the evolution of reactor sizes from pilot plant until the first years of large-scale production.

The Synthesis Gas

Hydrogen for the pilot plant tests came from the chlorine-alkali electrolysis. The stoichiometric ratio of nitrogen to hydrogen was adjusted by partial combustion of this hydrogen with air. At 0,4 tons/day of ammonia the capacity limit of the electrolysis unit was reached. Therefore, water gas was selected as an independent source of hydrogen. Two processes were brought up for discussion. One of them, the Messerschmidt process, worked on a cyclic principle. Iron oxide was first reduced to iron with water gas, and over the red-hot iron steam was then reduced to hydrogen, while the iron was reconverted to iron oxide. The second possibility consisted in the cryogenic separation according to the Linde-Frank-Caro process. In this process carbon monoxide is condensed out of the water gas at $-200^\circ C$ and 25 bar with liquid nitrogen, which is produced in a separate air separation plant. Both processes were already in industrial use to produce hydrogen for air ships and for fat hardening. However, no great demands were made on reliability. The decision was made in favor of the cryogenic process primarily because of its ability to remove catalyst poisons. The residual content of 1,5 % of carbon monoxide was removed by conversion to sodium formate in a washing step employing 10 % sodium hydroxide solution at the incredibly high temperature of $230^\circ C$ and a pressure of 200 bar. Scrubbing with copper solution had already been discovered as an alternative but could not be used, because the corrosiveness of the solution that was employed at first was excessive and the available steels were not sufficiently resistant. It is also of interest that CO_2 was removed before the cryogenic unit by washing with potash solution, which was regenerated by boiling.

The Technical Plant

As early as 1912 construction of a grass-roots plant with a capacity of 30 tons/day was started at Oppau, 5 km to the north of the existing BASF works. The Oppau plant involved total planning to an extent unknown until then. The individual units were erected on an area of 500 000 m^2 according to a uniform concept: transport of coal and coke, production of water gas and generator gas were included, together with gas separation, compression and ammonia synthesis, followed by further processing of ammonia to ammonium sulfate, storage in silos and shipping. In 1914 ammonia oxidation and production of nitric acid were added.

Fig. 9

Fig. 10

Fig. 9 shows a partial view of the construction site in 1912, Fig. 10 a total view of the Oppau works in 1914.

Production started in 1913 and reached full capacity in 1914. Not more than 4 1/2 years after the start of process development a commercial plant based on a radically new technology had been put on stream.

The introduction of high pressure techniques meant a revolution in chemical technology. Wherever one looked in the plants, one could discover new equipment and apparatus which had been completely unknown until then or had not yet been employed in chemical industry: high-pressure reactors with an inner diameter of 500 mm and, after 1916, of 800 mm, quick-acting magnetic shut-off valves, continuously functioning instruments to measure gas flows at high pressure, gas densities, gas compositions and oxygen concentration. New gaskets, joints and fittings, valves and other components were designed. The rapid growth of production to 120 tons/day of ammonia made it necessary to change the process for the production of pure hydrogen. The cryogenic Linde process was replaced by a CO-shift-conversion which Wild had discovered in BASF as early as 1912. This consists in a conversion of carbon monoxide with steam at 350 - 450°C over an iron-chromium catalyst to carbon dioxide and hydrogen, leaving a residual concentration of 2 % carbon monoxide. Carbon dioxide was removed by scrubbing with water at 25 bar, and the copper solution scrubbing process could now successfully replace the trouble-prone sodium formate process. At the end of 1916 daily production reached almost 250 tons of ammonia.

It was decided to build a further plant at another location. After a construction time of only 11 1/2 month a plant with a capacity of 36 000 tons/year was brought on stream at Leuna in close proximity to lignite mines. This capacity was rapidly raised to 160 000 tons/year and by the end of World War I construction work was in progress for further expansion to 240 000 tons/year.

By 1937 world-wide capacity had reached 760 000 tons of ammonia per year, of which 72 % were still concentrated in Oppau and Leuna. Fig. 11 shows a simplified process scheme of the plant as it finally operated in Oppau.

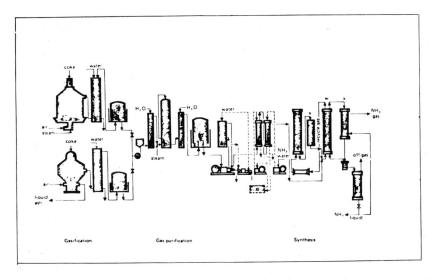

Fig. 11. Process flow diagram of Haber-Bosch process

The Chemical Engineering Aspects in the Work of Bosch

In the historical presentation of the development of the Haber-Bosch process, typical problems of a chemical engineering nature have already been pointed out to some extent: a systematic consideration will undoubtedly lead to the following conclusions:
1. This process development amounted to chemical engineering in a modern sense. 2. Bosch represented already the "Chemical Engineer" in today's meaning of the word if one considers his working methods and to some extent his professional training.
3. Chemical engineering does not necessarily presuppose the existence of a special "chemical engineering" profession, but can be practiced through team work involving individuals representing various disciplines, such as chemists, physicists, engineers, metallurgists and others.

Methodically the most suitable among the available processes – in some cases one should think of unit operation – were selected. Individual solutions for one process step were considered in conjunction with other steps, as in the connection between cryogenic separation and catalyst poisons. Rather than on expensive solutions involving apparatus or materials, reliance was placed on novel process design features (soft iron insert, "Bosch Holes", reverse flame). Instead of conditions which were chemically hard to realize (high conversion rates at low temperatures with an ideal catalyst), practical solutions were found through suitable process design (low conversion per pass at high temperatures with reasonable space-time yield and gas recycling). Finally the entire development starting in the laboratory always proceeded with regard to subsequent technical feasibility.

Not only the extensively described development of the process but also his own personal statements and the concept he had of his own activities indicate that Bosch was not only a chemist but above all an engineer. For him the uppermost principle was optimal technical realization with the aim of economical production. "Technology at its best is organized science", he once said. At another occasion he said: "In our projects I have frequently found that experiences gained on any occasion can be utilized in order to discover new approaches. One should not only look at chemistry and physics but also get acquainted with adjoining disciplines. The future lies especially in the borderline areas".

Further Development of High-Pressure Technology

The further development of high-pressure technology followed two different directions. In the methanol and the isobutyl alcohol processes the approach through selective catalysis was continued, which in due course added homogeneous to heterogeneous catalysis, as for example in the oxo-synthesis and "Reppe Chemistry".

The application of this technique in organic chemistry opened up a wide area which has decisively advanced the production of chemical intermediates. Many pure organic products are nowadays accessible on the basis of catalytic high-pressure reactions: Hydrogenations, aminations, ring closures and addition reactions. The focus of development in these cases was less on equipment, where the groundwork had been laid by the ammonia process, but rather on catalytic chemistry and reaction kinetics.

A different development was followed in the non-selective, destructive hydrogenation of coal. Here the main problem was the need to master the extreme conditions under which the equipment had to operate. Credit for having pushed outwards the boundaries of what is technically possible in this field belongs above all to Matthias Pier. Solid raw materials and non-selective iron catalysts were introduced into a system of extremely high pressures, 300 bar in the case of lignite and 700 bar in that of bituminous coal. The initial reaction consisted in a liquid phase hydrogenation to a medium-heavy oil, which was then further hydrogenated to gasoline in the gas phase at 300 bar over a fixed-bed catalyst. The mutual optimization of both steps was essential to reduce costs. Additionally temperature control was extremely critical. While at lower temperatures of 400 – 430°C the hydrogenation-dehydrogenation equilibrium favors hydrogenation, the rate of C – C bond breaking is too slow for satisfactory conversion. On the other hand at 470 – 490°C the conversion rate is good, but considerable dehydrogenation results in a large proportion of unsaturates and thus causes problems with polymerization. The temperature had to be controlled to within 10°C, which was partly achieved by the addition of cold hydrogen. Reaction heat values of $1,3 \cdot 10^3$ kJoule/m^3·h corresponded to conditions encountered in industrial furnaces. The residence time was just as critical. In 10 plants $3,5 \cdot 10^3$ tons of gasoline were produced during World War II, which were supplemented by $1 \cdot 10^3$ tons obtained from Fischer-Tropsch installations and $0.7 \cdot 10^3$ tons from the low-temperature pyrolysis of lignite.

All the processes were in the final analysis conditioned on the previous development of the Haber-Bosch process, and it would certainly be interesting to follow up the historical and personal connections between the various developments. Especially in the area of human resources a remarkable potential had been accumulated.

Within the individual companies like BASF, but also in the other IG Farben locations, a nucleus of highly qualified practical chemists had been trained, who according to American terminology could justly be called "Chemical Engineers".

It may appear strange that these enormous achievments in chemical technology, which also involved very considerable progress in chemical engineering know-how, were due to the efforts of traditionally trained chemists in cooperation with and assisted by engineers, physicists and metallurgists. It must not be forgotten, however, that in the previously mentioned processes the chemical aspects, the new chemical reaction or the new compound, usually were paramount. This was the case with the methanol-, oxo-, or Reppe-process. Interest was concentrated on the technical and commercial realization of the individual process. The transfer of the experience gained in a given process to a generalized system, that is, the elaboration of the scientific foundations with the purpose of making them generally applicable to other processes, did not constitute an immediate objective. The individual chemist or engineer with a special aptitude acquired for himself the art of chemical engineering through experience and from the literature. A large store of experience and numerous analogous examples were available within the large companies. On the other hand it is quite true that this experience rested to a large extent on a theoretical basis. For example in the late twenties the extensive results of Nusselt's studies on heat transfer were summarized in the form of parameter diagrams within the IG Farben concern as it existed then. These diagrams proved very useful in practice. The analogies between material and heat transport were already recognized. Nevertheless, this knowledge remained restricted to a small number of specialists and was not systematically developed to serve as the basis of a modern concept of unit operations and reaction engineering as was later done by Lewis, Colburn and others in the United States.

The Evolution of Chemical Engineering in the United States

The rapid development of the petroleum industry in the United States had brought about the accumulation of broad experience above all in the field of physical processes. Chemical problems frequently played only a subordinate role as, among other things, is shown by the fact that the first catalytic process was used there only in 1937, when Eugene Houdry introduced his moving-bed cracking process. Until then only thermal cracking had been applied. Hydrodesulfurization and high-pressure refining over cobalt-molybdenum catalysts was introduced into refinery practice only during the forties, although it had already been developed by Pier and his co-workers in conjunction with coal hydrogenation.

1923 is generally considered as the year when the science of Chemical Engineering was born. During that year the text book "Principles of Chemical Engineering" by Walker, Lewis and McAdams was published, which contains the first physical-chemical treatment of unit operations. This treatise inititated a tremendous evolution in the USA and contributed to a transformation of chemical engineering from an art into a science, which spread in a quite uniform fashion to almost all universities.

The basically physical concept of unit operations with its inherent possibilities for systematic and formal problem-solving dominated scientific chemical engineering for years. It implied a comprehension of chemical engineering as a combination of applied physics, applied mathematics and some physical chemistry.

Attempts to categorize inorganic and organic chemical reactions in analogy to unit operations found no resonance. Many students had to experience in practice that without a reasonable knowledge and appreciation of chemical properties and phenomena, innovation and practical problem-solving were greatly hampered. In the last 20 years there has been a clear shift in emphasis back to applied physical chemistry in the areas of professional training and research in chemical engineering.

To an ever larger extent general and chemical thermodynamics and reaction kinetics have contributed to unit operations and mechanical engineering aspects. Further development led to chemical reaction engineering as a concept providing some kind of synthesis of energy and mass transport phenomena, chemical kinetics and thermodynamics. The theoretical treatment of problems relating to reaction mechanisms is also becoming of greater interest to chemical engineers.

Two Development Trends in Germany

The development outlined above served as a pattern in all industrialized countries. In spite of a similar objective situation, the profession of "Chemical Engineer" has developed along two lines in Germany. One line of development, starting with the mechanical engineer, led to the so-called process engineer ("Verfahrensingenieur"), the other one added the special subject of technical chemistry to the basic chemistry curriculum. The process engineering variant puts the accent on the unit operation concept, while the technical chemistry variant rather emphasizes chemical reaction engineering. In practice, however, the traditionally trained chemist predominates in German chemical industry even nowadays. As before, technical problems are worked on by teams of specialists, mostly chemists, physicists and engineers, but there is an increasing tendency for the process engineer and the technical chemist to participate.

Modern chemical reaction engineering makes possible a mathematical formulation of the dependence of the chemical conversion on reaction rate and concentration, on the physical factors of mass and energy transport, on the flow conditions and on the spectrum of residence times. One can predict with a fair degree of certainty the optimal type, shape and dimensions of the reactor. The choice of the reactor also determines the subsequent processing steps and makes possible a cost calculation and, therefore, an economic prediction. In this initial design stage the chemical aspects predominate, since complex chemical reactions often have to be regarded with a view to the technical apparatus and, if necessary, have to be modified. Thus in the design stage of project planning it seems more logical and promising to give preference to the chemist with some chemical engineering experience. The "Verfahrens"-engineer, on the other hand, is more useful for detailed optimization in the sense of unit operations.

The Concept of the Total Plant

Yet one other aspect seems to be more essential than ever to process development and brings us back to the starting point of our considerations: in spite of all the optimization and calculation of individual steps and reactions one should never lose sight of the total process and the attendant aim of optimizing its economic performance. Proof of the correctness of this thesis lies in the development of the modern integrated ammonia and methanol plants empoying the total energy concept. A classical Haber-Bosch plant of the year 1940 using coke as raw material consumed 88 k J per ton of NH_3, while for a modern steam reforming plant operating on the basis of natural gas the corresponding figure is only 32.7 k J/ton NH_3.

More recent investigations show, that improvements based on systematic analysis can make the total process 10 - 15 % more efficient energetically on the basis of today's knowledge and using the classical catalyst. This means that the thermal efficiency of the ammonia process will have been increased from 22 % for the original Haber-Bosch design to 65 % at present.

More than ever the words that Carl Bosch spoke in a 1931 lecture apply to industrial chemistry today: "What is needed in engineering today is a comprehensive view of the state of science, full certainty regarding the objective, and a certain mature level of science and technology which lets the accomplishment of the given task appear promising".

REFERENCES

Appl, M.	A Brief History of Ammonia Production from the early Days to the Present; Nitrogen No. 100, page 47 (1976)
Aris, R.	Academic Chemical Engineering in a History Perspective, Ind. Eng. Chem., Fundam. Volume 16, No. 1, page 1 (1977)

Bartholomé, E.	Entwicklung der technischen Chemie in Deutschland, Chem.-Ing.Techn., Volume 48, page 913, (1976)
Bosch, C.	Über die Entwicklung der Chemischen Hochdrucktechnik bei dem Aufbau der neuen Ammoniakindustrie, Nobelvortrag, Stockholm 1932
	German Patent Specifications Nos. 235 421 and 252 275 (1908 and 1909) ("Circulation patents")
	German Patent Specification No. 11 3932 (1898) (Process for the production of sulfuric acid)
Grossmann	Die Bedeutung der chemischen Technik für das deutsche Wirtschaftsleben, Verlag Wilhelm Knapp, Halle, 1907
Hausbrand, E.	Die Wirkungsweise der Rectifizier- und Destillierapparate, Springer Verlag, Berlin, 1893
Hinrichsen, F.W.	Das Materialprüfungswesen, Ferdinand Enke, Stuttgart, 1912
Holdermann, K.	Der Stickstoff in der BASF, BASF company archives, 1940
Holdermann, K.	Carl Bosch, Econ Verlag, Düsseldorf, 1953
Knietsch, R.	Über die Schwefelsäure und ihre Fabrikation nach dem Contaktverfahren, Ber. dtsch. Chem. Ges., Volume 34, page 4069 (1901)
Mach, E.	Planung und Errichtung chemischer Fabriken, Verlag Sauerländer, Frankfurt/M., 1971
v. Nagel, A.	Stickstoff, BASF company archives, 1969
v. Nagel, A.	Fuchsin, Alizarin, Indigo, BASF company archives, 1968
Nagel, O.	The Transport of Gases, Liquids and Solids by means of Steam, compressed air and pressure water

THE EVOLUTION OF CHEMICAL ENGINEERING IN THE HEAVY INORGANIC

CHEMICAL INDUSTRY IN FRANCE DURING THE NINETEENTH CENTURY

André Thépot

Department of History
University of Paris X
92000 - Nanterre, France

INTRODUCTION

The managers of French chemical companies, while they did not revolutionize production techniques, did constantly try to improve them. In the nineteenth century, chemical engineering was still in its infancy; it nevertheless made great strides in the efficient utilisation of heat and in gas-liquid exchange processes. Up to the twentieth century, the French inorganic chemical industry had developed within a traditional framework of sulfuric acid and soda products manufacture. By a study of plant records and engineers' notebooks it is possible to show how the French chemical industry attempted to solve its problems.

GENERAL PROGRESS OF CHEMICAL ENGINEERING IN FRANCE

This study will concern mainly the years from 1810 to 1890, a period during which the inorganic chemical industry maintained a certain technical homogeneity since the industry had been founded largely on the production of two important basic products; sulfuric acid and artificial soda, and chemicals requiring these products in their production: nitric and hydrochloric acids, chloride of lime, etc. The LeBlanc process combined with lead chambers formed a system that would become increasingly important over the years. However, although the French chemical industry was relatively large, it lagged far behind that of Britain. Although only fragmentary information is available, it is estimated that, by about 1860, French production was 60,000 tonnes of sulfuric acid and 45,000 tonnes of soda products.

The economic context of the French industry during this period should be recalled. France was a country in which the Industrial Revolution was slow to become established. For many years, and at least until the 1860s, there co-existed in the same sectors truly modern industries and a still well developed craft industry. One of the main causes of this situation was that France in the first half of the nineteenth century did not constitute a viable domestic market; land transportation was infrequent and costly, and the railway network was still incomplete and left many areas isolated. Even when, after 1870, the network grew denser, rates for transportation of hazardous materials were so high that beyond one hundred kilometres there was no competition for acids between producers. Only in solid products with a high added value like soda salts was there a wide trade.

All this explains why the French inorganic chemical industry, although fairly large, was so widely dispersed. It was spread among a fairly large number of companies (about sixty soda manufacturers by 1860) and even after 1880, when mergers and commercial agreements kept the number down, the plants were still scattered throughout France. There were therefore no very large plants oriented to mass production of chemicals. For all these reasons it is easy to understand why until the end of the nineteenth century, plants remained dispersed in areas close to consumer industries, such as textiles, glass, and soap manufacturing.

Such a situation was clearly not always favorable to technological innovation. Because of the presence of abundant manpower, cheaper than in Britain, company managers had little incentive to develop chemical engineering procedures that would reduce materials handling. On the other hand, the scarcity of coal and its high price led to efforts at economy and the improvement of heat exchanges in chemical reactions. The same unequal development can be seen at the technical and scientific levels. Small firms limited to routine activity did continue to operate, particularly in the south of France, while the more dynamic companies kept in close touch with the latest scientific advances. In 1825 the Saint Gobain company began to make use of the expertise of genuine scientists as a matter of course, in turn, Clément-Desormes, Gay-Lussac and Jules Pelouze. Similarly, in 1857, when the company that was later to become the Pechiney firm was created, Balard became scientific adviser to its founder, Henri Merle. Frédéric Kuhlmann, creator of Manufacture des Produits Chimiques du Nord in 1842, had been a student of Vauquelin and professor of chemistry at Lille. Thus, unlike the situation in other industries, the French chemical industry did not, at least until about 1880, lag behind the British industry in technology.

CHEMICAL ENGINEERING IN FRANCE 57

HEAT UTILISATION

The processes for the production of sulfuric acid and LeBlanc soda were heat-intensive. In a country where coal was rare and expensive, the problems of the production, mastery and control of heat were important in achieving low production costs.

Furnaces

At first, sulfur was used for the production of sulfur dioxide. This was the case for a long time in some regions like the south of France, because of its ready access to Sicilian sources. At the Saint Gobain works, sulfur was used until the 1860s. The advantage of this material was that it was physically homogeneous and could be easily transported. It burned readily in ordinary hearth furnaces similar to those that were commonly used in metallurgy. However, when a government monopoly was instituted in Sicily, the price of sulfur rose and producers began to switch to pyrites. One producer, Clément-Desormes, had pointed out its usefulness as early as 1820, but had not furnished an industrial process for it. For roasting pyritic ores, the original inspiration for oven design came from the lime kilns in use in the south of France. Later, from 1838 on, these were greatly improved by Michel Perret and Jules Ollivier, owners of the principal French pyrites mines, those in Chessy and Saint Bel, near Lyon. They were in fact the first to develop multiple-hearth, stacked furnaces which could burn not only pyritic ores, but also powdered pyrites, without any exterior heat required to achieve ignition. The brothers Perret adopted and improved a type of furnace patented in 1851 by another talented engineer, Usiglio, who was manager first of the Rassuen and then of the Salindres works (future Pechiney plants), before being hired by Saint Gobain. The dual-purpose multi-hearth furnace of Perret was shown at the World Exposition of 1861. It was immediately adopted by Charles Kestner at Thann, and by the Salines Domaniales de l'Est at Dieuze. Construction of these ovens began later, about 1872, at the Saint Gobain plant in Aubervilliers close to Paris, in Frédéric Kuhlmann's plants, and in some plants situated in the Rouen area.

A steady rate of production of sulfur dioxide gas was ensured by grouping the furnaces into batteries, since one full roasting operation lasted for about six hours. The furnaces were at first arranged in groups of four, but more advanced batteries were developed: at the Saint Fons plant near Lyon in 1878, there were sixteen furnaces grouped around one central chimney. However, these facilities could not operate solely with powdered pyrites, although it was the most common and cheapest raw material available. The problem was solved about 1860 bu Malrtra in his plants at Rouen and Paris. He created a furnace with several superposed hearths. Some French works, particularly those at Saint Gobain and

Aubervilliers, adopted a furnace of another type called the Gesstenhaufer system: a furnace in the form of an upright cylinder with a triangular roasting device in which the powder could be distributed from a conical hopper. With these various systems, the French plants were able to utilize all the forms of pyrites available in France, and their furnaces were perfectly adapted to the needs of French industry, with its abundant, cheap labor. In such a case the reluctance to adopt automated furnaces, once they first appeared, can be readily understood. The first of these, the MacDougal type, required powerful steam engines that consumed a great deal of coal. They produced a great deal of dust (8 to 10 per cent of the load of pyrites) and presented overheating problems that were difficult to resolve.

Control of Furnace Temperature

One of the great difficulties with lead chambers was that it was difficult to control heat distribution in the various areas of the installations. The gases left the furnaces at $200°$ but the manufacturer had noticed that the temperature of the gases had to be lowered before entering the chambers for fear of rapid corrosion. The problem was all the more difficult to resolve because science and technology was of little help in this regard. At first it was attempted to use the excess heat to produce steam, and, to this end, boilers were placed on the furnaces. The task was not easy, however, because when the gases condensed on the boiler walls, they deposited acids which attacked the tubing and boilers. The result was serious problems when the corroded boilers leaked into the furnaces beneath. In the 1850s a way was found getting around this problem in the Perret plants in Lyon, Charles Kestner's works in Thann, and in the Salines Domaniales de l'Est in Dieuze. Beside the furnaces were placed tubular boilers composed of two separate parts: one in which the gas temperature was maintained at over $100°$ so there would be no condensation, and a second in which the temperature did not exceed $100°$, and in which the water used for supplying the preceding boiler was pre-heated.

Reduction in coal consumption was a goal sought by all the producers. A wide range of methods were employed for recovering part of the furnace heat. The boilers on top of the furnaces prevented the furnace heat from becoming excessive; for example, in installations that still used sulfur, excessive combustion caused volatilization of the sulfur and therefore a loss. Frédéric Kuhlmann, according to the report of the 1862 World Exposition, attempted to correct the situation by reducing the excessive heat consumption with cast iron boilers located on top of the sulfur furnaces. Apparently this allowed him to produce a large part of the steam needed for the lead chambers while achieving a substantial reduction in coal consumption.

The drawback in placing the boilers directly above the furnaces was that this system resulted in severe corrosion problems. There was therefore a shift in the 1860s toward the installation of coolers, which became standard in the principal French plants. The gas coming out of the furnaces was sent into a long lead pipe immersed in a large lead or sheet iron vat containing a constantly fresh supply of running water. To achieve the substantial temperature drop (from about $200°$ to $70°$) large units were required. The notebooks of Jules Ollivier show that as late as 1872 the Saint Gobain company installed a huge cooler of this type in its Chauny plant, one of the largest plants in France at the time. The cooler consisted of a long closed duct 1 m high and 1.25 m wide, containing a continuous current of water. This conduit was lined with refractory brick over a length of about 6 m at the furnace outlets to prevent excessive corrosion. Its total length was 20 m. This was not an exceptional length; according to Ollivier's notebooks, if the cooler was only 10 m long, the gas was still at a temperature of $92°$ when it reached the chambers. The tendency was therefore to lengthen the installations. In the Saint Gobain plant at Aubervilliers, near Paris, the coolers were 30 m long in 1875. More advanced heat exchangers appeared around this time. The Perret brothers appear to have been the first to use a shell-and-tube cooler; the gas passed through a series of pipes 10 cm in diameter that heated the water around them sufficiently to provide the steam required for the lead chambers. They thus saved a good deal of fuel while avoiding excessive corrosion in the chambers. In many plants, however, it was becoming increasingly common to place a small chamber (of less than 10 m^3), called "tambour", or "drum", upstream of the main chamber. This had a double advantage: it further reduced the temperature of the sulfurous gases when necessary by circulating cold water over the outside walls, and also served as a dust removal chamber and reduced the formation of sludge inside the chambers. This system was employed in the Kuhlmann plants at Lille and at the Saint Gobain plant in Aubervilliers (1872).

Soda Furnaces

In the nineteenth century, the furnace was the main tool for chemical transformation between two solids. For the manufacture of soda products, the furnaces were gradually modified on the basis of two needs: increased production capacity and the possibility of producing better soda by more continuous production. First the size of furnaces was gradually increased: the first furnace used by Nicolas LeBlanc in 1787 was only 2 m by 1.5 m. Payen's furnace, circa 1810, was double that size, and those built ten years later by Clément Desormes were sometimes 9 m by 3 m. As size increased, problems were encountered in effective manual stirring of the mixture of sulfate, chalk and coal. That is why, about 1860, the furnaces were most often no bigger than

6 m by 2 m. For continuous production over a longer time, the practice of stepped (multi-tiered) furnaces spread, especially in the south of France. The furnace hearth, formed of two tiers 10 cm apart, allowed two distinct operations to be carried out simultaneously. As soon as the furnace walls became red, the upper level was loaded with the mixture of sulfate, chalk and coal. After an initial mixing, the load was left for an hour and a half in this compartment. The mixture was then dropped onto the lower level, while a new load was placed on the upper tier. In some plants furnaces with three tiers were built.

The same concern for efficiency and optimal use of heat in a country where coal was costly prompted the use of a three-compartment furnace at Marseille; it was used simultaneously for both the preparation of sodium sulfate and for its decomposition into sodium carbonate by chalk and coal. The drawback of this practice was that the hydrochloric acid byproduct could not be collected, since the strong air draft necessary to operate such a furnace prevented the installation of a condenser; however, it did permit a substantial saving of fuel. All these furnaces required much handling. Their capacity was fairly small and they required considerable manpower. They nevertheless lasted until the end of the century and were replaced very slowly by automated furnaces.

GAS-LIQUID TRANSFERS

This is perhaps the area in which chemical engineering made the most progress, since producers faced their first large-scale problems here. The massive production of acid required the use of very large volumes of gas, air and water.

Lead Chambers

The proper operation of lead chambers, by oxidation of the sulfurous gases issuing from the furnaces, required a constant supply of oxygen. From the beginning of the nineteenth century, it was known that about 250 m^3 of air were needed for 100 kg of acid. The proper operation of the chamber therefore depended basically on efficient mixing of the gases and their proper circulation inside the equipment.

To this end, early in the nineteenth century, sulfur and potassium nitrate were burned in the same furnace to ensure good mixing of the gases, but the installations corroded rapidly. About 1835, Gay-Lussac came up with the idea of using nitric acid which, although more expensive, gave a higher yield. The acid was introduced inside the chambers where it was made to trickle down over miniature steps, or cascades. The manufacturers were attempting to promote oxidation by improving the gas-liquid contacting. For this purpose, Clément-Desormes had vertical glass plates placed in the first chambers of the Saint Gobain plant, in order to induce convection currents inside. The practice appears to have spread, as Anselme Payen, in the illustrations of his "Précis de Chimie Industrielle de 1849", showed lead partitions inside his installations. At the same time, the substitution of steam injection for simple spraying with water was also tried, although this method had the drawback of increasing the temperature inside the chambers.

The proper operation of the equipment depended also on the draft inside the chambers. For a long time, only the traditional means of operating hearths were available, ie, air and chimneys. As installations were expanded, the chimneys became taller and taller to provide sufficient draft. By 1863, the Perret plant at Saint Fons near Lyon had chimneys over 52 m high. This did not protect the chambers from changes in the weather. Only with installation of fans was it possible to truly regulate the operation of the chambers at both the inlet and the outlet, and this was achieved only at the end of the century when metallurgy was able to provide corrosion-resistant fans and when electric motors were available to operate them.

Despite all this progress, it is not difficult to see how far from uniform the design and development of lead chambers still was. Each producer combined chambers of various sizes in a different way, and the reasons for each choice are not always obvious. For example, about 1865, Kuhlmann had a plant of only modest size (1500 m^3) but composed of no fewer than eight consecutive chambers of different sizes. In contrast, the Perret firm had a larger plant (7000 m^3) that was arranged in two chambers of equal capacity. The Saint Gobain company used a four-chamber system. At the end of the century, about 1895-1900, the same diversity continued to prevail. The Saint Gobain company and the Maletra plants in Rouen stopped at three chambers whereas the Kuhlmann plant at Rouen had five.

Transportation of Acids

With the increase in the size of the chambers and especially the development of the Gay-Lussac and Glover towers the problem of the circulation of acids in the installation arose, especially with the Glover towers, which recycled up to sixty per cent of the acid leaving the chambers; this meant that large volumes had to be moved. The problem of circulation of acids had formerly been solved very simply by the use of gravity alone. The chamber floors were inclined slightly so that the acid could be drawn off. Later, pipe systems appeared in some plants. As early as 1863 the Perret plant at Saint Fons had a network of pipes linking the soda units as well as neighbouring works. This idea of using sloped pipes even led the Perret brothers, when they built their Chessy plant in the Monts du Lyonnais, to consider building a genuine lead pipeline running for several kilometres down the Lyon plain. In most cases, however, the acids were handled and transported only by means of hand-carried carboys. For this reason such basic problems as storage were complex. It was not until 1832 that Gay-Lussac, technical adviser at Saint Gobain, had the first acid tanks built of lead. Carboys were still used for carrying the acids. And it was not until 1873 that Frédéric Kuhlmann's son designed the first tanker for transporting sulfuric acid on the canals in the north of France.

The development of chemical cascades, beginning in 1842, meant that the acids had to be moved to heights of several metres in the supply tanks. Conventional pumps could not be used, because their iron or copper parts were quickly corroded. Instead a method that had long been used in the sugar industry, called "monte-jus" (because of the steam pressure, and literally meaning "juice pump") was used. There was difficulty in moving cold acid upwards in this manner since the steam condensed, causing a drop instead of a rise in pressure. Also, the temperature of the acids rose from the effect of the steam. This was not of great importance for the Gay-Lussac towers, but for the Glover towers it was preferable to have cold acids. That is why compressed air was increasingly used from the 1870s on. These monte-jus, which were used for the first time in France in 1832 by Charles Kestner at his Thann works, were developed into automatic returns by Cotelle in his Javel works in 1868.

Chemical Cascades and Gas Purification

The producers of sulfuric acid very quickly learned from experience that for complete reactions in the lead chambers, there had to be an excess of nitrogen oxides in the last chamber and the fumes issuing from the system had to be red in color. The check on whether the operation was going well was based on the brilliance of what was called at the time the "glowing red gases" coming out of the chimney. The result was excessive consumption of nitrates and the expulsion at the bottom of the chamber of strongly polluting fumes that produced great damage around the plants. All this was very costly both because the price of nitrates was very high and because the neighbours brought suit after suit against the producers. Thus the problem of industrial gas purification by the system of chemical cascades was first considered. In this area, Clément-Desormes seems to have been an innovator, for Payen reported in an article in the Revue d'Histoire des Sciences, back in 1819, that Clément-Desormes had developed an apparatus of that type for purifying illuminating gas and perhaps even for producing chlorine.

Despite Clément-Desorme's claim, genuine industrial use of

his process does not seem to have been achieved while, between 1825 and 1832, he was manager of the Saint Gobain company. A true industrial application of the chemical cascade was developed by Guy-Lussac between 1835 and 1842 in the Saint Gobain plants at Chauny for the recovery of nitrogen oxides. Not only were these gases recovered, but they were also recycled back into the chambers. The Gay-Lussac equipment was actually in two parts. The first was a true chemical cascade whose operation was based on the ability of nitrogen oxides to dissolve in concentrated sulfuric acid. A column lined with lead and stacked with coke was used. The gases were introduced at the bottom while the concentrated acid was trickled down from the top. The liquid recovered at the bottom of the column was used in the second part of the system, at the inlet of the first lead chamber. The acid, subjected to a strong injection of steam, released the nitrous gases which were passed back into the chamber.

The Guy-Lussac tower, despite its advantages, was slow to take hold in France. Saint Gobain, being very anxious to maintain its technical lead, assigned its patent very sparingly and only to companies too far away to be able to offer real competition. However, even when the Gay-Lussac patents lapsed into the public domain, many companies did not employ it because the tower required the use of concentrated sulfuric acid of which the production was costly and the scale small. Some competing companies attempted to use more traditional means; the famous Javel works, for example, instituted a system for the recovery of gases by arranging, at the chamber outlets, batteries of stoneware receptacles where the acid was concentrated. Among Saint Gobain's competitors, only Kuhlmann seemed to have found an original solution to the problem of nitrogen oxides. Unable to use the Gay-Lussac tower, Kuhlmann, known for his work on nitrification, tried to use the final gaseous products of the chambers to saturate aqueous ammonia flowing in coke-packed columns in order to obtain sal ammoniac. In the end, however, the Gay-Lussac tower won out. Kuhlmann himself adopted it about 1850. It appeared in 1859 at Montpellier. However, as late as 1879, the plants in the Marseille region still did not have it, whereas it had been proved that with this tower, the yield of the chambers was increased by one quarter. The Gay-Lussac tower was not really adopted until the Glover tower appeared and replaced the second stage of Gay-Lussac's system at the top of the chamber.

CONCLUSION

The decade from 1880 to 1890 saw the beginning of profound changes in the evolution of chemical engineering in France, resulting from technical developments in the chemical industry. The triumph of the Solvay process for soda, the rapid growth of the production of artificial fertilizers, and the birth of electrochemistry indeed marked a new stage. Another factor in the changes was the general growth of the industry which would henceforth be the driving force of the French economy. Metallurgy, the mechanical industries, and capital goods production in general were now capable of supplying what was needed for mass production. A new phase was therefore beginning: the twentieth century had arrived with its accelerated scientific and technical progress.

This paper, which was intended to present a kind of early history of chemical engineering, is concluded with a few remarks. The progress of chemical engineering was limited because in France the chemical industry developed in a fairly restricted technical framework, that of the production of acids and soda products by the LeBlanc process and lead chambers. Progress was limited also because the French industry of the nineteenth century did not master all of the techniques of the Industrial Revolution.

It would be wrong, however, to minimize the importance of that initial period in which the modern chemical industry was born. The French producers, at least the more clever among them, managed to find solutions well adapted to their technical means and their needs.

REFERENCES

Carnets inédits de Jules Ollivier
Archives de la Société SAINT GOBAIN-PONT-à-MOUSSON
Barreswill et Girard, "Dictionnaire de Chimie Industrielle", Paris, 1861-63
Dictionnaire Technologique: 1822-1835
Payen (Anselme), "Précis de Chimie Industrielle", 1849.

EARLY CHEMICAL ENGINEERING EDUCATION IN LONDON AND SCOTLAND

S. R. Tailby

Department of Chemical Engineering
University of Surrey
Guildford GU2 5XH England

ABSTRACT

Historical Background
Chemical Engineering Education in London

The work of Charles Graham and Watson Smith at University College 1878-1894 and the later development of chemical engineering at that college.

Professor H E Armstrong at the City and Guilds College (1880-1913). J W Hinchley at Battersea Polytechnic (now the University of Surrey) 1909-1917 and Imperial College (1910-1931) and the teaching of the subject at these colleges. The Diploma course at King's College, London (1928-1966).

Chemical Engineering Education in Scotland

The departments of Applied Chemistry and Chemical Engineering in the University of Glasgow and the Royal Technical College (now the University of Strathclyde).

The Chair of Technology in Edinburgh University (1855-1859).

The syllabi of these pre-1939 courses are collected together for the first time and given in appendices so that the narrative is not unduly interrupted.

HISTORICAL BACKGROUND

Let us begin by quoting a distinguished American chemical engineer, Warren K Lewis,[1].

> "*In 1887 Davis gave a course of lectures on chemical engineering at the Manchester Technical School; in 1901-1902 his lecture notes were expanded and published as 'A Handbook of Chemical Engineering'. Those Manchester Lectures, thirty years before the coining of the term, presented the essential concept of 'unit operations' and particularly an understanding of its value for education. Furthermore, insofar as the resources of the period allowed, the concept was developed quantatively.*"

The life and times of Davis, his attempts to form a Society of Chemical Engineers in 1880 and his secretaryship of the resulting Society of Chemical Industry in 1881 have been dealt with by Freshwater,[2] Underwood[3] and Johnson[4] and it is not proposed to detail them here.

Two questions spring to mind (i) Was there any significance in the date (1887) when Davis originated the idea of chemical engineering ? (ii) Why was the subject so slow in developing ? We cannot answer these questions or appreciate the significance of the early courses in London and Scotland unless we look briefly at the chemical industry and the state of knowledge of chemistry, engineering and education in the 1880's - the time of Davis. We must travel back 150 years or more and try to recapture the Zeitgeist - the spirit of the age.

The Chemical Industry

The industrial revolution dates from about 1760 with the rise of the cotton industry. In 1770 Arkwright's mill in Derbyshire employed 300, by 1820 there were 110,000 operatives in spinning mills[5]. The location of the mills was determined by the existence of water power and they were built in remote areas in Derbyshire, Nottinghamshire, Perth, Ayrshire and Wales. The gunpowder factory at Waltham Abbey in Essex was driven by water power (see photograph) in 1735. The real impetus for change was given by the invention of the steam engine by Newcomen (1712) and Watt (1781) and steam power was harnessed to a cotton mill at Papplewick, Notts, in 1785. By 1800 there were 89 steam engines in Lancashire, Yorkshire and Cheshire[6].

When a steam engine was attached to wooden machinery it shook it to pieces and an iron system was required. But the iron industry was ready. In 1740 annual output of pig iron was 17,350 tons

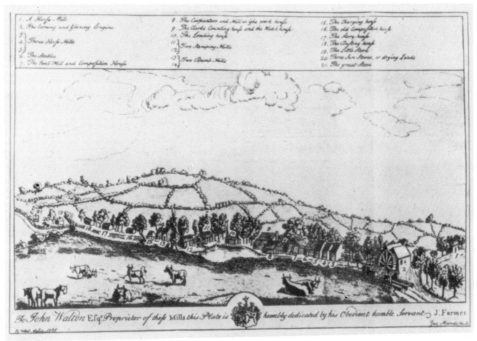

GENERAL VIEW OF THE FACTORY IN 1735
(FROM FARMERS "HISTORY")

Smoke over Leblanc Widnes

In 1850 Widnes was a rural village of 2,000 inhabitants. During the next few decades it became a leading centre for the manufacture of soda by the Leblanc process. At the peak of the development of the Leblanc industry more than a million tons of coal were consumed annually in the chemical works of the town.

produced by 59 furnaces (294 tons per furnace). By 1827 this had risen to 690,000 tons produced by 284 furnaces (2429 tons per furnace) and by 1850 a furnace could produce 6240 tons of pig iron per annum [7].

Now a chain reaction had begun. With steam driven iron machinery production could proceed apace but only as far as the bleaching process. This was the bottleneck. Bleaching was originally a cottage industry in which linen was spread out in the fields ('grassing') and treated with buttermilk ('souring') the treatment taking up to three months for cotton and six for linen. As production increased hundreds of acres were devoted to bleaching and the modern chemical industry was born in Britain in the wake of the mechanisation of cotton production, producing bleaching agents to do the job in factory production style in a matter of hours [8], and liberating land for agricultural use.

Sulphuric acid produced by Roebuck's lead chambers in Birmingham since 1746 [9], replaced buttermilk for 'souring' and by 1800 Britain was exporting 2000 tons of sulphuric acid yearly. Bleaching with chlorine was carried out in Glasgow in 1789 and Tennant absorbed the gas on lime to produce bleaching powder (1799) in his factory at St. Rollux. The textile industry required soap which had to be made from soda. This in turn was made from salt by the Le Blanc (Black Ash) process which released hydrochloric acid to pollute the air (see photograph) until the passing of the Alkali Act in 1863 [10], and overwhelmed acre after acre of the countryside with forlorn smelly hills of sulphide [11,12].

By the time of Davis there had been spectacular growth in the heavy chemical industry. In 1876 Britain exported 272,800 tons of alkali, worth more than £2m. Exports penetrated deep into Europe and even the U.S.A.[13]. Tennant's factory in Glasgow producing bleaching powder, sulphuric acid, soda and soap was one of the largest chemical works in Europe. It occupied ten acres and had a chimney 455 feet high. The Solvay (ammonia-soda) process was first operated by Mond in 1874 but electrolytic caustic did not appear until 1894. Here was the transition point from the old (Le Blanc) to current methods of manufacture.

In the years before 1850 the organic chemical industry could scarcely be said to exist [14]. The first public display of gas lighting was made in 1802, the Gas Light and Coke Company was established in 1812 and within ten years most cities had gas works. Mansfield obtained benzene from tar in 1849 and even in the 1860's benzene, toluene, etc. were looked upon as solvents rather than materials for the synthesis of new chemicals. In 1853 W H Perkin (Snr) accidentally discovered aniline purple ('mauve') and in 1857 set up a dye-works at the age of 18. Peter Griess discovered the diazonium compounds in 1858-64 developing a wide range of dyes by

1877. The greatest achievement was the synthesis of indigo by Baeyer in 1880. Unfortunately the dyestuffs industry passed later from England to Germany[15].

Chemical Knowledge

Dalton's atomic theory dates from 1807 but it was Berzelius who helped to get the theory accepted by determining the atomic weights of many elements from 1810 onwards. Even so, chemical ideas were so confused that even in 1865 the formula H_2O, for water instead of HO was by no means universal. The benzene ring was postulated by Kekulé as late as 1865. The chemistry of iron making was unknown before 1850 and none of the great London brewers knew the scientific nature of fermentation[17]. A Government committee in the 1860's found that the boards of chemical companies were uninterested in science and unable to think in scientific terms[18].

Engineering Knowledge

The Institution of Civil Engineers was formed in 1818 superceding the earlier Society of Civil Engineers formed in 1771. At that time there were only two classes of engineers, military and civil, the latter included people like Newcomen, Boulton and Watt who were concerned with steam engines[19]. The Bessemer process for making cheap steel was only developed in 1856 and so cast and wrought iron were the materials of construction. Steel production only got underway in the 1870's[20], (see graph 1). The electric motor and dynamo date from the 1830's and the I.E.E. was formed as late as 1871.

This was the great railway age. Stephenson patented his first engine in 1815. In 1832 there were 166 miles of track which increased to 6,802 miles by 1851 [21]. By 1850 the railways were the biggest single market for the iron industry and through that for coal. It was this combination of accessible coal and iron ore and engineering skill that gave Britain a lead over most other countries.

Victoria came to the throne in 1837 and married her cousin Albert of Saxe-Coburg in 1840. The Prince Consort was a science enthusiast and set up a Royal Commission with himself as president to launch the Great International Exhibition of 1851 in London to show British manufacturing supremacy to the world [22].

Education

The industrial revolution was not the deliberate outcome of an application of scientific knowledge nor a product of the formal educational system of the country but of techniques applied in an ad hoc manner. The Crystal Palace of the 1851 Exhibition was the idea of the head gardener of the Duke of Devonshire ! Spending on public education was almost zero [23] and in 1841 a third of the men and nearly half of the women signed their marriage register with a mark [24]. England was the most prosperous but the most illiterate nation in Europe !

Never was education more desired and sought after. For the common man the wonders of his world lay in the canals which climbed the Pennines, the flying coaches between London and the North, and the China clippers which covered 15,000 miles from Canton to the Channel in 109 days. It was a revolution that took place in the space of thirty or forty years, within the lifetime of thousands of people. There was an almost religious intensity of belief and faith that the accumulation of knowledge and scientific understanding would bring to the nation and the individual wealth, success and happiness. Mechanics in their Institutes, and the nobility at the Royal Institution, were moved by the same urgent sense of intellectual discovery, by a common feeling that they stood on the threshold of a new world. In these years science was woven into the fabric of the nation's life [6].

Oxford and Cambridge were *"places where the youth of the upper class prolong to a very great age their school education ... they were in fact still schools"*. The scientific revolution had occurred not through, but in spite of, the English Universities [24]. On all sides there were complaints about the *"wretched state of depression of the sciences and arts of England"* [25]. Science interests were catered for by societies such as the (Royal) Society of Arts formed in 1754, the British Association (1831) leading to the formation of the Chemical Society (1841), and a host of others [19]. The only widespread educational response coming in the wake of industry was the formation of the Mechanics Institute founded by George Birkbeck in 1824. They were self financing and within twenty-five years there were 610 institutes with more than 100,000 members [26]. They trained the NCO's of industry, the artisan anxious to improve himself into a foreman or master craftsman. In no sense were they the context from which might spring a technical university system as in continental countries. They withered away after the 1850's but many colleges of technology (e.g. Glasgow, Heriot-Watt, Manchester) rose from their ashes [24]. In Britain engineering was by tradition a craft, to be learned as an apprentice in the factory while in other European countries, and in Germany in particular, the engineer was a professional man with a long training in the basic sciences and technology behind him.

Empiricism cannot last forever and Nemesis was not long in arriving. Visitors to the Paris exhibition of 1867 found that Britain's industrial supremacy had vanished and that continental countries and the U.S.A. had become very serious competitors [27]. A famous letter to 'The Times' (see Appendix 1) written by Professor

EARLY EDUCATION IN BRITAIN

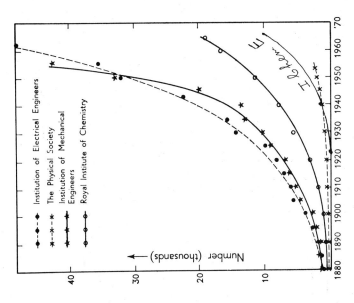

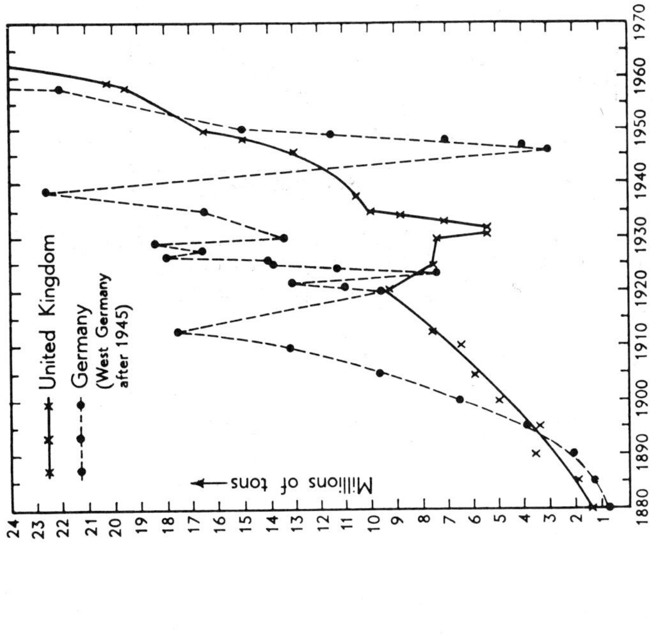

Graph 1. Production of steel ingots and castings in Great Britain and Germany, 1880–1965.

Graph 2. The rise of the professional in science and technology: membership of leading professional institutions, 1880–1965.

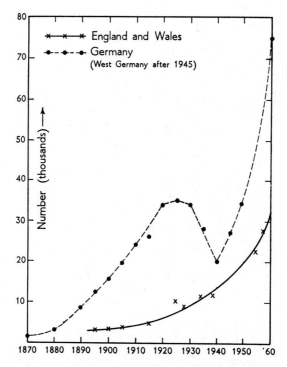

Graph 3. Students of science and technology at University: (a) England and Wales, and (b) Germany, 1870-1965.

(later Baron) Lyon Playfair [28], caused the Government to set up a select Committee to "*inquire into the Provisions for giving Instruction in Theoretical and Applied Science to the Industrial Classes*". A national crisis was discovered in the 1870's exactly parallel to the one which has caused the publication of the Finniston Report [29] in Great Britain in 1980. In 1872 only twelve persons were reading for the natural sciences tripos at Cambridge, most of them as part of a medical course. By then eleven technical universities and twenty other universities existed in Germany. The first grant of public money to support scientific education came in 1890 but by 1901 only £25,000 per annum was going from the Exchequer into British Universities. The very small numbers of scientific personnel at the time of Davis in 1887 is clearly shown in graphs 2 and 3. To quote from Finniston:

> "*Ours is the latest in a long line of official inquiries covering different aspects of the same remit. All of these reports made many practical recommendations for changes, yet none has succeeded in initially significant improvements in the engineering performance of industry. Many of the recommendations of these earlier reports ... even as far back as Balfour and Lord Playfair - are valid still ... and one speculates with a sense of regret what improvement in our economic situation might have been achieved if more effective action had been taken on implementing the recommendations of these earlier reports at the time.*" [29].

The ghost of 1870 is still with us.

SUMMARY

Our two questions can now be answered. Great ideas can only be brought to fruition if they are born at the right time and place, and we see that 1887 was about the earliest time that the basic ideas of chemical engineering could have evolved. At that time both the inorganic and organic chemical industries had simultaneously reached their transition points from the old order to the new and chemical theory had just stabilised. On the engineering side steel production only got under way in the 1870's (graph 1) and electrical engineering was so late in developing that the IEE was only formed in 1871.

The subject developed so slowly [3], because of the small number of professional scientific personnel (graph 2) and the poor opportunities for technological education compared with the Continent (graph 3). This slow development is clearly shown when we examine the growth of chemical engineering education in London and Scotland.

CHEMICAL ENGINEERING EDUCATION IN LONDON

University College London

In the early nineteenth century young men who wished to study at a university had to declare their religious faith according to the Church of England before they were admitted to Oxford or could take a degree at Cambridge. For those unable to declare that faith, there was no other university in England to which they could apply for admission. This situation disturbed an influential group of men in London, amongst whom was the philosopher Jeremy Bentham and the liberal (Edinburgh born and educated) Chancellor Lord Brougham. London was one of the few capital cities without a university, and a *"Council of the University of London was set up in 1826"* and in 1828 the building of what was then called 'The University of London', now known as University College (UCL), was begun. The *"Godless College of Gower Street"* met with considerable opposition since it would admit people of all faiths or none at all [30]. A second (Anglican) College – Kings – was established in the Strand in 1829 and in 1836 a Royal Charter set up the University of London. From 1836-1856 the University of London, comprising the affiliated colleges and medical schools, was in fact a paper university since it issued degrees but did not teach.

Chemical Technology or Applied Chemistry deals with the chemistry of substances, beer, cement, etc. Chemical engineering deals with processes. It is difficult to deal with the chemistry of substances without describing their manufacture and thus Applied Chemistry courses invariably lead to the inclusion of some chemical engineering topics. In almost every case university chemical engineering courses started in chemistry departments and transferred later to the engineering faculty. One of the exceptions being the present chemical engineering department at Strathclyde University, grown out of the mechanical engineering department.

The first Professor of Chemistry at UCL was Dr. Edward Turner followed by T Graham in 1837. The UCL Calendar of 1877 shows the chemistry department under Professor Williamson FRS, and Assistant Professor Charles Graham DSc.

The UCL Calendar of 1878 introduces Chemical Technology under Graham, it follows the chemistry section and lists four courses.

        A – Chemistry of Brewing
        B – Chemistry of the Alkali Trade
        C – Soap, Glass, Pottery, Cements
        D – Agricultural Chemistry

```
            Course A - Mondays   - 4-5 p.m.)
            Course B - Thursdays - 4-5 p.m.)  1st term
            Course C - Tuesdays  - 4-5 p.m.)
            Course D - Thursdays - 4-5 p.m.)  2nd term
```

Fee for each course £2-2 or £5-5 for the four courses.

<u>Laboratory</u> - The instruction in the laboratory in Chemical Technology under the direction of the Professor will consist of the examination and valuation of raw materials used, and of the final products obtained, in various manufacturing industries, and of experimental examination of the processes employed in the art of manufacture. The laboratory is supplied with apparatus and utensils for such investigation.

(note the last three lines).

Fee for session 25 guineas, 6 months 18 guineas, 3 months 10 guineas, 1 month 4 guineas, exclusive of the expense of materials.

The course continued in 1879 but the four subjects were now:

            A - Heating and lighting
            B - Dyeing and calico printing
            C - Chemistry of Brewing
            D - Metallurgy

In 1880 the course continued with the addition of Distilling, Vinegar-making, Bread and Biscuit making.

In the Calendar for 1880 we find the course separated from the chemistry section. It is now under the 'Department of Applied Science and Technology' along with Engineering and Architecture and by 1882 it had become a proper 3 year course. See Appendix 2.

Here is the chemical industry in watertight compartments and yet in the third year of each course the students are in the 'Laboratory of Chemical Technology' and this is capable of *"making investigations on manufacturing processes"*. This occurs in the 5th year of no 7 (Consulting Chemists).

The course was continued in 1883 and 1884 but in the 1885 calendar there is no mention of the three year course but only the words:

"The Professor will gladly advise with each student at the beginning of his college career as to the subjects of study essential for the following pursuits:- metallurgists, alkali, soap and manure manufacturers, manufacturers of

University College as it actually appeared at the time of its opening in 1828.

M. B. Donald entered the Royal College of Science from Felsted School but was then unable to finish the chemistry course owing to being commissioned as a signal officer in an Artillery brigade. He was subsequently awarded a Sir Alfred Yarrow scholarship to take a course in the newly founded chemical engineering practice school at M.I.T.

glass, cement, artificial stone, etc. Bleachers, Distillers, Vinegar manufacturers, sugar refiners, Glucose and Dextrin manufacturers, etc. Agriculturalists, Consulting Chemists and Public Analysts."

The same entry appears in the Calendar of 1886, 1887, 1888. Professor Charles Graham retired in the session 1888-9 and Watson Smith, lecturer in Technological Chemistry at Owen's College, Manchester, was appointed to the UCL chemistry staff. He had received his early training under Sir Henry Roscoe at Owens, proceeding later to Heidelberg and then to Zurich where he worked under Lunge. He had worked in industry before joining the staff at Owens. He was editor of the J. Soc. Chem. Ind. for 32 years (1882-1914). He did pioneer research on the coking of coal.

In the 1889 Calendar we have the heading 'Applied Chemistry' Lecturer Watson Smith FIC, FCS, and a brief syllabus similar to the one in 1878.

The next year (Calendar 1890) appeared the entry for a College Certificate in Chemical Technology (appendix 3) and the same course appeared in the Calendars of 1891, 1892 and 1893.

There is no mention of the course in 1894 and, in fact, Watson Smith left UCL in that year. Under A-Manufacture of Sulphuric acid, etc. it says that the *"general principles of Chemical Engineering will be illustrated"*. Watson Smith left Manchester in 1889 and Davis was giving his lectures in Manchester in 1887 - we do not know if they ever met. Note that the course was extremely expensive - 100 guineas - which is probably why it only ran for four years.

Chemical engineering (so called) died in 1893 and we have to leap forward thirty years before it returned.

During the 1914-18 war research work on the Haber process was carried out in the UCL chemistry department. This work demonstrated the detailed and prolonged research required to translate a laboratory reaction into a practical mass production process and showed that the country was short of men suitably trained for such work. This realisation led to the founding of the Institution of Chemical Engineers in 1922.

At this time a fund of £50,000 was raised to provide a memorial to Sir William Ramsay, who had occupied the Chemistry Chair at the college from 1887 to 1913. It was decided to devote £27,000 of the memorial fund to the endowment of the first Chair of Chemical Engineering in England. Professor E.C. Williams was appointed to the Chair in 1923 and the first two year course for graduates in

Chemistry or Engineering was run in 1924. An extract from the College Calendar gives details (Appendix 4).

This was a course in the true Davis tradition. It contained not only unit operations, but production and use of energy, materials of construction, heat transmission and fluids, all of which were dealt with in the Davis Handbook [2]. Professor Williams left the department in 1928 to become Research Director for Shell in California. He was succeeded by Professor W E Gibbs, who died in 1934.

In 1931 Mr. (later Professor) M B Donald joined the department at the age of 34 and stayed until his retirement as Head of Department in 1965. He had obtained his ARCS at Imperial College and followed this with an SM at MIT. He worked with the Chilean Nitrate Producers Association from 1925-29 and with Shell from 1929-31. According to Donald [31]:

"There was a very clear distinction between the early British teaching in chemical engineering and the new approach developed at MIT".

and Donald with his experience of MIT was able to bring the British and American ideas together. The course in 1933, the last in the lifetime of Professor Gibbs, is given in Appendix 5. It shows the original course somewhat expanded.

Gibbs was succeeded in 1934 by Professor H E Watson and in 1938 a full Degree course in chemical engineering was begun. The Diploma course was still available to graduates in chemistry or engineering, most of the lectures being common with those of the final year of the degree course. The syllabus of the first degree course at UCL is given in Appendix 6.

The department was badly damaged during the war and Donald, as well as doing war research, assisted with the course at Imperial College in 1942. He became Professor and Head of Department in 1951 until his retirement in 1965, when he was succeeded by the present holder of the Chair, Professor P N Rowe, well known for his fluidisation research at AERE Harwell. Space precludes the giving of details of post-war courses.

Imperial College London

The Imperial College of Science and Technology, constituted by Royal Charter in 1907, is a federation of three institutions – the Royal College of Science, the Royal School of Mines and the City and Guilds College. The three colleges are close together in South Kensington on land purchased from the profits of the 1851 Exhibition – as proposed by the Prince Consort. Appendix 7 is a map from the 1910 Imperial College Calendar showing the arrangement of the buildings at that time.

The Central College as it was in 1885. From a photograph taken in 1902.

The three colleges came to South Kensington from other locations. The Royal College of Science had started life as the Royal College of Chemistry opened in 1845 with the Prince Consort as President with premises in Hanover Square. The Royal School of Mines had started life in 1841 as a museum of Economic Geology, and the City and Guilds College had its beginnings at Finsbury in the north of London. The Royal College of Science moved to South Kensington in 1872, the Royal School of Mines completed its move in 1891 and the City and Guilds College in 1884. The three colleges issued Diplomas ARSM and ACGI. Not until the establishment of Imperial College in 1907 was the federation formally admitted as a School of the University of London in 1908.

Of great interest to chemical engineers is the work of Henry Edward Armstrong (1848-1937), FRS, FRSE, Hon. FCGI, LLD, DSc., Professor of Chemistry at the City and Guilds College until 1912. His life is the subject of a book by Vargas Eyre [15], two memorial lectures by Sir Harold Hartley [32] and his early education work has been dealt with by Brock [33].

Armstrong entered the Royal College of Chemistry in 1865 where he came under the influence of Frankland. He went to Leipzig in 1867 to work under Kolbe who was busy attacking Kekulés ring formula for benzene put forward in 1865 and returned to England in 1870. In that year he was appointed to the Chair of Chemistry at the London Institution, Finsbury Circus - a Mechanics Institute erected in 1819. The City and Guilds of London Institute which owed its inception to the City Livery Companies decided to build a Central Technical College and in 1879 Armstrong was appointed as lecturer in Chemistry and Ayrton as lecturer in Electrical Engineering at Finsbury Technical College which opened in 1883. All students took a common first year course consisting of chemistry, mathematics, mechanical drawing, electrical and mechanical engineering and French or German.

Meanwhile the Central Technical College (later the City and Guilds Engineering College) was being planned for South Kensington and Armstrong was appointed to the Chair. The College opened in 1885 with three departments, Mechanical Engineering, Electrical Engineering and Chemical Engineering.

The prospectus and time table of this 1885 course is given in Appendix 8, and it is quite clear that although the course was called Chemical Engineering it was not chemical engineering but a mixture of mechanical engineering and chemistry; in fact the Finsbury mixture given at university level. Perhaps it should have been called 'Engineering Chemistry'. Without doubt Armstrong was a pioneer in technical education, he was trying to produce 'work chemists' with a broad training but he had no idea of the unit operations concept and his products were not chemical engineers.

Two years later, in 1887, the Executive Committee of the Institute resolved that the Diploma award of the Central Institution should be in Chemistry and the term 'chemical engineering' was dropped from the Calendar. Armstrong's department was closed down in 1912 and his students transferred to the course in the Royal College of Science when it was decided to devote the City and Guilds College entirely to engineering studies.

In the Imperial College Calendar of 1910-11 Armstrong's course is in the City and Guilds section under the heading 'Chemistry' and there is no mention of chemical engineering.

In 1908 Armstrong's department, in its ordinary chemistry courses, was teaching such subjects as fuels, analysis of coal and producer gas, flue gas analysis, oils and lubricants and electro-chemistry. One would have expected Chemical Technology to have developed in the engineering college rather than in the Royal College of Science Chemistry department which was devoted to pure science but the planners decided otherwise.

Armstrong was 74 when the Institution of Chemical Engineers was formed in 1922 and he was wont to complain that Davis, rather than he, was recognised as the pioneer. In the writer's view Armstrong, although the doyen of British chemists in his day, never really understood chemical engineering and this is proved in an article in 'The Central' [34], which he wrote as late as 1928.

"Chemical engineers are much in demand. We may raise a few by striving to teach engineers to be chemists; a larger proportion perhaps by teaching chemists engineering because many chemists are constructive in their outlook as well as analytical: but in neither case will the hybrid be really competent in both subjects; if we are wise, we shall follow the German example and manacle chemist with engineer, attuning the two to work in sympathetic vibration."

Chemical engineering was not to return to the City and Guilds College until 27 years later.

To trace the development of Chemical Engineering at Imperial College, we must turn to the Royal College of Science (R.C.S.)[35]. The Calendar for 1908-9 shows the R.C.S. Chemistry Department under Professor W A Tilden, FRS. The fourth year of the course was devoted to research and short courses of which there were four.

(a) Rusting and Corrosion of Metals (a course for Engineers);
(b) The manufacture of azo dyes;
(c) Problems bearing on the Economy of Fuel;
(d) Town gas manufacture and use.

The last course was given by H G Colman PhD formerly Chemist to the Gas Committee of Birmingham Corporation.

The Calendar for 1909-10 shows Professor Tilden as Emeritus and Professor Sir Edward Thorpe FRS in charge. In the fourth year there was a special course of 16 lectures on 'Gaseous Fuel and Combustion' given by Professor W A Bone DSc., FRS of the University of Leeds.

The Calendar for 1910-11 gives the same syllabus for Professor Bone's lectures but has the entry

CHEMICAL ENGINEERING

Arrangements have been made for a special Fourth Year Course on 'The Design of Plant required for Chemical Engineering' the course will be given during session 1910-11, and full particulars will be announced as soon as possible. Apply to the Secretary for further information.

The Calendar for 1911-12 shows that Bone's lectures were continuing but we also have:

"A course of instruction including Lectures, Laboratory Work, and Drawing will be given on THE DESIGN AND ERECTION OF CHEMICAL PLANT by J W Hinchley ARSM, WhS, FCS (Consulting Engineer and Technical Advisor)."

The course will occupy two whole days each week, Tuesdays and Fridays, 10 - 5 p.m. October-Easter, one day per week in the summer term. Fee for course £5. The syllabus for the course is given in Appendix 9.

In 1912 Professor Sir Edward Thorpe was succeeded by Professor Brereton Baker FRS, and Professor Bone left Leeds and was appointed Professor of Fuel and Refractory Materials. In the 1912-13 Calendar the Department of Chemical Technology appears for the first time as a sub-department of the Chemistry Department.

The Calendar for 1913-14 shows Chemical Technology as a completely independent department with W A Bone as Professor, M G Christie (Otto Coke Oven Coy) as assistant Professor, and J W Hinchley as lecturer in Chemical Engineering. The latter was still a consultant and part-time lecturer. A DIC (Diploma of Imperial College) was introduced as a two year postgraduate course for students who had completed the ARCS course in Chemistry. This course was a mixture of fuel technology and chemical engineering, the syllabus for the latter being almost identical with that given in Appendix 9.

This course continued in 1914, 1915 and 1916 with J W Hinchley being appointed Assistant Professor in October 1917. His appointment as full Professor of Chemical Engineering did not take place

EARLY EDUCATION IN BRITAIN

John William Hinchley, one of the founders of the Institution of Chemical Engineers, was born on 21 January, 1871. This illustration is from a pencil and wash drawing made by his wife.

D. M. Newitt, a Past President (1949-50), is one of the two surviving members of the Provisional Committee which formed the Institution in 1922.

His first experience as a chemical engineer was in 1912 when he and Hugh Griffith were members of a research team which, under Sir Frederick Nathan and William Rintoul, was largely engaged in trouble-shooting at the Ayrshire Works of Nobel's Explosives Co.

During the 1914–18 war he served with the Indian Army in the North-West Frontier, Mesopotamia and Palestine campaigns. Returning to England in 1919 he graduated in Chemistry at the Imperial College of Science and later studied under J. W. Hinchley and W. A. Bone in the Department of Chemical Technology. He was elected Fellow of the Royal Society in 1942.

After the second world war during which he served with S.O.E., he was appointed to the newly established Courtauld Chair of Chemical Engineering, an appointment he held until his retirement in 1962.

as full Professor of Chemical Engineering did not take place until 1926.

From 1919 until Hinchley's death in 1931 the DIC course continued with much the same structure. The chemical engineering syllabus expanded to include the heating and ventilation of buildings, size reduction and mixing, the Factory Acts and Legislation connected with the chemical industry, safety and rescue work, factory records, book-keeping and balance sheets with the introduction of Chemical Engineering Research (Calendar 1921-22). Hinchley's fully developed syllabus for the course given in the last year of his life appears in Appendix 10.

When Hinchley died in 1931 at the age of 60 years the Governors decided not to fill the vacant chair until the retirement of Dr. Bone in 1936 and S G M Ure - who had joined the department in 1921 - was appointed to a Readership in Chemical Engineering in 1932. Bone retired in 1936 and died two years later.

During the twenty years that Bone and Hinchley had worked together they had had many quarrels and difficulties [35]. They were both north countrymen who by their own efforts and abilities had risen to distinction in their own fields of work. In bone's view the keystone of Chemical Technology was fuel, with chemical engineering and other subjects as appendages. Hinchley, on the other hand, regarded fuel technology as a subsidiary to chemical engineering.

Hinchley was the first secretary of the Institution of Chemical Engineers and was instrumental in forming that body. Whilst he had progressive ideas on the teaching of chemical engineering he tended to exaggerate the scope of the subject and expected his students, to carry out many of the tasks which are now properly considered to be the province of civil, mechanical and electrical engineers. His ideas did not die with him and the early degree syllabi of London University were clearly influenced by the course structure that he built up. Details of the lives of both Hinchley [35,36,37], and Bone [35,38], have been published.

After the retirement of Bone, A C Egerton (late Sir Alfred), Reader in Thermodynamics at Oxford was appointed to the Chair of Chemical Technology and Headship of the department. D M Newitt (later Professor of Chemical Engineering), who had entered the department as a postgraduate student in 1921 was made Reader and Assistant Professor. It was decided to commence an undergraduate course in chemical engineering. The first students to the BSc course were admitted in October 1937 and the first BSc degrees in Chemical Engineering were granted in 1940. Details of this course are given in Appendix 11. In retrospect this first syllabus had serious defects. The students spent their first two years studying chemistry and mechanical and electrical engineering, they heard little

BATTERSEA POLYTECHNIC 1910

UNIVERSITY OF SURREY 1980

MR. HUGH GRIFFITHS

DR. FRANK RUMFORD

William Charles Peck studied chemical engineering at Battersea Polytechnic. In the early days of the Institution he was senior Examiner, for which he was awarded the Osborne Reynolds Medal in 1946. He was elected to the Senate of the University of London on the Board of Studies in engineering, science, and medicine.

of Chemical Engineering and had no contact with the department. The same pattern applied to the early degrees at all the London Colleges (U.C.L., Battersea, West Ham). In 1956 the London syllabus was completely revised to emphasise the essential unity of the parent technology.

The early course was given by assistant Professor S G M Ure[39] and J M Coulson (appointed Professor of Chemical Engineering, Kings College, Newcastle-upon-Tyne in 1952). The new course came under the Engineering Faculty and was transferred to the City and Guilds College in 1939.

During the war the staff consisted of only three members, Sir Alfred Egerton, Assistant Professor Ure, with J M Coulson as lecturer. The other members of staff were away on War Service. When Ure died in 1941 Mr M B Donald (later Professor Donald of University College) was released part-time from war work to give the chemical engineering lectures and this he did until he was able to return to his own department (U.C.L.) so badly damaged by enemy action.

On returning to College at the end of the war Newitt was appointed to the new Courtaulds Chair of Chemical Engineering in 1945, and, on the retirement of Sir Alfred Egerton in 1952 he became Head of Department. He was elected Fellow of the Royal Society in 1942 and was a Founder Member of the Institution of Chemical Engineers[40] becoming its President in 1949 and 1950. He long held the record for the largest fish (110 lbs) caught in the Tigris. He retired in 1961 and died in March 1980[41] aged 85.

Battersea Polytechnic (University of Surrey)

The London Parochial Charities Act of 1883 allowed the Charity Commissioners to apply certain old City charities for educational purposes and they proposed the establishment of several 'Polytechnic Institutes' throughout London. Some two miles south of Imperial College, on the other side of the Thames, Battersea Polytechnic was opened by the late King Edward VII, then Prince of Wales, in 1874 [42].

During its early years it was a school by day and conducted classes for adults during the evening. In 1900 five members of the staff, including the Principal, became Recognised Teachers of London University and by the end of its first decade the Polytechnic was giving complete courses in preparation for the Science and Engineering degrees of the University of London.

The work of the chemistry department became well known internationally and at one time both the Principal, Sir Robert Pickard, and the Head of Department, Dr. Joseph Kenyon were Fellows of the Royal Society.

As well as undergraduate work the chemistry department gave evening lectures in Technical Chemistry, including such subjects as fuel, oils, fats, soaps and candles; gas analysis; paper making and bacteriology. It was therefore only a short step to Chemical Engineering.

In 1909 J W Hinchley delivered the first evening lectures in chemical engineering at the college (details in Appendix 12), one year before beginning his course at Imperial College and the subject has been taught at Battersea and Surrey without interruption since that date. Thus it can be said to be the oldest school of Chemical Engineering (under that specific title) in Britain [4,37], if not the world.

In the Calendar of 1915-16 we see the expansion of the course (Appendix 13). Of great interest is the last sentence of the introduction.

"He (the student) *will be able to do things, not merely know how*".

The sting was in the tail ! Clearly the course had a different slant from the one that Hinchley was giving at Imperial College at the same time.

Hinchley gave up lecturing at Battersea in 1917 when he was made assistant Professor at Imperial College and was followed by Hugh Griffiths who became President [43] of the Institution of Chemical Engineers 1945-46. At the age of 15 he scored a success which still remains a record, winning in free competition, with no age limit set, a National Scholarship in Chemistry. In the following three years he obtained his ARCS and BSc (First class honours) at the Royal College of Science and then worked for Nobels Explosives Company. Subsequently he ran his own consulting business until his death in 1954. He bequeathed his technical library to the Polytechnic.

The syllabus of Hugh Griffiths' lectures is given in Appendix 14. These were the only chemical engineering evening lectures available in London or anywhere else in England at the time, and Griffiths held the attention of successive 'generations' of young men coming from all parts after a day's work in the chemical industries of the South. Students travelled great distances, from Essex and the coast towns and some moved their jobs from the North to London so that they could prepare for the AMIChemE examination. His insistence on detailed flow sheets to cover everything that went into a process - materials, energy, labour and money - and, equally important, where everything came out, was characteristic of his methods. A recent tribute [44] to the work of Hugh Griffiths has appeared from Mr W C Peck and Professor F Rumford, at present both active 80 year olds. Both Peck[45] and Rumford[46], have given details of the early courses. In the 1920's Hinchley and Griffiths set up the laboratory with the practical help of C R Chesterman who had joined the Polytechnic in 1920 at the age of 14 and who retired in 1966.

Griffiths gave up his lecturing in 1934 when Frank Rumford, a Battersea chemistry graduate, then working for the South Metropolitan Gas Company, took the lectures until 1936 when he became lecturer at the Royal Technical College, Glasgow. The present writer attended his evening lectures in Glasgow in 1940.

Mr W C Peck gave the lectures from 1936 until 1947 when they were taken over by the writer and continued until 1968 when the college moved out of London. Mr Peck had graduated in chemistry at Battersea and had worked with Burgoyne Burbridges and Co. at East Ham subsequently setting up his own firm, the Apex Construction Company. Later Polytechnic Calendars do not contain syllabi of the courses given by Rumford and Peck.

In 1945 at the request of the Ministry of Education the Polytechnic agreed to conduct a one year full-time course in Chemical Engineering for returning ex-service chemistry and engineering graduates. The writer was appointed as the first full-time lecturer in the subject in 1946, to be joined by Dr W J Thomas in 1949. The first undergraduates were enrolled in 1947 and these students obtained the Internal BSc of London University in 1950. Students took London University degrees until 1966 when the college received the Royal Charter to become the University of Surrey. In 1968 it moved 30 miles away to Guildford being unable to expand on its original site.

King's College London

Chemical Engineering is first mentioned in the KCL Calendar of 1928-29 when a course covering one session to be taken by students in the year following their Final BSc Examination in either Chemistry or Engineering was announced.

The detailed syllabus was given in the Calendar of 1930-31 and is given in Appendix 15. Lectures in Chemistry were provided for the graduate engineers and lectures in the engineering department were provided for the chemists. The Chemical Engineering work was done by Mr H W Cremer, the Senior Lecturer in Inorganic Chemistry.

Cremer had been an undergraduate at KCL obtaining first class honours and the Daniell Research Prize. He was twenty-one in 1914 and became sub-manager of the TNT factory at Queensferry and later he edited for publication the Technical Records of the Department of Explosives Supply. He returned to lecture at his old college until 1939 when he left to assist in the Department of Explosives filling factories. After the war he established his own consulting practice still operating successfully under the name of Cremer, Warner and Train. Cremer was at one time Secretary, Vice-President, Treasurer and President [57] of the Institution of Chemical Engineers and later President (1951-53) of the Royal Institute of Chemistry.

After the war the course was restarted and transferred to the Faculty of Engineering, Mr S B Watkins MSc, FRIC, AMIChemE being appointed in 1947. The syllabus given in the calendar of 1953-54 is virtually the same as that of the 1930-31 course (Appendix 15). Chemical Engineering became a separate department in 1956 when Professor Watkins was assisted by Dr D R Morris.

The one year course never developed into a degree course and chemical engineering ceased to exist at King's College after the death of Professor Watkins in 1966.

CHEMICAL ENGINEERING EDUCATION IN SCOTLAND

Historical Introduction

England and Scotland were two completely independent countries until the union of the crowns in 1603 when James VI of Scotland became James I of England. Real political union was not, however, achieved until the Act of Union was passed in 1707 in the reign of Queen Anne.

The effects of such a separation can be seen today in that Scotland has its own National Church (Presbyterian rather than Anglican), its own legal system (Romano-Germanic), and its own Educational System. English children usually take six or more subjects at age sixteen (O-level) and three subjects, e.g. Mathematics, Physics and Chemistry at age eighteen (A-level). With this early specialisation they can proceed to a three year degree. Scottish children take a wide variety of subjects at age seventeen ('Highers') and because their knowledge of science subjects is therefore not as deep, the Scottish universities provide a four year degree.

Until the founding of University College, London in 1826 England had only two universities, Oxford (1249) and Cambridge (beginning of the 13th century). At this time Scotland had five universities, i.e. St. Andrews (1410); Glasgow (1451); Edinburgh (1582); King's College, Aberdeen (1495) and Marischal College, Aberdeen (1593). The last two were united by Royal Ordinance to become the University of Aberdeen in 1860. Since 1960 the four Scottish universities have grown to eight by the creation of Dundee from St. Andrews, Strathclyde from the Royal College of Science and Technology, Glasgow, formerly the Royal

Technical College, Heriot-Watt from Heriot-Watt College, Edinburgh, and the building of an entirely new university in Stirling in 1967.

Lord Ashby [24] has pointed out that the scientific revolution had occurred not through, but in spite of, the English Universities and that the Scottish universities were more sensitive to the spirit of the age. Edinburgh and Glasgow had flourishing medical schools and chemistry was one of the first scientific subjects to be taught, so far as it was required in medicine. One reason for the vitality of the Scottish universities was that throughout the eighteenth century they remained in touch with scientific thought on the Continent. Copland, a professor at Marischal College, had visited Geneva in 1787 and had a demonstration of cloth bleaching [8] by chlorine and this process, introduced to Scotland by Copland and Watt was of major importance in setting up the British chemical industry.

The only Scottish chemical teaching outside Edinburgh and Glasgow during the industrial revolution was at Aberdeen where the lectures were directed towards the application of chemistry to arts, manufacturers and agriculture. In this way they were different from those in Glasgow and Edinburgh.

The Royal Technical College, Glasgow (University of Strathclyde)

John Anderson, Professor of Natural Philosophy in the University of Glasgow from 1757-96, organised, in the University, courses of lectures for local mechanics and artisans to deal with problems that cropped up in their everyday experience. The garb and bearing of those who came to his well-attended lectures as well as the fact that Anderson was a fervid supporter of the French Revolution was doubtless offensive to the authorities and led to many quarrels.

When Anderson died in 1797 he bequeathed his property exclusively to the foundation of an independent institution for applied science instruction. This was the first technical college in Britain, and in the world second only to the Ecole Polytechnique founded in 1794. This 'Anderson's Institution' became in 1828, 'Anderson's University'. In 1886 other institutions concerned with technical education were brought in to form the 'Glasgow and West of Scotland Technical College'. In 1912 H.M. King George V directed that it should be called 'The Royal Technical College, Glasgow'. In 1913 it was affiliated with the University of Glasgow for the purpose of degree courses only. In 1956 it became the Royal College of Science and Technology, Glasglow. It received a Royal Charter in 1964 becoming the University of Strathclyde.

The history of the college and its chemical activities have been given by Sexton [47], Cranston [48], Wood [49], Richie [50] and Cumming [51]. Some details are also given in references [8] and [12].

The first professor of Natural Philosophy (including Chemistry) was Thomas Garnett (1796-1799) who went to the Royal Institution (London). He was followed by George Birkbeck (1799-1804) who went to London to found the college that bears his name today. After Andrew Ure (1804-1830) came the first Professor of Chemistry (instead of Natural Philosophy). Thomas Graham (1830-1837), who left to become professor at University College London.

The Royal Technical College, Glasgow has been unique in having two schools of chemical engineering, one in the Science Faculty under the title of Applied Chemistry, and the other in the Faculty of Engineering. These must be described separately.

The Applied Chemistry Department

One of Graham's students at Glasgow was James ('Parrafin') Young whose patent of 1852 for the distillation of Shale led to the formation of the Scottish Shale industry. In 1870 Young, then President of Anderson's University, gave 10,000 guineas to endow a Chair of Technical Chemistry and further sums for a building and fittings. The terms of this endowment leave no doubt that had the title been better known, Young would have called it a Chair of Chemical Engineering. A complete record of Young and each of the Young Professors has been given by Cumming [51]. The first Young Professor was W H Perkin (Sir W H Perkin of aniline dye fame) (1870-71) followed by G Bischof (1871-75), E G Mills (1875-1903), Thomas Gray (1903-32), W M Cumming, OBE (1933-49) and P D Richie (1950-1970). The Applied Chemistry Department awarded an Associateship (ARTC) at the end of a four year course and from 1915 a BSc in Applied Chemistry (Glasgow University). The syllabus of the course for 1888 given by Cumming [37ii] had the title 'Chemical Engineering' (Appendix 16).

Thomas Gray was essentially a fuel technologist and was working on the subject at the same time as Bone at Imperial College; it was during his time that the RTC laboratories really developed, because the Glasgow sugar manufacturers presented the college with a complete plant to set up a course in sugar manufacture [37ii]. From 1908 until the 1930's once a year a 5 cwt batch of sugar was worked using clay and charcoal filtration, a double effect evaporator and crystalliser and centrifuge. R.T.C. certainly had the best chemical engineering laboratories in pre-war days [52].

When Dr. Rumford left the South Metropolitan Gas Works and teaching at Battersea Polytechnic in 1936 he joined Professor Cumming and introduced a more formal training in chemical engineering as shown in the two books that he wrote for the course [53]. The Glasgow University Calendars of 1927-28 and 1937-38 contain identical Regulations for the degree in applied chemistry, which

EARLY EDUCATION IN BRITAIN

are given in Appendix 17. The course continued through the last war with never less than 10 students but the long vacations disappeared.

The BSc in Chemical Engineering (Glasgow)

In its list of courses for the degree of BSc in Engineering, the Glasgow University Calendar of 1924 mentions Chemical Engineering for the first time. Details of the course are given in Appendix 18.

The first mention of the course in the R.T.C. Calendar is in 1926-27 and the earliest teacher was W A Milnes (lectured 1920-30), to be followed up by A W Scott who had been on the staff of the Mechanical Engineering Department and who was appointed to the first Chair in Chemical Engineering in 1956.

The first degree examination was held in 1928, this was the first BSc degree in Chemical Engineering in Britain, the first graduate being John Watson Napier who was at one time a Member of Council of the Institution of Chemical Engineers [54].

At this time J W Hinchley attended a Chemical Engineering Conference in the U.S.A. [55] and he was extremely critical of the Glasgow course [3],

"... the teaching omits all of those special processes and unit operations that are essentially a feature of chemical engineering."

The writer has a copy of the paper in Chemical Engineering Design set on 27 March 1928, and it contains numerical questions on (1) a steam accumulator, (2) the calculation of the number of plates in a fractionating column, (3) the calculation of the number of bubble caps in a column, (4) the design of a centrifuge basket, (5) a double effect evaporator, and (6) a battery of solvent extraction towers. Hinchley's criticism would seem hard to understand.

Justification for the Two Courses

Surrounded as the College was by the manufacturers of both chemicals and chemical plant it was not surprising that two separate courses should evolve. Historically, in London, Birmingham, Manchester and other centres Chemical Engineering courses have always grown from Chemistry Departments and from 1937 onwards the Applied Chemistry Department had at least 30 students in its final year of whom only 3 or 4 came from the Engineering Faculty. Up to 1938 there were only three graduates from the Engineering Faculty [56]. There were also five who gained the College Associateship - the BSc Course plus a ten day design problem.

Since the war the position has reversed. With the retirement of Professor Cumming in 1949 and the departure of Dr Rumford to set up the Indian Institute of Technology in Delhi in 1960, the chemical engineering content of the Applied Chemistry course has declined and Chemical Engineering as a separate Department in 1948 under Professor Scott and under his successor, Professor G S G Beveridge, has flourished. It is clear that one cannot justify the existence of two schools in one Institution.

The University of Edinburgh

One of the earliest chairs of chemistry was set up in the University of Edinburgh in 1713. Dr William Cullen started extra-mural science lectures in Glasgow in 1744 and was appointed to the Edinburgh Chair in 1756 leaving his former pupil and assistant Joseph Black (1728-99) in charge at Glasgow. In 1766 Black succeeded Cullen at Edinburgh where his work on the alkalis led to the isolation of 'fixed air' (CO_2). In 1785 there was a Chemical Society of Edinburgh [58], and both Cullen and Black, coming from industrial Glasgow, recognised the importance of scientific (as apart from medical) chemistry and its application to manufacturers and agriculture [59]. A Chair in Technological Chemistry might have been expected to be an offshoot of such a famous chemistry department but this did not happen.

The Chair of Technology arose out of a movement by the Senate in 1852 to transfer the Natural History collections amassed by Professor Jameson, second only to those of the British Museum, to the nation. It was proposed that the Government should take them over and build a National Museum of Natural History and Technology on an adjoining site. This gave rise to the 'Industrial Museum', the foundation stone being laid by Prince Albert in 1860. In the meantime George Wilson had been appointed to the Chair of Technology in 1855 to be Director of the Industrial Museum.

Wilson had more than 40 students for his first course. His syllabus extended over three years, the first year being devoted to Mineral, the second to Vegetable and the third to Animal Technology. His course included lectures on Fuel, Building Materials, Glass, Pottery, the working of metals, and what was then known of Electrical Engineering. He had an audience not of candidates for degrees but of persons desiring practical information for the needs of life [60].

Unfortunately George Wilson died in 1859 and the Chair in Technology was abolished. It is interesting to note that Wilson's course was in the Faculty of Arts as there was no Faculty of Science until 1896. There was no continuous history of courses in chemical technology, either at the Watt Institution (now Heriot-Watt University) or the University of Edinburgh.

Nearly a century later, in 1955, Professor K G Denbigh was appointed to a Chair in Technology which was created in the former department of Industrial Chemistry (students taking the first three years of a four-year Honours Chemistry course). This was a joint Chair from 1955/56 with Heriot-Watt College, in which it was called the Chair of Chemical Engineering. In the University of Edinburgh it was not called that for a few years since Professor Arnold was Professor of Engineering (i.e. all branches) and only on his death did separate Chairs (including Professor Denbigh's Chair of Chemical Engineering) become established for a range of engineering disciplines.

The Present Position

The foregoing history covers the period up to the Second World War of 1939. With the post-war decision to build oil refineries in Britain, Chemical Engineering Education developed rapidly and there are now twenty-two Universities and four Polytechnics offering degree courses. The content of these courses was summarised by the writer [61] in recent years and they are accredited by the Institution of Chemical Engineers in accordance with strictly laid down requirements [62].

If we are satisfied with the present rate of technical progress, let us not forget the efforts of the pioneers - described in this paper - to whom we owe so much.

ACKNOWLEDGEMENTS

The writer would like to thank the Royal Society and the University of Surrey for a travel grant and the many people who have provided information for this paper. Special thanks are due to Professor Sargent of Imperial College, and Mrs Tingree of the Archives Department of that College; Professor Rowe and Miss Valerie Potter of University College; Professor Beveridge and Dr. Hendry of Strathclyde University; Professor Rumford for information about R.T.C. Glasgow; Mr. Alex Anderson of Heriot-Watt University Library, and Mr Peter McIntyre of the University of Edinburgh.

APPENDIX 1

Letter from Dr Lyon Playfair to Lord Taunton
published in 'The Times', Wednesday, 29th May 1867

This famous letter is reproduced in full in reference [24], appendix page 111.

APPENDIX 2

UNIVERSITY COLLEGE LONDON 1882

CHEMICAL TECHNOLOGY

Professor:- CHARLES GRAHAM, D. Sc.

Assistant:- C.J. WILSON

The Course of instruction in this Department is designed to afford to students who propose to devote themselves to industrial pursuits in which Chemistry plays an important part, or to prepare themselves for the profession of Consulting Chemist, the instruction essential for their success in their future line of work.

Assuming that the student enters for three years' study, the following will give an idea of the nature of the work during the period:-

No. 1 - Metallurgists

First Year - Junior Mathematics. Experimental Physics and Mechanics. Theoretical Chemistry (Perpetual Ticket) Mechanical Drawing. Mechanical Engineering. Chemical Laboratory daily (Summer Term).

Second Year - Theoretical Chemistry, repeated. Mechanical Engineering. Chemical Laboratory (daily). Lectures on Chemical Technology. Economic Geology.

Third Year - Lectures on Chemical Technology. Laboratory of Chemical Technology (daily) with special instruction in wet and dry assays and in the examination of the raw materials used, and of the metals obtained.

No. 2 - Alkali, Soap and Manure Manufacturers.

A similar course to No. 1 but the third year laboratory devoted to examination of the materials used and the products obtained in the respective manufactures.

No. 3 - Manufacturers of Glass, Cement, Artificial Stone, etc.

A similar course to No. 2.

No. 4 - Bleachers, Dyers, Calico Printers.

A similar course to No. 2.

No. 5 - Brewers, Distillers, Vinegar Manufacturers, Sugar Refiners, Glucose and Dextrin Manufacturers.

A similar course to No. 2.

No. 6 - Agriculturists.

A similar course to No. 2 but including Economic Geology and Land Surveying.

No. 7 - Consulting Chemists and Public Analysts.

This was a five year course. The first year was the same as course No. 1. The other years were as follows:-

Second Year - Theoretical Chemistry, repeated. Physical Laboratory (daily during First Term). Experimental Physics and Mechanics. Chemical Laboratory (daily during Second and Third terms). Lectures on Mineralogy and Geology.

Third Year - Theoretical Chemistry, repeated. Chemical Laboratory (daily throughout the Session). Lectures on Botany (Summer Term).

Fourth Year - Lectures on Chemical Technology. Laboratory (daily throughout the Session), with special instruction in the commercial analysis and valuation of raw materials. Lectures on Physiology.

Fifth Year - Lectures on Chemical Technology. Lectures on Materia Medica and Therapeutics. Laboratory of Chemical Technology (daily throughout the Session), with special instruction in making investigations on manufacturing processes and in drawing up professional reports.

APPENDIX 3

UNIVERSITY COLLEGE LONDON 1893

CHEMICAL TECHNOLOGY

Lecturer:- WATSON SMITH, F.I.C., F.C.S.

A – Manufacture of Sulphuric Acid, Alkali, etc.

Monday, 10 to 11 A.M.   *First and Second Terms.*

The Sulphuric Acid and Alkali Manufactures: Bleaching-powder; Potassium Chlorate; and, should time permit, Utilisation of the Burnt Cuprous Pyrites left as a by-product of the Sulphuric Acid Manufacture.

In this Course the general principles of Chemical Engineering will be illustrated:

Fee:-   £2.2s.

B – (I.)   Fuel and Gas Manufacture

Friday, 10 to 11 A.M.   *First and Second Terms.*

The Chemical Technology of Fuel, including the Manufacture of Coke with recovery of by-products.   Gaseous Fuel; Water-gas.   The Recovery of By-Products from Coal-fed Blast Furnaces.   The Manufacture of Illuminating-Gas.

(II.)   The Chemical Technology of Building Materials.

Friday, 10 to 11 A.M.   *Third Term.*

Building Stones and Bricks, Firebricks, Mortars, Clays and Cements, Wood, its preservative treatment, etc.

Fee:-   £2.12.6d. for either course;   £4.4s. for both courses.

(III.)   Methods for the Technical Chemical Analysis of Raw and Manufactured Products.

One Lecture per fortnight extending over the First and Second Terms, at a time to be arranged.   The subsequent carrying out of certain of these methods in the Laboratory is desirable.

Fee:-   £2.2s.

C – Coal-Tar Products and Colours

Tuesday, 9 to 10 and Wednesday, 11 to 12 A.M.   *First and Second Terms.*

The distillation of Coal-Tar and the purification of the Coal-Tar products.   Manufacture of the Intermediate Derivatives, with the Chemistry and Chemical processes involved in the preparation of the Coal-Tar Dyes, Colouring matters, etc.,   Fee:- £3.3s.

Applications of Chemistry to Engineering.

The requirements of Mechanical and Civil Engineers, and also of Architects, are especially considered in Course B (I.) and (II.), and those of Chemical Engineers in Course A and Course B (III.).

Methods of Technical Chemical Analysis.

The requirements of Chemical Engineers and Technologists are especially considered in a Course B (III).

Excursions

Occasional visits to Works will be arranged during the session.

Composition fee for the three year course - 100 guineas.

APPENDIX 4

UNIVERSITY COLLEGE LONDON 1924

(abbreviated syllabus)

CHEMICAL ENGINEERING

Ramsay Professor,   E C WILLIAMS, M.Sc., F.I.C., M.I.Chem.E.

Assistant Lecturer, BURROWS MOORE, M.Sc., Ph.D., F.I.C.

The Ramsay Department of Chemical Engineering has been instituted with the object of enabling young graduates in Chemistry and Engineering, who have already obtained a good training in the fundamental sciences of Chemistry, Physics, and Mathematics, to direct their studies and investigations towards the application of the principles of Physical Chemistry to the scientific design and operation of the apparatus and processes of Chemical industry in general.

The Course extends over two Sessions.

FIRST YEAR

A.1   (Professor WILLIAMS)   Industrial Chemical Calculations.

First Term: Tuesday at 2.

A.2   (Professor WILLIAMS and Dr MOORE)   Heat Transmission and the Dynamics of Fluids.

First Term: Monday and Thursday at 2.

A.3   (Dr MOORE)   Principles of Mechanical and Structural Engineering.

First Term: Friday at 2.

A.4   (Dr MOORE)   Production and Distribution of Energy in Works.

Second Term: Monday and Thursday at 2.

A.5   (Professor WILLIAMS)
               The Design and Operation of Unit Types of Chemical Plants.

Second and Third Terms: Tuesday and Friday at 2.

A.6   (Dr MOORE)   Materials of Construction.

Third Term: Monday at 2.

SECOND YEAR

This will be devoted entirely, with the exception of occasional lectures on special topics, to original research under the supervision of the Professor or other members of staff. The work may be carried out in the laboratories, or when circumstances require it, wholly or partly at industrial works.

Special lectures will be arranged dealing with the lay-out of plants and factories, factory administration and industrial economics.

COMPOSITION FEE

(for full course):   Session, 30 guineas;   Term, 11 guineas.

RESEARCH FEES:

               Registration 1½ guineas

               Laboratory (according to the work proposed)

APPENDIX 5
UNIVERSITY COLLEGE LONDON 1933
CHEMICAL ENGINEERING

Ramsay Professor:	W E GIBBS, D.Sc., M.I.Chem.E.
Lecturers:	J P MULLEN, M.Eng., A.M.I.Chem.E.
	M B DONALD, M.Sc., A.R.C.Sc., M.I.Chem.E.
Hon. Lecturer:	A J V UNDERWOOD, D.Sc., M.I.Chem.E.
Hon. Assistant:	H W RICHARDS, B.Sc.

The Course extends over two Sessions.

A College Diploma in Chemical Engineering may be obtained by students who have spent not less than three terms in the Department. The Diploma will be awarded on the results of an examination to which only those students will be admitted who have pursued the Course to the satisfaction of the Professor. The fee for the Diploma Examination is 5 guineas.

FIRST YEAR

A.1 (Professor GIBBS)   Materials of Construction.

First Term: Monday and Thursday at 10.

A.2 (Mr MULLEN)   Transport of Materials.

First Term: Tuesday at 10, and Friday at 11.15.

A.3 (Mr DONALD)   Heat Transmission.

First Term: Friday at 10

A.4 (Mr MULLEN)   Mechanical Construction and arrangement of Chemical Plant.

Second Term: Tuesday at 10, and Friday at 11.15.

A.5 (Mr DONALD)   Fuels and their utilisation in Industry.

Second Term: Friday at 10.

A.6 (Mr MULLEN)   Production and Distribution of Energy in Works.

Third Term: Tuesday at 10, and Friday at 11.15

A.7	(Professor GIBBS)	Design and Operation of Unit Types of Chemical Plant.

Second and Third Terms: Monday and Thursday at 10

A.8	(Dr UNDERWOOD)	A Special Course in the Mathematical Treatment of Experimental Data.

Second Term: Tuesday at 11.15

A.9	(Mr RICHARDS)	Industrial Uses of Electrical Energy.

Third Term: Tuesday at 11.15

PROBLEM CLASS

(Mr DONALD)

A problem class is held on Wednesday mornings throughout the session. Students are trained to solve problems in the design and operation of chemical plant.

APPENDIX 6

UNIVERSITY COLLEGE LONDON CALENDAR 1937

B.Sc. (Eng.) Degree Courses in Chemical Engineering

INTERMEDIATE OR FIRST YEAR COURSE

Students take either the Intermediate Course in Engineering with the following subjects:

Mathematics	Physics
Applied Mathematics and Mechanics	Chemistry
Engineering Drawing and Graphics	

Or the Intermediate Course in Science, with the following subjects:-

Mathematics	Physics
Applied Mathematics	Chemistry

Students taking the Intermediate Course in Science must, before the end of their second session, satisfy the Examiners in the First Year Engineering Drawing and Design and in Practical Applied Mathematics, in addition to taking the usual Second Year Examination.

A student who has already passed the Intermediate Science Examination in the above subjects may enter for the Second Year Course in Chemical Engineering, but at the end of his first session he must satisfy the Examiners in First Year Engineering Drawing and Design and in Practical Applied Mathematics, in addition to taking the usual Second Year Examination.

SECOND YEAR COURSE

Mathematics
Strength and Elasticity of Materials
Theory of Heat Engines
Hydraulics

Electrical Technology
Theory and Design of Structures
Physical Chemistry
Organic Chemistry

With appropriate Laboratory work and Workshop training.

THIRD YEAR COURSE

Theory and Design of Structures
Applied Thermodynamics
Electrical Engineering
Mechanical Engineering Drawing

Inorganic Chemistry
Chemical Engineering Problems
Metallography

With appropriate Laboratory work.

FOURTH YEAR COURSE

Materials of Construction
Transport of Materials
Heat Transmission
Plant Design
Fuels and their Utilisation

Distribution of Energy in Works
Unit types of Chemical Plant
Chemical Engineering Problems
Economics

With work in the Laboratory and Drawing Office. Visits to factories will also be included.

POSTGRADUATE COURSES

DIPLOMA COURSE

Students who have already taken a degree in Chemistry, Engineering or Physics may be admitted to the College Diploma Course in Chemical Engineering. The subjects of this course are substantially the same as those of the fourth year of the Degree Course, but modified according to the previous training of the student.

ADVANCED WORK AND RESEARCH

On completion of the Degree or Diploma Course a student may be permitted to undertake research work in the Laboratory. He may proceed to the degree of Master of Science at the end of one session, or to the degree of Doctor of Philosophy at the end of

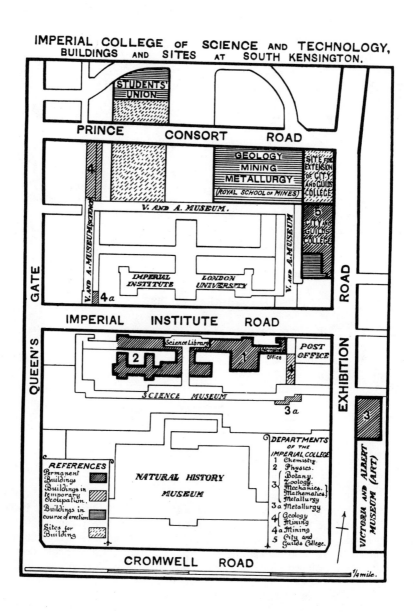

EARLY EDUCATION IN BRITAIN

two sessions under the appropriate regulations.

Senior students with adequate training, and experience, who are desirous of conducting original investigations may be admitted as research workers.

APPENDIX 8

The following is an extract from the original prospectus of the chemistry department, drawn up by Henry Armstrong and issued in January 1885.

> 'The course is arranged to suit the special requirements of those who will enter works where a knowledge of the principles, and the use of machinery, the strength of materials, building construction, etc., is of the greatest importance. Therefore a large amount of attention will be devoted during the first and second year to the study of engineering subjects.
>
> During the second year, students will be enabled to gain by their own work in the laboratory an intimate knowledge of chemical materials, and will become acquainted with the more important methods of analysis (mainly quantitative).
>
> In the third year, candidates for the diploma in Chemical Engineering, will acquire further practice as analysts, and will also devote attention to physical-chemical methods of enquiry, and they will attend one or more special courses on Applied Chemistry'.

The course during the first and second years is given as follows:

Chemical Engineering - *First Year*

Hour	Monday	Tuesday	Wednesday	Thursday	Friday
9.30-12	Mechanics Mathematics and Drawing	Mechanics and Mathematics		Mechanics and Mathematics	Mechanics Mathematics and Drawing
12-1		Chemistry and Physics	Chemistry and Physics	Chemistry and Physics	
1.30-2.30	Engineering Workshop or Laboratory of Mechanics				Engineering Workshop or Laboratory of Mechanics
2.30-4.00		Engineering Drawing		Engineering Drawing	
4-5	German				German

Chemical Engineering – *Second Year*

Hour	Monday	Tuesday	Wednesday	Thursday	Friday
10-12		Engineering Lecture	Mechanics and Mathematics		Engineering Lecture
12-1.30	Chemical Laboratory	Chemical Exercises F.C.L.	Chemical Lecture & Exercise	Chemical Laboratory	Chemical Lecture & Exercise
2-5		Chemical Laboratory	Engineering Laboratory		Engineering Design

APPENDIX 9

SYLLABUS OF HINCHLEY'S FIRST COURSE

at RCS. Calendar 1911-12

The course will deal with the economic aspect of designing plant for chemical work. Cost of production due to plant superintendence, materials and labour – Depreciation of plant – Essentials of good design; economic efficiency, ease of control, facilities for repairs, and low cost.

Preparation of specifications – estimation of costs. The testing of materials of construction from a chemical engineering point of view – Preparation of reports. The design of plant for the storage and transport of materials – The construction of tanks, bins, elevators, etc.

The arrangement of plant for crystallisation and filtration.

The design of plants for evaporation (a) by direct fire (b) by the circulation of hot fluids (c) by steam jackets (d) by steam coils – Calculations involved in the design of distilling, cooling and condensing plants – The design of multiple effect evaporators.

The design of an extraction plant of simple type.

The absorption and scrubbing of gases-scrubbers and mechanical washers.

The preparation and testing of cements and lutes for sealing and repairing chemical plant.

The study of the design of plants for typical chemical processes, the manufacture of acetone, nitric acid, sulphuric acid, hydrochloric acid and ammonia.

The design of a commercial plant from the experimental laboratory work of the student.

The planning of a simple factory.

The concurrent laboratory work will be based upon the lectures.

APPENDIX 10
(I.C. Calendar 1929-30)
HINCHLEY'S DEVELOPED COURSE
- the last in his lifetime

CHEMICAL ENGINEERING

Professor:- J W HINCHLEY, A.R.S.M., Wh.Sc., F.I.C. M.I.Chem.E.

Lecturer:- S G M URE, M.A., B.Sc., M.I.Chem.E.

Students who pass creditably through the first year course in Chemical Engineering, and who have had, subsequently, at least two years' practical experience of a kind satisfactory to the Institution of Chemical Engineers will be admitted to Associate Membership of that Institution.

CHEMICAL ENGINEERING, PART I

The lectures will be given on Mondays and Fridays at 10 a.m. throughout the session.

The *economic* aspects of chemical engineering. *Cost of production* due to plant, materials, superintendence and labour. Rent, interest on original cost and erection of plant, depreciation, repairs - items of increasing importance. Essentials of good design - economic efficiency, ease of control, facilities for repairs and low capital cost. Preparation of specifications. "General" and "complete" specifications. Commercial, legal and general clauses. Estimation of cost of plant.

Materials of construction. Iron and steel. Alloy steels and irons. Chemical cast iron. Lead, copper, tin, zinc, nickel, silver, etc., and alloys. Effect of heat on the physical and chemical properties of metals and alloys.

Timber. Varieties and properties of timber useful to the chemical engineer. Preparation and impregnation of timber for special purposes.

Cements and lutes. Preparation and testing of such materials for sealing and repairing chemical plant. Chemical pottery and stoneware. Brick construction. Stability of structures. Retaining walls. Foundations. Ferro-concrete construction.

Graphical methods of calculation. Heat transference calculations. Loss of heat from surfaces by radiation and air contact.

Estimation of the rate of transfer of heat through materials and surfaces. Co-efficients of heat transfer through metallic diaphragms from gas to liquids, steam to liquids, liquids to liquids, etc.

Heat insulation. Calculation of heat losses through composite diaphragms.

Size-reduction of *solid materials*. Crushing, grinding and disintegrating. Work done in size-reduction.

Estimation of output, horse-power required, etc.,

Wet and dry grinding.

Sifting machinery. A "sifting action". Estimation of size of sifting and dressing plant.

Mixing operations. Handling of plastic compositions, celluloid, rubber, margarine, etc.

The transport of solid materials. Elevators, conveyors, etc.,

Estimation of capacity, etc.

The transport of liquids and gases; pumps, acid elevators, etc.,

CHEMICAL ENGINEERING PART II

The lectures will be given on Wednesdays at 10 a.m. throughout the session.

Workshop practice and plant maintenance.

Design of "wet" and "dry" stills. Plant for evaporating, condensing and cooling. Estimation of sizes of heating elements, steam coils, etc.

Area of evaporating surfaces. Prevention of entrainment. Design of condensers.

Furnace and chimney design. Calculations relating to output, loss of heat, available temperatures, sizes of beds, muffles, flues, chimneys, etc. Friction in flues and conduits.

Dry distillation of materials. Heat transference.

Stirring gear. Dust and tar extraction.

Filtration. Water purification and filtration for chemical works. Slow, multiple, and rapid methods. Loose, compact, woven, and felted materials as filtering media. Types of filtering plant. The filter press-chamber and frame types. Relative utility of different types of filter. Leaf filters - latest types. Continuous mechanical filtering plant. Theory of filtration. Rate of filtration and washing. Filtration factors.

Pneumatic transport of materials.

Drying plants. The drying of gases on a large scale by refrigeration and chemical reagents.

Design of high pressure autoclaves, cylinders and plant.

Problems of power transmission.

CHEMICAL ENGINEERING PART III

This course of lectures will be given during the second year on Tuesdays and Thursdays at 10 a.m. throughout the session.

Thermodynamics.

The design of extraction plant. The design of crystallising plant.

Multiple effect evaporation. Corrosion.

The design of jet apparatus for the transport of gases, liquids and solids. The design of absorption plants. Refrigeration plants.

The theory of fractional distillation and design of column apparatus.

The planning and design of factories. Heating and ventilation of buildings.

Factory accounts. Factory records and book-keeping. Balance sheets. Flow sheets.

The Factory and Alkali Acts. Leglislation connected with the Chemical Industry.

DRAWING OFFICE and LABORATORY WORK

First-year Course. From the third week in November daily throughout the Session.

Second-year Course. Daily throughout the Session.

Determination of the co-efficients of heat transmission in chemical plant. Loss of heat by air contact and radiation.

Condensing and cooling factors.

Determination of rate of evaporation in crystallising and other plants.

Experimental study of units of plant. Reaction towers, fractionating columns, compressors, filters, grinding mills, sifters, dressers, etc.

Wet and dry still. Multiple evaporator, flow of gases and liquids through granular beds. Transport of gases, liquids and solids by jet apparatus.

Designs of simple units of plant. Estimation of cost and output. Energy balance sheets.

The working out of designs of commercial plant from the student's own notes and experimental work; drawing up of plans and specifications, bills of quantities and costs of erection, etc. The planning of factories.

CHEMICAL ENGINEERING RESEARCH

Courses of Instructions and Research in the Chemical Engineering of the oil, varnish, pigment, paint, rubber, plastic material, paper, sugar, starch and other industries will be arranged when convenient.

A P P E N D I X   11

First Undergraduate Course at Imperial College

(College Calendar 1939-40)

First Year Students would enter with the Inter B.Sc. (Eng) i.e. with Mathematics, Physics, and Engineering Drawing and Workshop.

Second Year Students take courses in Machine Drawing and

Workshop, Strength of Materials, Heat Engines, Mechanism and Hydraulics in the Department of Mechanical Engineering, in Chemistry in the Department of Chemistry in the Royal College of Science, and Mathematics and Mechanics in the Mathematical Department.

These courses consist of lectures and tutorial classes together with experimental work in the laboratories of the Department.

Third Year Students take advanced courses in the subjects they have studied in their second year and in addition attend a course of lectures, tutorial and laboratory work in Electrical Engineering in the Electrical Engineering Department. A course in Metallurgy is taken in the Metallurgical Department in the Royal School of Mines.

Fourth Year Students take courses in Fuel and Combustion, Applied Thermodynamics, Applied Electro-Chemistry, Materials of Construction (Non-Metals), Chemical Plant Design and Construction, Economics, and General Chemical Engineering in the Department of Chemical Technology. They also attend a course of Heat Transmission in the Department of Mechanical Engineering and a short course on the setting out of works in the Department of Civil Engineering.

The courses consist of lectures and tutorial classes together with experimental work in the laboratories of the Departments.

The syllabus in chemical engineering was similar to that given in Appendix 10.

A P P E N D I X   12

Entry in Battersea Polytechnic Calendar 1909

Hinchley's original lecture syllabus

(10) - Chemical Engineering

J. W. HINCHLEY, A.R.S.M., Wh. Sc.

LECTURES - TUESDAY, 7.15 to 8.30, commencing September 28th.

FEE - 10s.6d.

SYLLABUS - Twenty-Five Lectures - The Physical Treatment of Materials (Crushing, Grinding, Separating, Concentrating, etc.) Handling and Storing Solid Materials. The Flow of Liquids through Pipes. Storage of Liquids; Pumping and Elevating Liquids. Evaporation, Distillation, Fractionation, and Condensation of Liquids. Filtration and Crystallisation. The Flow of Gases through Pipes, Chimneys, Flues, Injectors, Exhausters, Compressors, Scrubbers. Autoclaves, Boilers; Crucible, Blast, and Reverberatory Furnaces. Fuels. The application of Electricity to Chemical operations.

The Properties of Materials (as Brickwork, Metals, Rubber, Wood, etc.) in relation to their use in the construction of Plant for Chemical Processes. The Designing and Planning of Works and Plant.

The Course will be found useful to Works Chemists, to Students taking the "Technological Chemistry" paper at the Institute of Chemistry Examination, and to industrial Chemists in general.

A P P E N D I X   13

Entry in the Battersea Polytechnic Calendar 1915-16

Hinchley's expanded evening syllabus

(1) - Chemical Engineering

J W HINCHLEY

FIRST YEAR COURSE

LECTURES - FRIDAY, 7.30 to 8.30
PRACTICAL WORK - THURSDAY, 7.0 to 10.0

SECOND YEAR COURSE

LECTURES - FRIDAY, 8.45 to 9.45
PRACTICAL WORK - THURSDAY, 7.0 to 10.0

Fee 15s. for each Course (Lectures with Laboratory Work) and a Deposit Fee of 5s. for Apparatus

These Courses will be found useful to those who are or may become Assistant Managers, Works Chemists, and Assistants to Industrial Chemists in general, and to Students taking the Technological Chemistry Paper at the Institute of Chemistry Final Examination.

During the first year the Student will determine co-efficients and factors of design; in the second year he will undertake a Course of practical design of Chemical Plant.

The Courses are designed to supplement the usual scientific training, and to develop that sound judgment, wider experience, and sense of proportion in practical matters which is essential to industrial success. In this Course, *e.g.* the Student becomes familiar by experiment with simple plant, with the range of accuracy of his calculations when designing factory plant. He will also develop an initiative for experimental work with crude apparatus (which in most cases is all that can be obtained in a factory), enabling him to tackle and solve problems which otherwise appear beyond his power. He will be able to *do* things, not merely know *how*.

SYLLABUS - *First Year* - Materials of Construction. Metals, Timber, Brick-work, Stoneware, Porcelain, Cements, Lutes, etc.

Plant for Crushing, Grinding, Sifting, Mixing, etc., Filtration and Crystallisation.

The Design of Evaporating, Distilling, and Condensing Plant. Fractional Distillation.

LABORATORY WORK (Mr HINCHLEY and Mr WOOD) - The Testing of Materials. Determination of Efficiency in Plant of Different Descriptions. Grinding Plant, Evaporating Plant, etc.

Determination of the Co-efficients relating to the Design of Plants. The Transmission of Heat. The Loss of Heat from Hot Surfaces by Air Conduction and Radiation. Rate of Evaporation of Liquids under different conditions. Filtration of liquids and gases. Experimental Testing of Insulating Materials.

Second Year - The Economics Aspect of Designing Plant for Chemical Factories. The Principles of Factory Organisation. Cost of Production due to Plant, Superintendence, Materials, and Labour. Essentials of Good Design.

Erection of Plant. Influence of Site, Climate, etc., Design of Foundations.

Plant for the Storage and Transport of Materials. Tanks, Bins, Elevators, etc., Determination of the Sizes of Pipework for the Transport of Liquids and Gases.

The absorption and Scrubbing of Gases. The Design of Scrubbers and Mechanical Washers.

Plants for Drying Materials.

The Design of Plant for Typical Chemical Processes. The Manufacture of Acids, Ammonia, Acetone, etc.

The Preparation of Specifications and Estimates.

The Principles of the Planning of Chemical Factories.

DRAWING OFFICE WORK - The design of simple parts of Plants. Tanks of Wood, Iron, Stone, Slate, Stoneware, etc., Evaporating Pans, Stills, etc. Furnaces and Chimneys. The Design of Plant from Data obtained by Laboratory Experiment.

APPENDIX 14

Entry in Battersea Polytechnic Calendar 1919-20

(I) - Chemical Engineering

H. GRIFFITHS

FIRST YEAR COURSE

LECTURES - FRIDAY, 7.15 to 8.15   -   September to Easter
PRACTICAL WORK - WEDNESDAY, 7.15 to 9.45   -   Commencing Wednesday, October 1st

SECOND YEAR COURSE

LECTURES - FRIDAY, 8.30 to 9.30   -   September to Easter
PRACTICAL WORK - WEDNESDAY, 7.15 to 9.45   -   Commencing Wednesday, October 1st.

Fee 15s. for each Course (Lectures and Laboratory Work), and a Deposit of 5s. for Apparatus

The Courses are intended for those who are or may become Assistant Managers or Works Chemists, but will also be found useful to Industrial Research Chemists.

SYLLABUS

A - LECTURE COURSE

The Lectures will deal with the principles involved in the Design, Erection, and Working of Modern Chemical Plant, and the Economic Aspect of Chemical Engineering.

Examples from practice will be given and discussed in details, as far as time will permit. The Lectures will be illustrated by diagrams and drawings of plant in actual use.

Selections will be made from the following, depending upon the requirements and experience of the students:-

1. Materials of Construction - Review of the Mechanical, Physical and Chemical Properties of the Constructional Materials used in Chemical Engineering.

2. General principles of Design. Limitations of Materials of Construction. Principles of Design. Determination and use of Experimental Data. Estimating in Chemical Engineering. Calculation of Working Costs.

3. Measurement. Methods of Measurement of Solids, Liquids, and Gases. Physical Measurement in Chemical Manufacture.

4. Fuel and Fuel Economy.

5. Water Supply.

6. Steam Raising.

7. Power Generation, Transmission, and Control.

8. The Application of Heat. Sources of Heat. Utilisation of Waste Heat. Application of Heat in Relation to the Construction of Furnaces, Dryers, Stills, Evaporators, etc.,

9. The removal of Heat. Refrigeration, Cooling, and Condensation.

10. Mechanical handling of Materials. Storage, Transport, and Control of Solids, Liquids, and Gases in Chemical Works.

11. Preparation of Materials. Grinding, Mixing, Agitating and Kneading.

12. Methods of effecting reactions. Discussion of the methods employed in effecting reactions between Materials by subjecting to various Physical conditions. Methods of Control.

13. Methods of separation: (a) Solids from Solids. Solution, Vaporisation, Crystallisation, etc., Screening, Sieving, etc. Levigation and Elutriation. Magnetic separation. (b) Solids from Liquids. Filtering, Pressing, Decanting, etc., (c) Solids from Gases. Gas Washing, Fume Precipitation, etc., (d) Liquids from Liquids. Settling, Centrifugal separation, etc., (e) Liquids from Gases. Drying of Gases. Tar removal, etc., (f) Gases from Gases. Absorption, Condensation, etc.,

14. Applications of Electricity. Electrical Furnaces and Electrolytic Plant. Electro-Chemical Calculations.

15. Buildings. Construction for Chemical Work, Heating and Ventilation.

16. Package. Methods of packing Chemical Products. Shipping in bulk.

17. Organisation. Sampling and testing in control of manufacture. Manufacturing books and reports. Costing systems, Stores systems, etc., Labour, Methods of payment. Accident prevention. Insurance. Welfare of workers.

Practical work. Students will determine the simpler types of Chemical Engineering Data, and will be required to derive these from experimental results in such a manner that an experience in Chemical Engineering calculation is acquired.

Certain portions of this work will be conducted in the Mechanical and Electrical Engineering Laboratories.

The advanced Course will deal with a selection of subjects from the Syllabus of the First Year Course, each being treated in a more comprehensive manner. A knowledge of Physics and higher Mathematics will be assumed, and students will be expected to be familiar with Engineering units.

Students who have not taken the First Year Course will only be accepted for these Lectures if they can show that they already possess sufficient knowledge to benefit by the Course. Lecture Course. The exact choice of subject matter for advanced treatment will depend upon the requirements of students who present themselves, but the following gives an indication of the scope of the Lectures. The subjects will in certain cases be taken as examples from practice and discussed in full detail.

1. The work of the Chemical Engineer. Cost of production in relation to Chemical Engineering: Estimating and Planning of work: Constructional Costs, etc.,

2. Power in Chemical Works. Special requirements of Chemical Works in respect of Power and Steam: Discussion of Steam-Raising Systems.

3. Examples of application of Chemical Engineering principles. Discussion of Design of various classes of Chemical Plant in detail.

PRACTICAL WORK. Practical Designing of Chemical Plant. Chemical Engineering calculation and Drawing Office Work. Application of Experimental data to Plant Design.

APPENDIX 15

KING'S COLLEGE, LONDON. CALENDAR 1930-31

For Chemistry Graduates

S1 Materials of Construction and Structural Design.

20 lectures and laboratory.

S2   Mechanics of Fluids and Theory of Machines.

        30 lectures and laboratory.

S3   Transmission of Heat and Heat Engines.

        30 lectures and laboratory.

S4   Electrical Technology.   Lectures and laboratory 40 hours.

S5   Engineering Drawing.   7 hours per week throughout the session.

S6   Workshop Practice.   3 week vacation course.

For Engineering Graduates

S7   General Theoretical Chemistry.

        40 lectures and laboratory.

S8   Principles of Organic Chemistry.

        30 lectures and laboratory.

S9   Chemistry of Metals.   20 lectures and laboratory.

For Chemistry and Engineering Graduates

S10 Physical and Reaction Treatment of Materials. (Mr H W Cremer). The principles governing the Design, Layout, and Operation of common types of Chemical Plant are considered. Examples in the design of typical units of plant are given, involving application of the principles dealt with in the other parts of the course, and including the preparation of quantitative Flow Sheets of Materials, Energy, and Time.

Conveyance and Storage of Materials - Weighting - Size Reduction; crushing, grinding, grading, and screening - Mechanical and Electrical Separation Processes - Filtration; theory of filtration; classification of filters; choice of filters; filter calculations; water purification and filtration - Leaching and dissolving - Basic principles of Vaporisation Processes - Evaporation; classification of evaporation processes; single and multiple effect vaporation; condensing and cooling factors; heating elements, entrainment; calculations for evaporator design - Drying; methods of drying; general principles of drying by vaporisation; drying apparatus and auxiliary equipment - Distillation; distillation processes; design of stills; destructive distillation - Characteristics of Crystallisation

Plant – Furnace and Kiln Design – Vapour and Gas Absorption Processes; scrubbing towers; solvent recovery; solid adsorbents – Electrolysis – Catalytic processes – Design of apparatus to withstand high pressures – Nitration, Sulphonation, Reduction, etc., mixing and agitation; heating and cooling factors; preparation of acid mixtures; spent acid treatment. (60 Lectures).

Laboratory – Determination of co-efficients of heat transmission – Determination of rate of evaporation – Quantitative study of units of Chemical Plant including crushing and grinding plant; mechanical separators; filters; stills and evaporators; reaction towers; nitrators, etc.

S11 General Principles of Organisation and Management in Chemical Works. (Mr H W Cremer)

Factory records and book-keeping – Cost of production due to plant, materials, labour, superintendence – Cost and Flow Sheets – Plant maintenance and depreciation – Health and safety regulations. (10 Lectures)

APPENDIX 16
Syllabus of the Glasgow and West of Scotland Technical College for 1888

Chemical Engineering

The Construction and Use of Chemical Plant

Steam. Water Purification

Refrigerators

Open and Closed Evaporation

Stills

Storing, Mixing, Grinding. Squeezing, Crushing, Precipitation, Floating, Washing, Filtering

Sublimation

Materials of Construction – Lectures illustrated by reference to Tar Distillation, Brewing, Spirit Manufacture, Gas Manufacture, Aerated Waters, Alkali Manufacture Purification, Boiling, and Storage of Vegetable and Animal Oils.

Thirty-six years later we find this syllabus in a more developed form.

R.T.C. Calendar 1924-5 (Technical Chemistry)

The Manufacturing Operations - Crushing and grinding. Methods of mixing and separating solids, solution and lixivation. Separation of solids from liquids by settling, filtration, vacuum filtration, and filter pressing; comparison of efficiencies of the three methods; hydraulic presses. Methods of mixing and separating liquids; different forms of mechanical stirrers; agitation by compressed air. Separation of gases; gas washers and scrubbers, construction and efficiency. Drying: types of hydro-extractor, drying by hot air and steam. Evaporation: overheat and underheat; evaporation by steam; multiple-effect evaporation. Distillation: different types of still. Crystallisation.

Materials of Construction - Timber, stone, bricks, clay and stoneware, glass, silica-ware, porcelain, mortars and cements, Portland cement. The metals: iron and steel, copper, lead, tin, zinc, nickel, aluminium and platinum.

Water for chemical purposes. Purification of water.

APPENDIX 17

B.Sc. in Applied Chemistry (Glasgow)
Abstract of Regulations 1927-28 (1937-38)

Candidates must attend at least <u>nine</u> courses (listed below). <u>Five</u> of the nine must be taken in the University of Glasgow, the other four could be taken in R.T.C.

1. Mathematics

2. Natural Philosophy (Applied Physics)

3. Chemistry (3 terms)

4. Advanced Inorganic and Physical Chemistry (3 terms)

5. Organic Chemistry (3 terms)

6. General Technical Chemistry or Chemical Engineering (1 term 30 hr/week)

7. Engineering Drawing (2 terms 6 hrs/week)

8. Advanced Practical Physics

9. A half course in Engineering (50 lectures General Engineering). A half course in <u>one</u> of (a) Fuels (b) Dyes (c) Oils and Fats (d) Sugar (e) Biochemistry (f) Technological Mycology (g) Metallurgical Chemistry (h) Coal Tar and Intermediate Products.

Suggested Course of Study

First Year — Mathematics; Natural Philosophy; Chemistry

Second Year — Organic Chemistry; Advanced Practical Physics; Engineering; Engineering Drawing

Third Year — Advanced Inorganic and Physical Chemistry; * Technical Chemistry I; Engineering Drawing

Fourth Year* — General Technical Chemistry with Special Courses.

The whole of the Fourth Year was spent at R.T.C. the students being in the Chemical Engineering Laboratory every day. They carried out a research project, but not a Design Project.

Professor Cumming kept the title "Applied Chemistry" with the Course in the Science Faculty so that his students could obtain the Associateship of the Institute of Chemistry (then A.I.C. now A.R.I.C.).

Students from the Chemical Engineering Degree Course (Appendix 18) came to the Applied Chemistry Course in the Fourth Year.

(Private Communication from Professor Rumford, May 1980)

A P P E N D I X   18

University of Glasgow (Ordinance No: 30)

(R.T.C. Calendar 1924-25)

DEGREES IN SCIENCE IN ENGINEERING

Subjects of study ... shall include the following:-

A (for all Branches)

---

* Held at R.T.C.

(i) Mathematics
Natural Philosophy (with practical work)
Chemistry (with practical work)

(ii) Higher Mathematics
General Engineering and Drawing

B (for particular Branches)

Branch (a) Civil Engineering .. .. subjects
" (b) Mechanical Engineering.. .. .. .. "
" (c) Electrical Engineering.. .. .. .. "
" (d) Naval Architecture .. .. "
" (e) Mining Engineering .. .. "
" (f) Chemical Engineering .. .. "

Higher Natural Philosophy

Applied Mechanics

Engineering Laboratory Work

Technical Chemistry with Laboratory Work

Engineering Design (including Design of Chemical Plant)

Heat Engines (Subsidiary Subject)

C (additional subjects, one at least to be selected)

Civil Engineering	Metallurgy
Heat Engines	Fuels
Electrical Engineering	Aeronautics
Naval Architecture	Advanced Mathematics
Astronomy and Geodesy	Engineering Economics
Geology and Mineralogy	Engineering Production

Technical Chemistry

The R.T.C. Calendar of 1936-37 contains the syllabus for Chemical Engineering Design

CHEMICAL ENGINEERING DESIGN

The Course deals with the design of plant in common use in the chemical industry. Heat transmission, application of theory to design of heat exchangers. Multiple-effect evaporation. Air humidity, rotary and spray drying plants. Distillation fractionating columns. Absorption towers. Centrifugal dryers. Steam accumulation. Autoclaves. Combined pressure and temperature stresses in cracking stills. Crushing and grinding

REFERENCES

1. W. K. Lewis, "Chemical Engineering Progress Symposium," Series 55, no. 26, p.1. (1959).

2. D. C. Freshwater, "George E. Davis, Norman Swindin and the empirical tradition in chemical engineering," Part 1 of this Symposium. Honolulu, (1979).

3. A. J. V. Underwood, "Chemical Engineering - Reflections and Recollections," Trans. Inst. Chem. Eng. 43, 302T (1965).

4. A. H. O. Johnson, "History and Development of Process Engineering," Chemistry and Industry 4th August (1979) p. 496.

5. P. Mathias, "The First Industrial Nation - an economic history of Britain 1700 - 1914," Methuen & Co. London (1969) p. 129.

6. J. H. Plumb, "England in the Eighteenth Century 1714 - 1815," Pelican Books (1950) p. 145.

7. G. Beecroft, "Companion to the Iron Trade," J. Y. Knight, 3rd edn. (1851).

8. For the development of bleaching see:-

    A. Clow and N. Clow, "The Chemical Revolution" Batchworth Press, London, (1952). Chapter IX.

    D. W. F. Hardie and J. Davidson Pratt, "A History of the Modern British Chemical Industry," Pergamon (1966) p. 30.

9. T. I. Williams, "The Chemical Industry," Pelican Books, (1953) Chapter 2.

10. J. E. Colehan, "The Centenary of the Alkali Act," The Chemical Engineer CE15 February 1964. I Chem E.

11. For the alkali industry see:-

    Reader, "Imperial Chemical Industries - a History," Vol. I Oxford University Press (1970), and 8 and 12.

12. L. F. Haber, "The Chemical Industry during the Nineteenth Century," Oxford University Press (1958).

13. Ref. 12 above p. 55.

14. F. Sherwood Taylor, "A History of Industrial Chemistry," Heinemann, London (1957). Chapter 16.

15. J. Vargas Eyre, "Henry Edward Armstrong 1848 - 1937," Butterworth (1958), p. 20 - 21 deals with the dyestuffs industry. On Perkin see H. F. Mark, Chem. and Ind. 5th July (1980) p. 533.

16. See Ref. 15 p. 30 and Ref. 14 p. 179.

17. See Ref. 5 p. 137.

18. See Ref. 12 p. 189.

19. Sir James Taylor, "The Scientific Community," Oxford University Press, (1973). "Sources of History Series," Macmillan, (1967).
    G. W. Roderick, "The Emergence of a Scientific Society in England 1800 - 1965."

20. For steel production statistics see Ref. 5 p. 410.

21. See Ref. 5 p. 280.

22. Much information about this exhibition is contained in the library of the Royal Society of Arts, John Adam St., London, W2N 6EZ.

23. See Ref. 5 p. 463. In 1840 - 49 expenditure was £0.3 m. representing 1% of net public expenditure, it reached 10% by the end of the century.

24. Lord Eric Ashby, "Technology and the Academics" Macmillan (1958) p. 50.

25. Charles Babbage FRS and Sir David Brewster in 1830 quoted in Ref. 12 p. 33.

26. This figure is given in 12 p. 34. 200 institutes with over 50,000 members in 20 years is given in Refs. 5 and 24.

27. On the Paris Exhibition see Refs. 24 p. 57 and 12 p. 74.

28. Playfair Lyon (1818 - 1898) had studied chemistry in Germany and was a follower of Liebig. In 1843 he was appointed a member of the Royal Commission on the Health of Towns. He was chemist to the Geological Survey in 1845 and to the Royal School of Mines in the 1850's. He was a member of the Prince Consort's organising committee for the 1851 Exhibition and Professor of Chemistry at Edinburgh in 1858. He became an MP in 1868 and was Post-Master General in 1873 - 74. Had the Prince Consort not have died in 1861 Playfair and his colleagues might have had more success.

29. "Engineering Our Future," Report of the Committee of Inquiry into the Engineering Profession. Chairman Sir Montague Finniston FRS, HMSO January, (1980).

30. See for example J. H. (later Cardinal) Newman's letters on "The Tamworth Reading Room," reprinted in J. H. Newman's "Discussions and Arguments" Pickering (1873) p. 254.
A pictorial history, "The World of University College, London 1828 - 1978," by Harte and North has recently been published and cannot be too highly recommended.

31. Letters written to F. R. Whitt on 28th August 1967 and 2nd January 1968 (Private communication).

32. "H. E. Armstrong and the Development of Organic Chemistry," Sir Harold Hartley. Chemistry and Industry, December 22nd and 29th, (1945) p. 398 and 406.

33. "An Experiment in Technical Education," W. H. Brook, New Scientist, 22nd November (1979), p.622.

34. "The Central," - the Journal of the Old Students of the Central Institution of the City and Guilds. Bound copies may be referred to in the Imperial College Archives. The passage occurs in the Armstrong Memorial Number Vol. 35, June (1938) p. 55. See also Ref. 3.

35. "History of the Department of Chemical Engineering and Chemical Technology 1912 - 1939," Compiled by Mrs M. de Reuch (née Gratwich). Imperial College Archives.

36. Hinchley's Obituary. Nature Vol. 128, 402, (1931).

37. F. R. Whitt, "Early Teachers and Teaching of Chemical Engineering,"
    (i) The Chemical Engineer CE 356 October, (1969), and
    (ii) The Chemical Engineer No. 254 p. 373 October, (1971).

38. Obituary. Gas Journal January, (1959) p. 100.

39. Ure died in 1941. Obituary. Nature Vol. 149, p. 133, (1942).

40. "The Foundations of the Institution," The Chemical Engineer, p. 135, April, (1972).
J. F. Davidson, "The Institution of Chemical Engineers," Chemistry and Industry p. 879, 4th July, (1970).

41. D. M. Newitt, Obituary, "The Chemical Engineer," p. 287 May, (1980).

42. A complete history of the Polytechnic is given in "Pioneering in Education for the Technologies," by H. Arrowsmith Published by the University of Surrey. See also D. M. A. Leggett, "Battersea College of Technology," Chemistry and Industry, 6th October, (1962) p. 1730 - 39.

43. The President. Trans. Inst. Chem. Eng. 23, XVIII, (1945).

44. W. C. Peck, and F. Rumford, Chemistry and Industry, 3rd November (1979), p. 748.

45. W. C. Peck, "Early Chemical Engineering," Chemistry and Industry, p. 511, 2nd June (1973).

46. F. Rumford, "Chemical Engineering at Battersea Polytechnic 1924 - 36," The Chemical Engineer CE 263, December (1967).

47. A. H. Sexton, "The First Technical College," Chapman and Hall, (1894), 176 pages.

48. J. A. Cranston, "The Royal Technical College, Glasgow," Journal of the Royal Institute of Chemistry, March (1954) p. 116 - 124.

49. C. G. Wood, "The Royal College of Science and Technology, Glasgow," Chemistry and Industry, 27th July, (1959) p. 758 - 762.

50. P. D. Ritchie, "Chemistry, Applied Chemistry and Chemical Engineering," in RCSTG. Ref. 49 p. 762 - 764.

51. W. M. Cumming, "James Young and the Young Chair," The Young Centenary Lecture given before the Oil, Shale and Cannel Coal Conference, Glasgow, 1950. Published as "Oil, Shale and Cannel Coal," Vol. II, by the Institute of Petroleum 1951. "Early Chemical Engineering Teaching," The Chemical Engineer CE 33, March (1968).

52. A. H. Loveless, "An experimental still and High Pressure Installation in the Technical Chemistry Dept, RTC Glasgow," Industrial Chemist Vol. II, 441, 500, (1935).

53. F. Rumford, "Chemical Engineering Operations," Constable & Co. (1951).
    F. Rumford, "Chemical Engineering Materials," Constable & Co. (1954).

54. G. S. G. Beveridge, "Chemistry and Industry," 15th December 1979, letter p. 863.

55. Chemical Engineering Education. Trans. Amer. Inst. Chem. Eng. Vol. 21, 80, (1928).

56. W. Cullen, Trans. Inst. Chem. Eng. Vol. 16, 10, (1938).

57. The President. Trans. Inst. Chem. Eng. Vol. 25, XXII (1947).

58. J. Kendall, "Some Eighteenth Century Chemical Societies," Endeavour Vol. 1, 106 - 9, (1943).

59. See Ref. 8, p. 587 - 596.

60. S. Grant, "Story of the University of Edinburgh," (1884) p. 354 - 361.

61. S. R. Tailby, "Chemical Engineering Education Today," Chemistry and Industry, 20th January, (1973).

62. "A scheme for a Degree Course in Chemical Engineering," Institution of Chemical Engineers, 12 Gayfere Street, London, SW1P 3HP.

FOUR SCORE AND SEVEN YEARS OF CHEMICAL ENGINEERING

AT THE UNIVERSITY OF PENNSYLVANIA

A. Norman Hixson and Alan L. Myers

Department of Chemical Engineering
University of Pennsylvania
Philadelphia, Pennsylvania   19104

IN THE BEGINNING

Benjamin Franklin founded the Charity School of Philadelphia in the year 1740. Franklin's Charity School grew up to become the University of Pennsylvania, but a century passed before the field of engineering was started. The Department of Mines, Arts, and Manufacture was formed with four professors in 1855 and within a few years there were twenty-two students enrolled. The school year started on the first day of December and lasted until the end of March. Each course consisted of thirty lectures and the fee was five dollars for the term. This fee was collected by the janitor at the start of the term and was the only salary for the professors.

The period from 1861 to 1869 was one of slow growth for engineering because of the Civil War. In 1869, ground on the present campus across the Schuylkill River in West Philadelphia was purchased for a scientific school. College Hall on the new campus was dedicated on Friday, October 11, 1872. There was now a faculty of twelve professors and four instructors in engineering. For admission a student had to be fourteen years old and he was required to pass examinations in ancient and modern geography, English grammar, arithmetic, and algebra as far as quadratic equations. The school session of three terms started in September and ended in June. In 1875, John Henry Towne, a University trustee, left an estate to endow salaries in the science department. This gift was the largest ever received by the University and the school was named the Towne Scientific School. By 1883, engineering degrees were granted in civil, mining and mechanical engineering. A full four-year program leading to a B.S. degree in chemistry was established in 1892. In 1893, a four-year program for a B.S. degree in chemical engineering was started. Each year since 1893 a curriculum has been taught in chemical engineering at the University of Pennsylvania. This continuity confers upon the Penn chemical engineering program the distinction of being the first program in continuous operation in the United States.

CHEMICAL INDUSTRY IN PHILADELPHIA IN THE 19TH CENTURY

The need for these early courses in chemistry and in chemical engineering is closely related to the history of chemical industry in Philadelphia in the nineteenth century. Three American colleges had offered courses in chemistry before the Revolution: Columbia in 1767, Penn in 1769, and William and Mary in 1774. A chemical society was organized in 1792 at the University of Pennsylvania, which was followed in 1811 by the Columbia Chemical Society organized in Philadelphia. The formation of the Association for the Advancement of Science in Philadelphia in 1847 arose from the need and desire for communication in the scientific field. A quotation from *The American Chemical Industry - A History* by Haynes explains: "No doubt the chemical famine during the War of 1812 encouraged other adventuresome apothecaries to become chemical manufacturers. Most of them must have been wiped out by the flood of postwar imports. At all events, the chronicles of the enterprises that eventually established the industry firmly in this country was written first in Philadelphia, then the qualified chemical headquaters, and later in New York, Baltimore and Boston."

Several chemical manufacturing plants were established early in the nineteenth century in Philadelphia, and many of them have descendants that exist today. One of these was founded by John Harrison in 1793, who built the first lead chamber plant that produced sulfuric acid in America. In 1814 he introduced another innovation: a platinum still weighing seven-hundred ounces with a twenty-five gallon capacity for concentration of the acid. Later, because of competition, he expanded his manufactures to salts and particularly to the field of paint pigments. The plant, which is located across the Schuylkill River from the University on Grays Ferry Road, remained the site of this company until it was purchased in 1917 by the duPont Company. duPont still manufactures paint at this location. A grandson of John Harrison, Charles C. Harrison, became provost of the University of Pennsylvania. During his fifteen year tenure as provost, Provost Harrison was very active in making the arrangements for the construction in 1904 of a new facility on the campus for teaching and research in chemistry. This facility was named The John Harrison Laboratory of Chemistry. Competition on a large scale in the sulfuric acid business was created by Charles Lennig in 1829 when he built the largest lead chamber plant in America, in Bridesburg near Philadelphia. It was also equipped with platinum stills and discharged a steady stream of concentrated acid. By 1859, this plant became one of the most important heavy chemical operations in the country and it remained a prominent producer in Philadelphia until the mid nineteen-thirties.

Quoting again from Haynes, "The real foundation of fine chemical manufacture in the United States was laid down in Philadelphia between 1818 and 1822 by five young men, all of them foreign: two Swiss, a German, a Hollander and an Englishman." They founded two firms: Powers and Weightman, and Rosengarten and Co. Although primarily in the acid business, they built plants that manufactured quinine sulfate, mercurials, morphine salts, and other medicinal chemicals that later became the real base of the business. These two companies later combined to become Powers, Weightman and Rosengarten until, finally, they merged with Merck in 1927.

The first white lead plant in this country was built at Broad and Chestnut Streets in Philadelphia in 1804. Those acquainted with Philadelphia would recognize this as the choicest location in the city, now occupied mostly by banks. This plant burned down, but in 1811 a new plant was started in a different location and the manufacture of white lead by corroding lead buckles became very successful. This operation was taken over by merchants who organized the National Lead Co. In 1938 and 1939, senior classes at Penn visited plants in Philadelphia that were still corroding lead buckles by the Dutch Boy process.

An interesting and very durable chemical manufacturer is the Philadelphia Quartz Company, founded here in 1831. It is remarkable that this modest-sized company making soap and soluble silicates has operated so long in this field and is still in business. When questioned about their lasting powers, Dr. James Vail, who was president of the company in the nineteen-thirties and nineteen-forties stated laconically, "There are several of the big boys who could push us out of business if they wanted; it's just up to us not to get big enough to make it worth their while. As we are now, it isn't and they don't." James Vail was a prominent Philadelphia Quaker, a member of the Friends Service Committee, and served as president of the American Institute of Chemical Engineers.

Although not in Philadelphia, the duPont Company was founded in 1802 in neighboring Wilmington, just thirty miles south of Philadelphia. They have, of course, exerted a great influence upon the chemical industry and chemical engineering education in the Delaware valley.

George T. Lewis, who was very prominent in the booming lead pigment business in Philadelphia, was a leader in founding the Pennsylvania Salt Co. in 1850. Today, after a series of acquisitions, they are known as the Pennwalt Co. Another early chemical company that is of particular interest to the University of

Pennsylvania is the Henry Bower Chemical Manufacturing Co., which in 1858 produced ammonium sulfate from the condenser and washer liquors then running to waste from the Philadelphia Gas Works next door. In 1892, three sons of Henry Bower were in charge of the company. George, the president, had an Arts degree; William, a vice-president, majored in chemistry; and Frank, also a vice-president, held a bachelors degree in Mechanical Engineering, all from the University of Pennsylvania. They are still in operation in Philadelphia.

There were many chemical operations in existence in Philadelphia during the nineteenth century, far too many to list here. A few are especially notable. The Moro-Philips and the United States Works of the General Chemical Co. were located across the Delaware River in Camden, where they made heavy chemicals. The Barrett Manufacturing Co. in northeast Philadelphia (Bridesburg) produced benzol and coal tars. These two concerns were brought together as part of the formation of Allied Chemical and Dye Corporation in 1920. The Barrett division is still quite active. There were several gas manufacturing operations and one of the earliest was the United Gas Improvements Corp., which is still active. Among the drug companies, the Philadelphia directory for 1844 lists Smith, Kline and French as a manufacturer of cosmetics and well as medicines.

Several soap manufacturing concerns were active in Philadelphia but among those who persisted in this area for a long time were Charles W. Young and Fels and Co. Fels and Co. was originally a manufacturer of fancy soaps, but most people will remember it for production of Fels naphtha, a brown rosin soap that was a very good antidote for poison ivy.

Although Edwin Drake drilled the first oil well in Titusville, Pennsylvania, in 1859, it was not until later that refineries began to appear in the Philadelphia area. In 1882 the Philadelphia directory listed the Atlantic Refining Co. located at Point Breeze on the Schuylkill River in Philadelphia. This is still the location of their big refinery. In 1885, Atlantic Refining Co. put into operation the first continuous stills and, ten years later, the first packed tower. The packing was made of stone and concentrates were removed at three points.

This sketchy history of chemical industry in Philadelphia indicates the lively interest and activity in the field of chemistry. There was a need for new products made locally. The opportunities for new undertakings in Philadelphia attracted talent from our growing country and from abroad. This combination produced a thirst for new knowledge and methods and set the stage for close cooperation between the fledgling industries and the developing departments of engineering at the University of Pennsylvania.

UNIVERSITY OF PENNSYLVANIA

HISTORY OF CHEMICAL ENGINEERING AT PENN

The first separate chair for chemistry was established in 1769 in the Medical School, and was occupied by Benjamin Rush, who was widely featured in the scientific and medical fields of his time. In 1795, James Woodhouse became the first Professor of Chemistry at the University. This chair was not in the Medical School and was entirely devoted to chemistry. The position was offered to Woodhouse after it had been declined by Joseph Priestley, the discoverer of oxygen.

During the last three decades of the nineteenth century, there were three professors of chemistry who were responsible for the growth of chemistry and chemical engineering at Penn: F.A. Genth, Samuel Sadtler and Edgar Fahs Smith. All three obtained their PhD's in Germany. Genth was president of the American Chemical Society in 1880, Sadtler was the first president of the American Institute of Chemical Engineers in 1908, and Smith was president of the American Chemical Society in 1895, 1920 and 1921.

Genth established a consulting firm in Philadelphia in 1850. He was an outstanding inorganic chemist and mineralogist and was quite successful. In addition to this business, he was Professor of Analytical Chemistry from 1872-74, and Professor of Chemistry and Mineralogy from 1874-1888. Smith was his assistant from 1876 to 1881. Sadtler taught at Penn from 1874 to 1891. His last position was as Professor of Organic and Industrial Chemistry and he, too, ran a successful consulting business.

Both Genth and Sadtler were influential in the early development of academic and industrial chemistry. Sadtler was deeply involved in the organization of the American Institute of Chemical Engineers. However, it was Smith who organized the four-year B.S. program in Chemistry in 1892, and in chemical engineering in 1893. The department was named Chemistry and Chemical Engineering and was in the Towne Scientific School. This organization lasted until 1951, when Melvin (Mike) Molstad became Chairman of the School of Chemical Engineering. Thus, Chemistry existed in a school in which all of the other curricula were engineering, whereas all of the other sciences at Penn were in the College of Arts and Sciences. This close association of chemistry and chemical engineering affected both curricula. For example, an extensive course taken together by seniors of both groups was Industrial Chemistry, a laboratory course which included plant trips to local industries. This course called for ten hours per week of experimental studies in inorganic and organic chemistry. The plant trips were made at least every fortnight until the second world war, which forced a temporary discontinuance of the activity. A senior course in chemical engineering called Chemical Processes grew out of this early course in Industrial Chemistry.

The Catalogue and Announcements of the University of Pennsylvania for 1892-93 contains the following description of the course in Chemical Engineering: "This course has been arranged with the view of enabling chemical students to familiarize themselves with mechanical subjects to such a degree that they will be able to overcome the many difficulties which are constantly presenting themselves to those who are engaged in extending the applications of chemistry. The chemical studies introduced into this course will not only give the student a thorough acquaintance with the fundamental principles of chemical science, but will also afford him a complete drill in analysis, and in the preparation of inorganic and organic products. Instruction in technical analysis and applied chemistry is reserved until the last year. This has been purposely so arranged. It permits of the previous preparation in chemistry and mechanics, so necessary for the intelligent comprehension of the mechanisms involved in the applications of chemistry. The course aims to be practical. Laboratory methods will be preferred for instruction. Frequent excursions will be made to adjacent plants for the purpose of studying practical processes in operation and examining in detail the mechanical appliances that are used.

Table 1. Subjects for Junior Class, from the Four-Year Course in Chemical Engineering Description in the Penn Catalogue for 1892-93.

Chemistry 4.-Analytical Chemistry. Laboratory Practice. Lectures and recitations in Gravimetric and Volumetric Analysis. Assaying. *Ten hours (One Term)*. Professor Smith and Dr. Keith.

Chemistry 6.-Organic Chemistry. Lectures. Laboratory work in making Organic Preparations. *Ten hours (One Term)*. Professor Smith and Dr. Keith.

Physics 9.-Experimental Physics. *Three hours*. Professor and Assistant Professor Goodspeed.

Physics 8.-Laboratory work. *Three hours*. Professor Barker.

Metallurgy 1.-Metallurgical processes. *One hour*. Mr. Brown.

Mechanical Engineering 1.-Statics. *Two hours*. Mr. Huffington.

Mechanical Engineering 2.-Hydrostatics and Hydraulics. *Two hours*. Professor Spangler.

Mechanical Engineering 9a.-Steam boilers. *Two hours (One Term)*. Mr. Huffington.

Mechanical Engineering 11.-Electricity. *Two hours*.

Mechanical Engineering 15a,21a.-Mechanical and Electrical Laboratory. *Three hours*.

Mathematics 16.-Calculus. *Three hours*. Assistant Professor Crawley.

UNIVERSITY OF PENNSYLVANIA

The degree conferred upon graduates of this course will be Bachelor of Science in Chemical Engineering. Three years after graduation those bachelors of science who have shown marked progress in their professions and who submit a satisfactory thesis, may be granted the degree of Chemical Engineer (Ch.E.)."

Looking at the junior year of the first program in Table 1, it is apparent that this was a curriculum in chemistry with mechanical engineering (including mechanical drawing) superimposed. There was a heavy emphasis on laboratory work and the load required five and one-half days per week of class activity. There was a much greater emphasis on inorganic than organic chemistry, particularly in the qualitative and quantitative procedures where the student was given the dreaded "unknowns." The work load earned for chemical engineering the reputation as the hardest engineering program.

Smith was Chairman of the Department until 1898, when he became Vice-Provost and, in 1911, Provost. This was the University's highest office and removed him considerably from operations in chemistry. He retired in 1920. Although the chemistry professors, including Smith while he was Provost, carried on research, there was no consideration of a graduate program in chemical engineering.

The primary changes in the courses at Penn up to 1920 were a clearer definition of the various fields of chemistry, a considerable increase of emphasis on organic chemistry, pure and applied, and the introduction of electrochemistry because of Smith's prominence in the field. During this period came the introduction of a continuing puzzle for chemical engineers: learning thermodynamics from physical chemists and then from mechanical engineers. It was not until 1927 that two undergraduate courses labeled ChE were introduced. One was a course in Materials of Construction (corrosion), and the other was a start in unit operations. By 1930 there were three assistant professors of chemical engineering: Russel Heuer, Edward Fennimore and A. Kenneth Graham.

During the late nineteen-twenties and early nineteen-thirties the size of the chemical engineering classes increased until in 1936 the senior class ranged from 30 to 45 students. This was considerably larger than the corresponding classes in the chemistry program and created a problem of lack of space.

The Towne Building was designed to house Civil and Mechanical Engineering. When it was dedicated in 1906, it received considerable acclaim. The comment from Engineering News was "It is with little doubt the finest, largest, and best equipped structure devoted to instruction in engineering in the United States, if not the world." It contained 128,000 square feet of floor space and cost $730,000. Gradually during the nineteen-twenties, chemical engineering began to acquire space and finally became completely housed in the Towne

Fig. 1. The Towne Building of the University of Pennsylvania, dedicated in 1906.

Building shown on Fig. 1. This meant some rather drastic and difficult alterations of space assignments. There was less space for drafting rooms and machine shops, and a foundry was eliminated. The building was well-constructed and remains an excellent facility today.

With the arrival of Norman Krase in 1936, Harding Bliss in 1937, and Norman Hixson in 1938, a completely revised and modern undergraduate program was developed and a graduate program leading to the PhD was inaugurated. A refusal of the university administration to create a separate chemical engineering department led

UNIVERSITY OF PENNSYLVANIA

to the departure of Krase and Bliss in 1939 and the arrival of Melvin Molstad and Erwin Amick who, along with Hixson, were in charge of the undergraduate and graduate functions of the department. During the wartime years the undergraduate civilian student body fell off sharply, but these were replaced by a naval V-12 training program and the army ASTP program. Seventy-five "volunteers" were assigned to Penn for training, but their backgrounds varied from holders of trade school diplomas to one with a masters degree in chemistry. The group was finally reduced to 17 excellent and highly motivated students. Unfortunately the army cancelled the entire program nationwide just before this group were finished. Having been warned of this impending disaster, Penn managed to transfer fifteen of the seventeen for active duty on the Manhattan Project at Columbia University.

For a period during the war, there was a subcontract of the Manhattan Project operated in the Chemistry and Chemical Engineering Department at Penn. Because of the draft deferment policy the graduate work, although diminished, continued at a reasonable level and there was considerable activity in the masters program with chemical engineers employed locally. Courses for these part-time students on evenings and Saturday morning were quite successful. The Department has continued this tradition of providing graduate courses taught by the regular faculty during the evenings for part-time students from industry.

Shortly after the start of the war, Amick left to work in industry and Hixson and Molstad were left to run a busy two-man show. In 1943, they were joined by Ralph Thompson as an instructor and later as an assistant professor. The influence of John Goff, a mechanical engineer who became Dean of the Towne Scientific School in 1938 and continued during the wartime period, should be mentioned. He pushed for the development of graduate programs while teaching his special brand of thermodynamics by the Caratheodory approach. His graduate course in thermodynamics was required for chemical engineers. He also had a Naval project for a Thermodynamics Research Laboratory for the development of portable equipment to produce liquid air and oxygen. Prominent among those involved in this project was Barnet F. Dodge of Yale, who wrote one of the early thermodynamic textbooks designed specifically for chemical engineers.

During the period of Harold Stassen's presidency of the university, in 1951, the Department of Chemical Engineering was made independent of Chemistry. This came as a complete but pleasant surporise to everyone. As mentioned previously, Molstad became the first Chairman and held this office until 1962. Following this change there occurred a series of important events which made possible the expansion of the chemical engineering staff and its graduate and undergraduate programs. The first of these was the appointment as president of the University of Gaylord Harnwell, a physicist who backed engineering with enthusiasm. Next an office of vice-president in charge of engineering was created, and this post was filled by Carl C. Chambers, an electrical engineer. Under him were two assistant vice-presidents, Reid Warren for undergraduate affairs and Norman Hixson for graduate affairs and research. One of the results of this administrative reorganization was that it unified all of the engineering faculties into one common faculty which met often and could make decisions on an engineering-wide basis. It made possible the undertaking of bold steps to realize what the faculty considered to be their full potential.

In 1963 the Ford Foundation granted the University three million dollars for the expansion of its full-time graduate program leading to the PhD. A portion of these funds combined with matching funds from the National Science Foundation enabled us to convert our buildings into first-rate research facilities. A vigorous program for obtaining new talent for our engineering staff was begun. Particularly important in this recruitment are Arthur Humphrey, who is now at Lehigh University, Lee Eagleton, now at Penn State, Alan Myers, Daniel Perlmutter, John Quinn and Stuart Churchill. All of these came to Penn during the nineteen-fifties and nineteen-sixties. Many of them shared in the direction of the department: Humphrey from 1962 to 1972, Perlmutter from 1972 to 1977, and Myers from 1977 to date. Arthur Humphrey was mostly responsible for the development of a biochemical and biomedical engineering group within engineering and particularly in chemical engineering.

In summary, after being in the business of teaching chemical engineering for four score and seven years at Penn, from 1893 to 1980, we now have a staff of thirteen which include Professors Stuart Churchill, William Forsman, Norman Hixson, Arthur Humphrey, Mitchell Litt, Alan Myers, Daniel Perlmutter and John Quinn; Associate Professors Elizabeth Dussan V., David Graves and Warren Seider; and Assistant Professors Eduardo Glandt and Douglas Lauffenburger. The undergraduate enrollment is approximately 180, and a full-time graduate student body of 55 produces an average of eight PhD's each year. A recent survey showed that there are thirty seven chemical engineering degree holders from the University of Pennsylvania actively teaching at the collegiate level in this country and abroad. Consistent with its tradition over these 87 years of existence, the department still maintains an excellent rapport with the chemical industry in the Philadelphia area. We provide evening graduate courses so that part-time students can start their graduate work while fully employed. Many of our best graduate students began in this way. Furthermore, the industries in Philadelphia have been very generous in cooperating with us by sending design engineers to work with the senior class and the faculty in the chemical plant design course. For the past several years, an average of seven engineers from local industry have volunteered to spend one afternoon every other week during the

Spring term working with our senior students. This strong coupling of industry and academia is an essential ingredient of our program. Indeed, the phenomenal growth of chemical industry in the Philadelphia area stimulated the development of the Department of Chemical Engineering at Penn during the last four score and seven years.

HISTORY OF CHEMICAL ENGINEERING AT THE UNIVERSITY OF MICHIGAN

Donald L. Katz, James O. Wilkes, and Edwin H. Young

Department of Chemical Engineering
The University of Michigan
Ann Arbor, Michigan 48109

ABSTRACT

The chemical engineering program at the University of Michigan dates from 1898[1,2]. Discussion of the first 26 years shows the transition from an affiliation with chemistry to the new East Engineering Building with a pilot-scale laboratory. Involvement by chemical engineering faculty in professional activities is seen to include professional societies, research projects, and consultation.

Research developed quickly in the East Engineering Building; some programs of long-standing duration are mentioned. The Ford Foundation Computer Project is taken as an example of an activity that had a national impact by including faculty from a large number of engineering schools.

In recent years, research interests and the curriculum have broadened into mathematical and biological areas. The Herbert H. Dow Building under construction will provide appropriate new facilities.

The paper enumerates some highlights of the past 82 years.

THE FIRST 25 YEARS, 1898 - 1923

Engineering at Michigan was established in 1853, within the College of Literature, Science and the Arts. Early emphasis was on civil engineering, followed by mining engineering (discontinued in 1896), mechanical engineering (1868), and electrical engineering in 1889. The Department of Engineering was granted independent status in 1895, and became the College of Engineering (along with Architecture until 1931) in 1913, the same year as the building of Hill Auditorium.

The development of chemical engineering was anticipated, at least in part by President Henry Tappan, during his inaugural address of 1852 in which he said:

> "To this end, we propose to establish a scientific course parallel to the classical course. . .There will be comprised in it, besides other branches, Civil Engineering, Astronomy with the use of an Observatory, and the application of Chemistry and other Sciences to Agriculture and the industrial arts, generally."

However, it was not until April 1898 that a course of studies in chemical engineering was approved by the Board of Regents. Professor Edward DeMille Campbell of the Chemistry Department was placed in charge; he was appointed Professor of Chemical Engineering and Analytical Chemistry in 1902, and Director of the Chemical Laboratory in 1905.

Until 1909, our laboratories were housed within the old Chemistry Building and part of what is now the Economics Building. Spacious new quarters were then available in the new Chemistry Building from 1909 until 1923, when we moved to the newly completed East Engineering Building (Plate 1). Thus, when we move to the Herbert H. Dow Building (currently under construction on the North Campus), we shall leave behind us a home of 60 years.

The other early prime mover in chemical engineering, and a person who was to bring the department to a stage of preeminence over some 45 years, was Alfred Holmes White, who was appointed Instructor in Chemical Technology in 1897, having just spent a year at the Federal Polytechnicum in Zurich. Teaching in the early years was roughly divided into the areas of chemical technology (White) and metallurgy (Campbell). Administrative pressures in both chemistry and chemical engineering led Campbell to resign as Professor of Chemical Engineering in 1914, and A.H. White then became chairman of the department.

Other prominent faculty appointments of the period were A.E. White (1911), W.L. Badger (1912), C. Upthegrove (1916), J.C. Brier (1917), E.H. Leslie (1919), and G.G. Brown (1920).

Substantially increased enrollments after the First World War were straining both the staff and facilities, which were much relieved when the move was made to East Engineering in 1923. Plate 1 was photographed in 1980. The 4th floor of this building was devoted

Plate 1. East Engineering Building (1923), and New Wing (1946).

to a foundry and to the metallurgy and fuels laboratories of the department, and there was adequate space, extending over three floors, for the evaporator laboratory, newly developed by W.L. Badger.

Thus, the end of the first 25 years came to see chemical engineering firmly established as a discipline, and permanently housed in proper accommodations.

DEVELOPMENT OF THE CURRICULUM

The overall requirements for the bachelor's degree during the past 70 years are summarized in Table 1. Throughout, the first year (at least) has been common to all disciplines - civil, mechanical, electrical, chemical, and marine engineering, etc.

Early, in 1912, the chemistry courses were dominated by the industrial chemistry needs of qualitative and quantitative analysis,

Table 1. Distribution of Credit Hours in Undergraduate Curriculum.

Subject	1912	1931	1957	1980
Chemical Engineering	19	25	34	33
Chemistry	27	28	26	22
Physics	10	10	10	8
Mathematics	18	16	16	16
English	5	6	10	12
Modern Language	16	16	--	--
Metal Shop	4	2	2	--
Mechanical Engineering and Mechanics	14	14	7	4
Surveying	2	--	--	--
Economics & Accounting	--	3	6	3
Electives	17	8	14	15
Electrical Engineering	4	4	4	4
Digital Computing	--	--	--	2
Humanities & Social Science	--	--	--	9
Drawing & Design	4	8	6	--
TOTALS	140	140	135	128

which together accounted for 18 of the required 27 hours. Mathematics consisted of algebra and analytical geometry, followed by a substantial portion of calculus, the latter part of which was devoted to particle and rigid-body kinetics. Physics included mechanics, sound, light, heat, electricity, and magnetism. A firm grounding in two modern languages was an important feature, requiring at least 16 hours, and as much as 24 hours for a student admitted with no previous language experience.

The general flavor of the early chemical engineering part of the curriculum is shown in Table 2, which lists all courses offered in the department for 1911-12. With only one graduate student among 130 undergraduates, the graduate program was scarcely underway in 1912. Nominally, however, the Master of Chemical Engineering degree was available to qualified students who wished to spend another 36 hours, with a minimum of 10 hours in chemical engineering.

With its further 15 hours of free electives, the current program allows remarkable flexibility for the student who wishes to pursue other areas in depth. Along these lines, the biochemical option in applied microbiology has been popular and successful over the past 25 years.

UNIVERSITY OF MICHIGAN

Table 2. List of Chemical Engineering Courses Offered, 1911-12.

1. Fuels and Refractory Materials, Iron and Steel
2. Chemical Technology
3. Chemical Technology of Carbon Compounds
4. Metallurgy of the Non- Ferrous Metals
5. Micro-Metallurgy
6. Technical Examination of Gas and Fuel Laboratory
7. Methods of Assaying Gold and Silver Ores
8. Chemical Technology Laboratory
9. Evaporation, Distillation, Filtration, and Transportation of Liquids on the Manufacturing Scale
10. Machinery and Processes for Conveying, Drying, Calcinating, and Grinding

Table 3. Chemical Engineering Courses Required for the BSE, 1980

Thermodynamics I & II	Gases, Liquids, Solids, and Surfaces
Rate Processes I & II	Engineering Materials in Design
Chemical Engineering Laboratories I & II	Principles of Chemical Engineering Design
Senior Elective in Chemical Engineering	

Table 4. Chemical Engineering Graduates at University of Michigan

Total Degrees Granted in Period

Decade	B.S.	M.S.	Ph.D.
1898-1908	72	0	0
1908-1918	283	40	1
1918-1928	365	68	14
1928-1938	367	218	77
1938-1948	712	316	47
1948-1958	747	432	108
1958-1968	491	293	152
1968-1978	529	214	72
Total at end of 1979: 3678		1609	479

Plate 2. TEACHING STAFF, DEPARTMENT OF CHEMICAL ENGINEERING 1959. (See also Table 5).

Table 5. Teaching Staff in Chemical, Metallurgical and
          Materials Engineering

                   Department Booklet 1959-60

1.	Donald LaVerne Katz	20.	Kenneth F. Gordon
2.	John Crowe Brier*	21.	Edward E. Hucke
3.	Lloyd Earl Brownell*	22.	Donald William McCready*
5.	Stuart Winston Churchill	24.	Guiseppe Parravano*
6.	Richard A. Flinn	25.	David Vincent Ragone
7.	Lloyd L. Kempe	26.	Richard Emory Townsend*
9.	Richard Schneidewind*	27.	Keith H. Coats
11.	Maurice J. Sinnott	29.	Walter B. Pierce*
12.	Lars Thomassen*	31.	William A. Spindler
14.	G. Brymer Williams	32.	Mehmet R. Tek
15.	James Wright Freeman*	33.	Richard Earl Balshizer
16.	Jesse Louis York	34.	James A. Ford
17.	Edwin H. Young	35.	Richard A. Hummel
18.	Lawrence H. VanVlack	36.	Peter B. Lederman
19.	Wilbur C. Bigelow	37.	Robert L. Norman
	38. Paul K. Trojan		*Deceased

Professors Carrick, Mason, Siebert and White were absent
when the picture was taken. Dr. Rudd and Pehlke joined
the staff in February, 1960. Professor J.J. Martin was
on leave, 1959-60.

Table 6. Chairmen, Chemical Engineering Department,
          University of Michigan

1898-1914	Edward DeMille Campbell
1914-1942	Alfred Holmes White
1942-1951	George Granger Brown
1951-1962	Donald LaVerne Katz
1962-1967	Stuart Winston Churchill
1967-1971	Lawrence H. VanVlack
1971-1977	James Oscroft Wilkes
1977-	Jerome S. Schultz

In 1931, 50 courses were listed in the department, new directions being represented by heat and material balances, thermodynamics, unit operations, petroleum refinery engineering, X-ray studies, and electrochemistry. With 31 MSE and 20 doctoral students the graduate program had become well entrenched by then. The same as today, the MSE degree requirements were one year of coursework beyond the BSE, and included a master's thesis.

Over the past 70 years, the effort devoted to mathematics, chemistry, and physics has changed little. However, mathematics now includes linear algebra and differential equations; chemistry has emphasized the physical and organic aspects; and, mechanics, electricity, and magnetism have come to dominate the first two physics courses. Modern languages have essentially been replaced with humanities and the social sciences, and further English.

In the same period, the number of required hours in chemical engineering has increased, and here the trend has been away from chemical technology towards unit operations (now somewhat in decline) and chemical engineering principles. The current (1980) required courses are listed in Table 3. Equipment and chemical engineering process design in the senior year have continued to provide an important vehicle for applying the principles of earlier courses towards solving problems of industrial significance.

Although most of our students have, of course, come from Michigan, and we have also attracted goodly numbers from other states, notably New York and Ohio. The number of degrees granted for each decade since our founding appears in Table 4. The names of the department chairmen over the same period are listed in Table 6.

The faculty photograph (Plate 2) on the steps of East Engineering Building in 1959 was during the period when the title was Department of Chemical and Metallurgical Engineering.

A MAJOR EDUCATIONAL CONTRIBUTION

Of the many instructional enterprises in our department, we select one as representing the very best in terms of its positive and enduring effect on engineering education. 1959 saw the beginning of a three-year project, "The Use of Computers in Engineering Education", with major funding from The Ford Foundation. The broad goal of its Director, D.L. Katz, was to explore how the then newly emerging digital (and analog) computers could benefit instruction in all avenues of engineering. By bringing together faculty from all over the country (and indeed internationally) for workshops, seminars, and summer sessions in Ann Arbor, a substantial number of ideas were developed and interchanged. Thanks also to the efforts of Assistant Directors E.I. Organick, S.O. Navarro, and B.

Carnahan, the resulting publications, showing how computers could advance engineering education both generally and in each of its individual disciplines, attracted widespread interest and circulation. Their effect in catalyzing the use of computers at an early stage throughout the nation is well remembered and appreciated 20 years later.

PROFESSIONAL INVOLVEMENT OF FACULTY AND GRADUATES

The eldest co-author was impressed as an undergraduate to find that his Organic Chemistry professor, Moses Gomberg, was President of the American Chemical Society (1929), and the Chairman of the Chemical Engineering Department, A.H. White, was President of the American Institute of Chemical Engineers (1929-30). It was the practice for undergraduates to attend the A.I.Ch.E. Student Chapter Meetings; at one meeting, with the full faculty in attendance, students mimicked by lectures the eccentricities of their teachers in turn.

With a large-scale (pilot-plant) laboratory maintained by W.L. Badger and others for research in evaporation, the Dowtherm boiler, and heat transfer equipment, the students were impressed with the link between academic studies and the industrial sector. Problems given to the students were recognized as coming from the experiences of their teachers as consultants.

Brief comments will be made on involvement in the professional areas of national societies, consultation, and industrially related research.

Professional Societies. Engineering faculty find meeting with others through engineering societies to be broadening and to afford additional contacts with industry and government. A.H. White initiated activity with the A.I.Ch.E. during 1915-1920. In 1922, he was responsible for forming at the University of Michigan the first Student Chapter of the American Institute of Chemical Engineers. Faculty members E.M. Baker, D.L. Katz and Brymer Williams have served as Chairman of the Student Chapters Committee. The first prize for the Student Contest Problem is the A. McLaren White Award, named after the son of the A.H. Whites. Both faculty and graduates have given leadership to many engineering societies. Table 7 lists A.I.Ch.E. Presidents; another 9 faculty and graduates have served on Council.

In the American Society for Engineering Education, A.H. White served as President in 1942 and J.J. Martin in 1971. E.H. Young was President of the National Society of Professional Engineers in 1968-69; J.J. Martin was President of the Engineers Joint Council in 1973-75. It is beyond the scope of this paper to note the many involvements of the faculty, but a few more will be mentioned.

Table 7. Faculty and Graduates Who Have Served As President of the American Institute of Chemical Engineers

1929-30	A.H. White
1944	G.G. Brown
1950	W.L. McCabe
1954	G.G. Kirkbride
1958	G.E. Holbrook
1959	D.L. Katz
1962	J.J. McKetta
1966	S.W. Churchill
1971	J.J. Martin
1975	K.E. Coulter
1980	J.G. Knudsen

In the American Chemical Society, J.J. Martin has served as Chairman, Division of Nuclear Chemistry and Technology, and on the Executive Committee of the Division of Industrial and Engineering Chemistry. Memberships on advisory boards to journals and award committees have been held by J.J. Martin and D.L. Katz, and R.L. Kadlec has served as editor of the A.I.Ch.E. Journal.

Among the activities is the accreditation of curricula in chemical engineering via the joint efforts of the A.I.Ch.E. and the Engineers Council for Professional Development. A review of the various committee members in the E.C.P.D. shows a continuity of between one and four faculty members from 1936 to date: A.H. White, G.G. Brown, W.L. Badger, D.L. Katz, R.R. White, S.W. Churchill, G.B. Williams, J.E. Powers, R.E. Balzhiser, and J.J. Martin. They served E.C.P.D. and A.I.Ch.E. as members of Visiting Committee, A.I.Ch.E. Representatives, Education and Accrediting Committee, and E.C.P.D. Executive Committee.

The first national meeting devoted to the peaceful uses of Nuclear Energy was held at the University of Michigan in 1954. It was organized as part of the formation of the Nuclear Division of the A.I.Ch.E.; D.L. Katz was chairman, R.R. White was General Chairman, and most of the faculty and the Detroit Section had part in handling the meeting, with 1,200 in attendance.

Engineering Consultation and Research. Today, younger staff members accept the challenge to master some field of engineering science as part of their development. In due time, applications of their expertise are called for in solving problems of industrial

ALFRED HOLMES WHITE
1873-1953

GEORGE GRANGER BROWN
1896-1957

DONALD LAVERNE KATZ
1907-

Plate 3. Three Department Chairmen.

interest. In this way, faculty members gain an insight into the essence of their specialty.

The College of Engineering established a policy in 1910, which stated in part:

> "Teachers of engineering. . .not only should be permitted to engage in professional work outside of their University work, but should be encouraged. . .in so far as it can add to the effectiveness of their work as teachers, and does not impair their services to the University, and where such work related to problems of public interest the special kinds of skill to be found in such a teaching staff should, as far as possible, be made available to the public. . ."

With this policy, it becomes important for faculty to develop their personal code which always places foremost the interest of students at the University. Research came naturally to the young chemical engineering staff since Prof. E.D. Campbell of the chemistry faculty was active in studies of steel by 1900 and passed on the satisfaction which comes from pursuing new frontiers.

A.H. White early in the century worked with the local coal-gas company, and published papers on gas analysis as well as a book in 1912 on Technical Gas and Fuel Analyses. W.H. Badger worked with the Swenson Evaporator Company, and included their pilot-sized evaporators in the unit operations laboratory of 1923. G.G. Brown carried out assessment of natural gasoline as automotive fuels, and had an engine laboratory in the department laboratories. This led him into the application of thermodynamics to the petroleum and natural-gas industries. D.L. Katz, based on experiences in reservoir engineering in industry, joined in developing underground storage of natural gas. His books on Natural Gas Engineering and Underground Storage of Fluids were prepared in good part from experience in industry.[3]

The Michigan Gas Association initiated a fellowship in chemical engineering in 1900 for a student to work with A.H. White. This fellowship program has had continuous support for 80 years, with some lapse in appointees during World War II. In the 1940's and 50's the program emphasized synthetic gas, under R.R. White; since 1960, it has supported work in natural gas storage and more recently in coal gasification. Later, the initiated projects were supported more fully by the Pipeline Research Committee of the American Gas Association. The University has prepared four monographs covering this work since 1960 under the guidance of Katz and Tek.

The Engineering Summer Conferences afford an excellent example

UNIVERSITY OF MICHIGAN

of faculty interacting with industry. A formal program that started in 1953 for the College of Engineering, with 6 courses initially, has grown to some 40 courses annually, with from 3-4 being offered by chemical engineering faculty. The Chrysler Center, devoted to continuing engineering education, was constructed in 1967. The course in underground storage of natural gas (by D.L. Katz and M.R. Tek) has been given 15 times since 1957, and that in Applied Numerical Methods (by B. Carnahan and J.O. Wilkes) 17 times since 1964. Other courses include Corrosion and Electrochemical Engineering (both by F.M. Donahue) and Polymer Processing (by G.S.Y. Yeh).

RESEARCH

The Engineering Research Institute. The College of Engineering was a pioneer in formal contract research for industry and government under Albert E. White of the chemical engineering staff[4]. In 1920 he established mechanisms for carrying out small-and large-scale projects in university laboratories under staff direction, with most workmen drawn from student ranks. The arrangements included compensation for staff for the extra load carried in research efforts, and provided formal reporting services and approvals for use of facilities; overhead charges returned funds to departments for equipment and travel.

A.E. White retired as Director of the Engineering Research Institute in 1953, being replaced by Richard G. Folsom. In 1958, the Engineering Research Department was expanded to become the university contract agency under a Vice President for Research.

Today, sponsored research for supporting graduate programs is commonplace. In the 1920's and 1930's it was considered pioneering to involve students in the current problems of industry and government. Participation in research instills a desire to carry on throughout a career with a spirit of inquiry.

Selecting specific research programs worthy of mention here is difficult because of the voluminous nature of some 479 doctorate theses and a multitude of formal research projects. The following are a representative sample of sustained efforts.

Distillation and Fractionation. E.H. Leslie brought his interest in distillation to the university and early studies were of vapor-liquid equilibria with petroleum fractions. Brown and Souders pursued the acquisition of operating data of fractionators and developed design procedures. Brymer Williams continued with studies of tray efficiencies.

Vapor-Liquid Equilibria/Physical Properties. Phase behavior of hydrocarbon mixtures under pressure has received much attention in several categories. Beyond Brown's work, Katz directed studies measuring equilibrium constants, surface tension, viscosities and gas hydrates. Critical phenomena and retrograde condensation were photographed in glass-windowed pressure cells. R.R. White, B. Williams, and J.H. Hand with their graduate students had excursions into these areas.

Finned-Tube Heat Transfer. The Wolverine Tube Division of Calumet and Hecla sponsored research in 1940 to explore heat-transfer performance of various integral-wall finned tubes. Over a period of 36 years, until 1976, continuous project and fellowship support was given for research by graduate students under A.S. Foust, D.L. Katz, and E.H. Young. Sixty-five research reports and 45 technical papers provided heat-transfer data for the process and refrigeration industries; 8 doctorate theses were included.

Thermodynamic Properties Under Pressure. G.G. Brown's interest centered in applied thermodynamics and he initiated experimental programs for doctorate research into enthalpy measurements for hydrocarbons. J.J. Martin's program included P-V-T and specific-heat measurements with calculated thermodynamic properties of refrigerants. Katz initiated a program for calorimetric measurements on natural gas constituents at low temperature. Direction was assumed by J.E. Powers for gaseous mixtures and new calorimetric techniques were devised. In parallel with early research was W.L. McCabe's work with enthalpy of caustic solutions.

Light Scattering, Particle Size, Sprays. C.M. Sliepcevich developed a method for determining droplet sizes and distribution in atomized sprays based on light scattering. In parallel, J.L. York carried out research in a spray laboratory.

Applied Microbiology/Biochemical Research. Renewed interest in applied microbiology resulted from fermentation processes developed during both World Wars; this involved production of ethanol, intermediates for explosives, biological warfare agents and penicillin in commercial amounts. The technology developed for these processes led to applications in large scale food processing, and more recently to developments in medicine. L.L. Kempe developed the process for sterilizing bone fragments used in repair of vertebra. This work was accompanied by the establishment of standards for γ radiation sterilization of surgical equipment and non-acid canned foods, where botulism is the criterion. J.S. Schultz has been working with basic fermentation studies and studying diffusion through membranes as models for transport phenomena in biological systems. He is applying this work to the use of membranes in separations. Collaborative work with physicians in the medical school is directed towards understanding blood coagulation on artificial surfaces as in dialyzers and heart pumps.

Other Research Programs of Interest. Less extensive research programs are given in Table 8. Also, some topics under investigation are included. The authors are keenly aware that this review, by space limitations, has omitted the names of their students who carried out the research.

Table 8. Research Areas and Current Interest.

Area	Topic	Faculty
Paint, varnish, vegetable oils	Drying, weathering, Processing	L.L. Carrick J.C. Brier
Filtration	Flow through porous media, drying	L.E. Brownell
Gamma Radiation	Food & transplant preservation, chemical reactions	L.E. Brownell L.L. Kempe* J.J. Martin*
Synthetic Fuels	Adsorption, catalysis, reaction rates	R.R. White
Catalysis	Mechanisms, chemisorption, new concepts	G. Parravano J. Schwank*
Heat Transfer	Mathematical solutions, conduction, convection, liquid metals, boiling mercury, correlations, polymers in boiling	S.W. Churchill J.O. Wilkes* R.E. Balzhiser D.L. Katz* D.E. Briggs* J.H. Hand*
Coal and shale conversion	Nature of residues, filtration, application of microwaves	D.E. Briggs
Polymers	Mechanical Properties	G.S.Y. Yeh*
Two-phase flow	Pressure gradients in wells, pipelines	M.R. Tek*
Waste disposal	Use of wetlands	R.H. Kadlec*

Table 8. Research Areas and Current Interest. Continued.

Area	Topic	Faculty
Reactor performance	Control	R.H. Kadlec
Computer applications	Process simulation & optimization, graphical techniques	B. Carnahan
Mathematical applications	Numerical methods	J.O. Wilkes B. Carnahan
Mass Transfer	Mixing of dispersions, diffusion	R.L. Curl
Solid-liquid reactions	Reactivity of solid constituents in rock, acidization in wells, sonic stimulation	H.S. Fogler
Transport properties	Diffusion, viscosity, light scattering	E. Gulari*
Electrochemistry	Corrosion, plating	F.M. Donohue*

* Denotes current (1980) faculty member. J.S. Schultz, J.E. Powers, and E.H. Young are not listed here but have been included in paragraphs above. H. Wang has also recently been added to the staff.

TRANSFER TO THE NORTH CAMPUS

The large-scale pilot laboratory in East Engineering had become of diminished interest by 1950. In planning the eventual move of the Engineering College to the North Campus, the G.G. Brown Building was constructed in 1957 and included new but limited large-scale facilities. Since then, the faculty and students have been handicapped by having offices and classrooms in East Engineering and many research facilities in the G.G. Brown Building. Our move to the North Campus will be completed in 1982, when the Herbert H. Dow Building, currently under construction, is finished.

UNIVERSITY OF MICHIGAN

Table 9. General Description of Herbert H. Dow Building

Gross Area: 105,000 square feet
Net Assignable Area: 63,200 square feet
Types of Space: laboratories, classrooms, faculty offices, student service areas
Range in Classroom Capacity: 40 to 200 stations
Estimated Construction Time: 24 months
Typical Instructional Laboratories:
    Chemical Engineering
        Polymer Preparation and Testing
        Chemical Engineering Process Design
        Biochemistry Engineering Process Design
    Materials and Metallurgical Engineering
        Analytical
        Physical Measurement
        Mechanical Testing
        Heat Treatment
        Process Metallurgy
        Physical Ceramics
Graduate Laboratories:
    Chemical Engineering
        Energy Processes
        Bioengineering
        Polymers
        Thermodynamics
        Environmental Processes
    Materials & Metallurgical Engineering
        High Temperature
        Wet Chemical
        Ultrastructures
        Electron Microscopy
        X-ray Diffraction

NOTE: The Dow Building will have three levels. On Level II there will be a service connector to the existing G.G. Brown Building and the 30,000 square feet gross (24,000 square feet net) in that building which are being renovated.

Two magnificent gifts, from the Herbert H. and Grace A. Dow Foundation and the Harry A. and Margaret D. Towsley Foundation, have made possible the construction of the Herbert H. Dow Building, which will house the Departments of Chemical Engineering and Materials and Metallurgical Engineering on North Campus. The combined Dow/Towsley gifts are the largest ever given to the College, and one of the largest foundation gifts in the University's history.

The total cost of the Dow building will be $10.5 million, of which $4 million was given by the Dow Foundation and $1.5 million by the Towsley Foundation. The balance comes from construction and undesignated gifts to the College's recent Capital Campaign.

Named for the founder of the Dow Chemical Company and his wife, the Herbert H. & Grace A. Dow Foundation was established in 1936 and dedicated to "the public benefaction of the inhabitants of the city of Midland and the people of the State of Michigan". It concentrates its efforts on educational and civic projects. Groundbreaking ceremonies were held on April 22, 1980, with completion of the building scheduled for 1982. An architect's sketch of the North Campus Engineering Complex is shown in Plate 4. A general description of the Dow Building is given in Table 9.

With their resolution to name the Dow Building as they did, the U-M Regents have formally linked the name of one of the most distinguished figures in the history of the American chemical industry with an engineering college which has, over the years, achieved an international reputation for academic excellence. The association is fitting, for the ties between the Dow Chemical Company and the U-M College of Engineering are strong. Both grew to positions of prominence in the state and nation at roughly the same time, and over the years the College and the Company have enjoyed a mutually beneficial relationship.

THE FUTURE

Some indication of future developments may be gleaned from a historical perspective. After the faculty are settled on the North Campus in the Dow Building, one would expect a burst of new activities. The challenge is to maintain freedom for each faculty member to investigate any area which has great need where chemical engineering expertise will assure progress. At the same time, formal classroom instruction in the central disciplinary sciences of mathematics, chemistry, physics, biology, and chemical engineering science can provide a common background.

The needs are great in providing our energy supplies, in controlling the unwanted changes in our environment, in applying chemical engineering discipline to new areas like biology, while remem-

Plate 4. THE HERBERT H. DOW BUILDING - NORTH CAMPUS
CHEMICAL ENGINEERING / MATERIALS AND METALLURGICAL ENGINEERING
UNDER CONSTRUCTION 1980

bering that the process industries provide the most significant vehicle for chemical engineers to serve society.

REFERENCES

1. The University of Michigan: An Encyclopedic Survey, Volume III, University of Michigan Press, Ann Arbor, 1953. "The College of Engineering", pp. 1161-1180, and "The Department of Chemical and Metallurgical Engineering", pp. 1190-1200.
2. D.L. Katz, "Development of Chemical Engineering at the University of Michigan", Chemical Engineering Progress Symposium Series, Vol. 55, No. 26, pp. 9-15, 1959.
3. D.L. Katz, "Practice What You Teach", Phillips Lecture, Oklahoma State University, 1978.
4. B.A. Uhlendorf, "The Engineering Research Institute (University of Michigan)", Michigan Alumnus Quarterly Review, Vol. LIX, No. 21, August 8, 1953.

CHEMICAL ENGINEERING AT THE UNIVERSITY OF WISCONSIN: THE

EARLY YEARS

        Edward E. Daub

        General Engineering Department
        University of Wisconsin-Madison

    Historians of science are known to say that a bad scientific theory is better than none because, despite its errors, it offers a point of departure for thought. That observation seems also to apply to the account that Charles Burgess's biographer has given of the origins of chemical engineering at the University of Wisconsin. McQueen opened his discussion as follows:

> One of Burgess's good friends was John Butler Johnson, who was called to Wisconsin in 1899 to assume the newly-created office of Dean of the College of Mechanics and Engineering.... After taking office he lived only three years; but they were years of notable service to the college, and to the department which had been entrusted to Professor Burgess. He sympathized with Burgess's theories of education, and promised to aid him in his efforts to establish a strong modern course in chemical engineering.[1]

    According to McQueen, Burgess hoped to enlarge the applied electrochemistry degree program that he already headed so as to include instruction in chemical engineering and thereby serve the needs of all chemical industries. The new Dean, the story continues, lent Burgess his aid.

> Dean Johnson had some good ideas about publicity. To help gain general support for the proposed new department, he enlisted the aid of Magnus Swenson '80, one of Wisconsin's most famous citizens, renowned as educator, inventor, chemical engineer and industrialist. Swenson responded generously,

and gave several talks. The first of these,
delivered in Madison early in 1900, was printed
as Bulletin No. 39 of the University of Wisconsin,
and read throughout the State.... He had seen
chemical engineering work wonders ... and called
on the University of Wisconsin to aid the good work
... by setting up a separate department of Chemical
Engineering. Burgess was delighted. He knew that
this was the best possible type of propaganda...[2]

McQueen's account, thus, provides us with a point of departure
for studying the beginnings of chemical engineering education at
the University of Wisconsin, alerting us to three major actors and
a scenario. The actual events at the turn of the century, however,
show that Dean Johnson and not Burgess played the key role and that
he proposed to the University President in 1900 that Magnus Swenson
chair the new degree course.[3] Swenson though was busily pursuing
the many commercial ventures spawned by his various patents and,
with President Adams taking leave of his office due to illness
and then resigning in 1901, Dean Johnson's aggressive moves failed
to bring formal Regent action and the necessary funding.

The only visible fruit of his effort was the addition in
1902 of instruction in chemical machinery under Burgess's direction
in the Applied Electrochemistry Course that he chaired. Johnson
died that summer in a freak accident before he could complete his
plans for this new venture in engineering education. The seed that
he had sown, however, took firm root in Burgess's ambitions,
culminating in the establishment of an eight course sequence and
a chemical engineering degree in 1905.

Thus, chemical engineering at the University of Wisconsin
bears the unique feature of emerging from the previous program in
applied electrochemistry, and not from industrial chemistry as
was wont to be the case. In fact, the Chemistry Department at
Wisconsin under Louis Kahlenberg's new chairmanship initiated an
industrial chemistry degree course in 1908 that rivalled chemical
engineering and threatened its survival, a dramatic reversal from
the cooperative days that Burgess and Kahlenberg had enjoyed in
applied electrochemistry. Burgess left in 1913 and not until the
1920's did chemical engineering under Otto Kowalke clearly out-
distance its rival.

THE APPLIED ELECTROCHEMISTRY COURSE

In the spring of 1899 the College of Engineering established
a new degree course in applied electrochemistry with young
Charles Burgess as its head. It was to be a part of the electrical
engineering program with the first two years in common with the

traditional electrical engineering course. That course had only recently been established in 1891 with the appointment of Dugald C. Jackson, a former division engineer with the Edison Company. Under D.C. Jackson's leadership, electrochemistry became a significant emphasis in the new electrical engineering curriculum. Thus, an 1899 study of eighteen electrical engineering curricula showed Wisconsin offering several times as many credit hours in electrometallurgy and electrochemistry as other institutions and, in addition, ranking among the leaders in chemistry hours as well. Some institutions did not even offer electrochemistry.[4]

In a paper read in 1894 D.C. Jackson indicated that three laboratories were needed to teach electrical engineering, namely, laboratories of dynamo, electrolysis and alternating current equipment. With regard to electrolysis he said:

> The electrolysis laboratory has received comparatively little attention in American engineering schools, but its importance makes it deserve an allotment of its full share of equipment and instructional force. There seems to be a time coming when the electrolysis equipment will be made equal in importance to the dynamo equipment in all first-class engineering schools.[5]

And D.C. Jackson set about developing such a laboratory at the University of Wisconsin.

Instruction in applied electrochemistry originally fell to Professor John E. Davies, a physicist who had served as professor of electrical engineering prior to D.C. Jackson's coming,[6] but Davies' mind apparently had too theoretical a cast to suit the more practical task of preparing engineers for industry that Jackson had in mind. In speaking about electrical engineering laboratories in 1894 Jackson made it quite clear that a professor of electrical engineering must be in charge.

> The practice, at present common in many colleges, of placing the electrical engineering course with its classroom and laboratory instruction under the direction of the professor of physics is just as absurd as it would be to devolve the instruction in advanced thermodynamics and the design and construction of steam engines under that professor. It is equally absurd for the professor of electrical engineering to attempt to teach physics.[7]

Samuel Fortenbaugh, a Cornell graduate with industrial experience, was appointed assistant professor in electrical engineering in 1894 and assigned the role of teaching applied electrochemistry and developing the electrochemical laboratory.[8] Davies' original description for the course, Electrolysis and Electrometallurgy, was immediately changed to sound a more practical tone, for example, topics such as Grotthus chains and the velocity of ions were no longer mentioned. Moreover, a footnote stressed that the approach in all applied electrochemistry courses was thoroughly practical and that those students who wished to enter teaching should take appropriate courses in physics.[9]

Burgess entered the scene when he graduated with the first class from the new electrical engineering course in 1895 and was appointed as assistant to Fortenbaugh. By 1898 he had gained a master's degree and attained the position of assistant professor with full responsibility for the applied electrochemistry program. Some indication of the progress made in developing the electrolysis laboratory during those intervening years may be gained from his brief article in the Western Electrician on the electrolytic equipment at the University of Wisconsin.

> ... [electric furnaces] three in number, are available for investigation involving the use of the high heat of the electric arc and the electrolysis of compounds fused by the heat of the electric arc and the electrolysis of compounds fused by the heat of the electric current or by heat externally applied. The one in the middle [of the figure] is a Moissan furnace and large magnets for directing the arc which is maintained by a current of 100 amperes at from 40 to 60 volts. The electrodes are kept from excessive heating by circulating water. At left a Borchers apparatus for the electrolysis of melted substances is shown. A carbon rod encased in a porcelain cylinder serves as anode, while the iron crucible serves as cathode. The heat for melting the substance is applied by means of a Perrot furnace and a Fletcher burner. The apparatus is designed for a current of about 50 amperes and intended for the separation of such metals as magnesium, lithium, etc. from the double chlorides. The other, a Borchers apparatus for the electrolysis of electrothermally melted substances, is constructed for currents from 100 to 150 amperes, and is available for use in the separation of the melted oxide of aluminum.[10]

And in another article we find Burgess elaborating what may have been one rationale for this investment in energy intensive electrochemical equipment, a rationale that sounds remarkably contemporary, namely, how to raise "the central station load factor... to a value which will enable the equipment to be utilized to its fullest extent."[11] Methods proposed to solve the load problem via energy storage, according to Burgess, involve energy losses in the transformations for and from storage as well as capital losses due to interest payments for and depreciation of the costly equipment required. The solution to the load equalization problem, therefore, might be better found in electrochemical and electrometallurgical processes, provided that such processes can be operated intermittently and that the amount of power required can be varied in response to other load demands. Unfortunately, Burgess lamented, few electrochemical processes could presently come to the electrical engineer's aid.

He summarized their pros and cons. Electroplating requires only small amounts of power and that power is but a small fraction of the cost. The refining of metals entails large capital investment in anodes and cathodes and therefore the equipment could not be profitably left idle during the periods of the day already high in electric load. Electric furnace processes, such as the manufacture of calcium carbide, might prove promising because the capital costs are not so high and the processes themselves operate intermittently.[12]

Then again, Burgess said, decomposition of salts might be the best possibility. He speculated about producing sodium hypochlorite in a sodium chloride decomposition cell with the diaphragm removed to permit diffusion and reaction of the dissolved chlorine and sodium hudroxide.

> The electrolytic apparatus necessary for this process is of the simplest possible kind, consisting of an insoluble anode, a solution of salt, and a cathode of almost any metal. The resulting solution might be utilized in the neighborhood of the plant or might be distributed around the city to hospitals, etc., for disinfecting and other purposes...[13]

But here too there were problems to solve: to find an anode that is cheap and not readily disintegrated; to devise a method to maintain the bleaching strength of the solution; to avoid the loss of undecomposed salt remaining in solution. Thus, Burgess could not yet propose a satisfactory electrochemical answer to the load problem encountered by electrical engineers in electric power generation, apparently a major reason for attempts to link chemistry and electrical engineering.

His efforts to bring together the chemist and the electrical engineer, however, did lead in 1899 to the formation of the applied electrochemistry degree as an alternative course under electrical engineering. In his article in the <u>Western Electrician</u> he had commented on the communication gap between the two professions.

> It is an unfortunate fact that electrical engineers
> have devoted little time and attention to chemical
> questions, and that chemists have been equally
> negligent of the technical side of electrical engi-
> neering, and as a result the former have been telling
> of the disinfecting properties of electrolyzed salt
> water and the latter persist in describing an electric
> current as having a pressure of so many Smee cells and
> a quantity of so many volumes of gas liberated per
> second.[14]

In launching the new course he had found a cooperative and competent chemist, Louis Kahlenberg, of whose help he wrote,

> In establishing this...course the Electrical Engineering
> department is especially fortunate in having the
> co-operation of the departments of Chemistry and
> Physical Chemistry. The latter department, which has
> charge of the instruction in theoretical electro-
> chemistry, is under the direction of Dr. Louis
> Kahlenberg, who has worked with some of the noted
> chemists of Germany, and who, since coming to this
> university, has accomplished results of such note as
> to place him in the front rank among those in the
> field of Physical Chemistry.[15]

Burgess seems to have been particularly enthused over Kahlenberg's success in the electrolytic separation of lithium and other metals from their electrolytes at room temperature, circumventing fused electrolysis, and he speculated that the same might now prove true for aluminum and even make aluminum electroplating possible.[16]

In any case Kahlenberg's support must have helped immensely in gaining approval of the new engineering course for, after a brilliant record in chemistry at the University of Wisconsin in the early 1890's, he had gone to Leipzig to study physical chemistry with Ostwald, receiving his doctorate in 1895 summa cum laude. Upon returning to the University of Wisconsin, he found the position in physical chemistry occupied but managed to find a temporary appointment as instructor in the School of Pharmacy. The position in the Chemistry Department soon opened to him, and he rose quickly to full professor by 1900.[17]

Both Kahlenberg and Burgess gained expanded roles in the new degree course. Students in the applied electrochemistry course were to take a full year of electrochemistry and a semester of physical chemistry with Kahlenberg, and a quartet of courses with Burgess -- the long established pair, Primary and Secondary Batteries and Electrochemistry and Electrometallurgy, plus a new pair, Industrial Electrochemistry and Electricity Applied to the

Treatment of Metal Surfaces.[18] When the university newspaper, The Daily Cardinal, announced the Regents' formal approval of the new engineering course in 1899, it probably expressed the hopes of both young professors in saying, "it is expected that the enrollment will be very large owing to the large demands at present for just this kind of engineering work."[19] Unfortunately, no class was ever to number more than three, and when the new Dean of the College arrived in that same year of 1899, he came with a proposal to link chemistry with engineering that ultimately proved more successful and led to the chemical engineering degree.

JOHNSON'S PROPOSAL AND SWENSON'S SUPPORT FOR CHEMICAL ENGINEERING

John Butler Johnson became the first Dean of the College of Mechanics and Engineering in 1899. Although instruction in engineering dated back to the Morrill Act, a concerted effort to develop engineering education at the University of Wisconsin came much later. Upon the recommendation of a special study committee, the Regents formally constituted the College of Mechanics and Engineering in 1889 and authorized a number of new ventures, including the establishment of the electrical engineering course to be accomplished by D.C. Jackson.[20]

The expansion of the College and the growing need for a new engineering building made it necessary to find an engineering educator to head the program. The Regents delegated that task to President Charles K. Adams and, after considerable search, the suggestions that Adams received directed his attention to Johnson, then Professor of Civil Engineering at Washington University in St. Louis. He was appointed in January 1899 and, after the Legislature had voted the funds for the new building, he spent the next half year visiting technical schools in Europe.[21]

Returning in September, Johnson collected his thoughts for his inaugural address entitled, "Some Unrecognized Functions of Our State Universities," to be given in conjunction with the meeting of the Western Society of Engineers in Madison. President Adams made it a special point to invite Magnus Swenson, an alumnus who had distinguished himself as a chemical engineer in the sugar, meat packing, and textile industries and who now designed and manufactured special machinery primarily for chemical industries.[22] Adams asked Swenson to offer some remarks.

> I should be much gratified if you would respond and speak to those who are present on some feature not foreign to an occasion of this kind. I understand Dean Johnson's address will be on the general subject of the organization and scope of a modern College of Engineering. I am not sure as to the title of his

> paper, but in a general way I think I understand the trend of his thought.[23]

And one of the major trends of his thought was to recommend the establishment of chemical engineering.

Johnson viewed chemical engineering as the combination of the talents of the chemist and the mechanical engineer. His visit to Germany included schools where both industrial chemistry and mechanical engineering were taught enabling the students to become chemical engineers, and he noted that movements were afoot in this direction in America. The present industrial practice, Johnson said, of having chemists diagnose the problems and mechanical engineers attempt the remedy is analogous to having one physician diagnose an illness and another prescribe treatment without understanding the disease.[24]

Both talents must be joined in one individual, for great mechanical ingenuity is needed to solve many of the problems that the chemist confronts in industrial practice and "failure is certain unless the chemist is also an expert mechanical engineer."

> Whenever these two kinds of ability have been embodied in the same individual, and these coupled with a reasonable amount of executive power, the fortunate possessor has not only attained to wealth and fame, but he has proved a blessing to all the industries with which he has been connected.... Furthermore, there are untold realms remaining to be conquered by this potential giant, the chemical engineer.[25]

To prepare students for those realms, Johnson envisioned courses in a department of chemical engineering which, in addition to traditional metallurgical engineering, would cover

> ... the applications of chemistry and mechanical engineering to the manufacture and testing of various products such as sugar, glucose, starch, glue, soda, soap, wood-pulp, paper, portland cement, leather, paints, varnishes, dyes, tobacco, beer, spirits, chemicals, illuminating gas, and a hundred other commodities of daily use.[26]

Johnson's catalogue of applications may have suggested to some of his listeners that courses in chemistry might suffice, so he reiterated his earlier point that pure chemistry lacks an emphasis on how chemical knowledge is applied in industry and that chemists lack the mechanical and inventive skills that the teaching of industrial chemistry require. The graduate from this new course should be an engineer and not a chemist.

We have no record of the remarks that Magnus Swenson may have made on this occasion, but he must have responded most positively because Johnson was soon to remind Swenson by letter of their discussion regarding a lecture he would give "before our engineering faculty and students, which we can have printed as a bulletin and circulate throughout the state as a sort of campaign document in favor of the establishment of a strong course in chemical engineering."[27] Johnson asked that Swenson reinforce what he had said, demonstrating from personal experience the advantages of the combination of chemistry and mechanical engineering, punctuated by actual cases "of the benefits on the one hand, and the financial loss and sometimes ruin, on the other, which would make the case so strong that there would seem to be but one side left to the question."[28] Johnson reiterated that Swenson's audience would be the people of the state and, thinking ahead perhaps to the budgetary increase that he would soon be asking for, he added, "including the members of the incoming legislature."

Swenson did not disappoint. In his opening paragraph he called the chemical engineer the joint bearer of two professions, "chemist as well as mechanical engineer;" and in the final paragraph he eloquently appealed to the university and to the people of the state:

> Let the University of Wisconsin take hold of this matter in earnest, and I predict for the department of Chemical Engineering a career that will reflect credit on the University and redound greatly to the prosperity of the state.[29]

Moreover, in his discussion of the sugar industry as a case in point, he italicized every specific reinforcement of Johnson's proposal.

Chemists, Swenson said, had not been employed in sugar manufacture until costs became crucial in the price squeeze of the eighties. Only then did chemists identify various losses in sugar manufacture, such as entrainment of sugar solution in the vapor from the evaporators, but these problems were not readily solved.

> This was due largely to the fact that <u>the men who were able to point out the losses were unable to apply the remedy</u>. They were simply analytical chemists whose sole duty seemed to be to make the planters as unhappy and miserable as possible by pointing out to them the enormous losses without being able to suggest a remedy for them. Moreover, <u>the remedy had to be applied largely by engineers</u>

who could not grasp the situation owing to lack of technical knowledge of the subject, and the contempt which many of these practical gentlemen had for the opinions of the theoretical and apparently helpless chemist.30

But, he added, sugar yields have now increased far beyond any expectations because of "the gradual evolution of the chemical engineer who has both the necessary chemical knowledge and the technical training."

Swenson then touched upon a number of chemical industries already active in the state that could benefit if they developed higher grade products with improved profit margins by using the scientific skills of the chemical engineer. In the paper industry he pointed out that the United States was almost wholly dependent for photographic paper upon France and Germany. Such paper, he said, must not contain any chemicals that react with silver salts and, given the increasing market created by the rise of the amateur photographer, whoever solves the problem will be richly rewarded.31

He also recommended the manufacture of Portland Cement by making use of the extensive deposits of marl in Wisconsin. Natural cement made directly from Wisconsin marl brought only one-third the price that artificial or Portland Cement could claim on the Chicago market. The artificial variety, of course, required a careful blending of limestone with the right proportion of clay. Keeping those proportions within narrow limits called for chemical vigilance, and "the thorough grinding, sizing, and blending of the ingredients" provided an additional field for the chemical engineer. Swenson again indicated that America could not yet produce enough to meet national needs and judged this lack to be "a strong commentary on the backwardness of our chemical industry."32

Thus, Swenson dealt with two industries from the many that Johnson had catalogued as subject matter for the chemical engineering program, contenting himself with brief mention of the others -- "glue, beer, soap, glycerine, zinc white, wood alcohol, acetic acid, chemical fertilizers, matches, ammonia from gas liquors" -- that would also benefit greatly from the presence of the chemical engineer. He then devoted the last half of his talk to the inviting prospects for the development of a beet sugar industry in Wisconsin.

He did so for a variety of reasons. He had made his mark as a chemical engineer in the sugar industry and knew the need for reliable machinery, since with any delays the sugar in dilute solution would irreversibly ferment or invert. The economic forecasts were good; Michigan had recently built beet sugar processing plants with excellent results, and the Wisconsin

Agricultural Experiment Station had demonstrated that Wisconsin's climate and soil were equally right for growing sugar beets. It was also an industry where Swenson foresaw opportunities for American inventive ingenuity in contrast to the conservatism of Europeans who then were operating most of the plants.[33]

But more importantly, especially for us as we search for the roots of chemical engineering education, Swenson gave this more narrow focus because the territory of chemical engineering was too broad to cover in a new department.

> It will no doubt seem to you that I have contracted my subject to a narrow limit; but it is as impossible for me to treat of the subject of chemical engineering broadly and impress with its importance as it would in my opinion be imprudent and unwise to start a department of chemical engineering on too broad lines to begin with. Better teach a few branches thoroughly and well than a dozen indifferently. I have selected the one that to me seems by far the most important and promising for the state.[34]

Swenson envisioned a small experimental sugar beet processing plant where the students would gain practical experience and opportunity to experiment.

> ... such a plant would yield results that would be in every way practical, and one thoroughly familiar with its operation would with very little experience be fully competent to design and take charge of a large plant. Moreover, such a plant would be easily susceptible of changes, and processes that would suggest themselves to the students could be readily tried.[35]

Swenson was wrestling here with the problem that plagued the early attempts to formulate a curriculum for chemical engineering, a problem that Edward Orton, Dean of the College of Engineering at Ohio State, was to articulate so very well in 1902 when he discussed the question of how to subdivide the field of chemical engineering.[36] It was the conflict between the industrial tradition that expected an engineer to have sufficient practice in his field so that he could immediately contribute to solving real industrial problems and the fact that chemical industry was so diverse that no curriculum could possibly give a student the necessary depth of practical knowledge and experience to cover that diversity.

Chemistry students, Orton said, receive a thorough grounding in the fundamental principles in all branches of chemistry but without industrial knowledge. He compared them with graduates from an ordinary academic course who find themselves "at the threshold of life with well-trained tastes, keen perceptions and good reasoning power, but of no special utility anywhere."[37]

In engineering, by contrast, the field is subdivided and narrowed so that the ground to be covered is reduced and the student can gain competence. "All who bear the name engineer," Orton said,

> ... have been familiar with this feeling. It is a sort of cult among engineers that they shall not only know how to reason and how to learn, but shall know where and how to start for the attainment of any given goal.[38]

A primary condition for awarding the degree of chemical engineer, therefore, should be that the graduate is equally competent for his field as other engineering graduates are for theirs. Or, as Orton succintly stated it, "hold up the value of the word engineer and make it mean something to carry it."[39]

The problem was that, apart from the fields of metallurgy and ceramics, into which Orton collected the clay, glass and cement industries and for which Ohio State had developed a course already in 1894, he could find no unifying theme among the chemical industries.

> There are large industries, such as the manufacture of acids, alkalies, salts, chemical reagents, etc., in which the work is strictly inorganic. But there are dozens of others in which organic work is exceedingly varied and important, as in the manufacture of aniline colors, perfumes, soaps, glycerine, explosives, etc. It seems, therefore, that no general or far-reaching affinities can be shown to exist between the chemical industries which are neither metallurgical nor ceramic.[40]

In conclusion Orton maintained that successful training for this more diffuse field of chemical engineering would be possible with a five year course that placed a strong emphasis on diverse chemical operations. Orton's suggestion seems to point toward unit operations, the direction that chemical engineering education was later to take.

> Personally, the writer believes that, notwithstanding the great diversity of industries represented in the field of general chemical engineering, it is possible to train men in five years so that they could enter any of these industries fairly well equipped. The

> operations of the chemical plant are merely the
> operations of the laboratory on a large scale.
> If the student be given a good drill on quanti-
> tative chemical preparations and has access to
> apparatus for performing the ordinary processes
> of heating, cooling, calcination, reduction,
> oxidation, filtration, lixiviation, etc., in the
> same general way they would be done in the works,
> there is no reason why he should not give a good
> account of himself even in an industry of which
> he knows practically nothing at the start.[41]

Thus, it would seem that Swenson, in focusing on the potential for a sugar beet industry in Wisconsin and the prospects for making an experimental processing plant a part of the education of chemical engineers, was trying to define a field that was commercially significant but of sufficiently reduced scope so as to assure the competence expected of engineers. When Johnson submitted his report to President Adams at the close of his first academic year, he incorporated Swenson's idea into his request for funds to found a chemical engineering degree, and suggested that Magnus Swenson head the program.

> I desire ... to obtain a grant from the next
> legislature to establish a course in <u>Chemical
> Engineering</u>. This would require a suitable
> person (like Mr. Magnus Swenson), to direct the
> work of the course, and two or three assistants
> in industrial chemistry and in the mechanical
> engineering for chemical operations. It would
> involve an addition to the Engineering Building
> and the constructing therein of a small beet-
> sugar experimental plant. (See Mr. Swenson's
> lecture on this subject.) ... With this state aid
> we could at once inaugurate a most successful
> Chemical Engineering Course, and thus develop in
> this state a great beet-sugar industry.[42]

THE CHEMICAL ENGINEERING COURSE ESTABLISHED UNDER BURGESS

But fate dictated otherwise. President Adams took leave of his office due to illness and resigned in 1901; Dean Johnson suffered a fatal accident in 1902. Nonetheless some progress was made. A new course was added to the applied electrochemistry curriculum in the spring semester 1902, Chemical Machinery and Appliances, under the direction of Charles Burgess. The <u>Daily Cardinal</u> reported that

> ... the remarkable progress in industrial chemistry

and electrochemistry in this country has been due
most largely to the improvements in machinery and
appliances warrants a study of the economical
handling and treatment of raw materials; machinery
for crushing and grinding; apparatus for drying,
crystallizing, and evaporating; and the properties
of materials for constructing such apparatus.[43]

Moreover, at the next meeting of the Society for the Promotion of Engineering Education, Burgess outlined another new direction towards chemical engineering that the applied electrochemistry curriculum had taken, a course in chemical preparations and industrial chemistry that he described as follows:

Each student makes on a fairly good scale many common
chemical products. He is supplied with the necessary
raw materials which are charged up to him at current
market prices, and the finished produce is weighed,
analyzed for purity, and the value of this same is
credited to his account, at the market rates of this
material. He is required to make a report giving a
statement of the efficiency of the process, materials
consumed, materials produced, waste products, methods
of procedure on a large scale, the machinery and
apparatus necessary, and an estimate of labor and
power.[44]

Thus, in the absence of official action by President Adams, the Regents, and the Legislature in support of the strong recommendations that Dean Johnson had made, the applied electrochemistry curriculum became the province for several new courses in chemical engineering; and the mantle for a department of chemical engineering intended for Magnus Swenson later came to rest on Burgess's shoulders.

It is not certain, of course, that official inaction alone can account for the fact that Magnus Swenson did not assume that role. His discussion of the sugar beet industry reflected the interests of the inventor-entrepreneur that he was more than the educator that he might have been. Between 1886 and 1896 Swenson had secured patents that significantly improved many of the processes and machinery used in the sugar industry, from cane-cutters and the automated conveyor chains to move the crushed cane to horizontal tube multiple effect evaporators.

Swenson evaporators made their way to Australia, France, Germany, Austria, Algeria, Cuba, Hawaii and Scotland.[45] He began this work after successfully developing a process for refining sugar from sorghum cane at a major installation in Fort Scott, Kansas. His switch to the design and construction of machinery led to the formation of the Walburn-Swenson Company headquartered in Chicago where he reigned as Vice President from 1891.

Thus, the prospects for a beet sugar industry whet his appetite for inventing new machinery and for selling it. By the close of 1902 he already was doing so. In a letter to the President of the Wisconsin Sugar Company he wrote,

> I heard some time ago that you intended to erect a couple of factories in Wisconsin,... If this is the case I would be very glad to see you with reference to the machinery. As I previously wrote you, we have furnished practically all of the special machinery to three large plants during the past season. All of them are running very successfully and the one at Sebewaing is considered to be the finest factory in the state of Michigan.[46]

And additional correspondence during the years that Dean Johnson held office reveal that Swenson had a project of even greater scale in mind.

The letters suggest that he was working on a combination of processes that would produce the following: cotton fibers for paper making, oil from the cotton seed, alcohol from sorghum, sugar from sugar beets, and then, indicative of his motto to "save the waste," a cattle feed consisting of dried cotton seed hulls, sorghum cane and sugar beet refuse. He had an extensive correspondence on the topic with Harvey W. Wiley, Chief of the Bureau of Chemistry at the United States Department of Agriculture, because Wiley held patents for making alcohol and for drying vegetable matter.

The two had met when they were "just plain Sorghum Cranks," to use Swenson's phrase, as he later wrote on the occasion of the 25th Anniversary dinner in honor of Wiley in 1908. His account of their meeting is priceless.

> It must have been about 1889 when he and his department were marooned out in the extreme western part of Kansas, a place as dry as the Lord and prohibition could make it. After the most strenuous efforts, covering many weeks of time, I succeeded not only in keeping them alive but floated them back to civilization. It was not a question then of bottles or kegs - but barrels. Anyone at all intimate with Dr. Wiley in those days could have easily predicted his coming greatness as a food expert, as he had, even then, a very strong developed taste...[47]

At the turn of the century Wiley occasionally wrote Swenson

about potential projects for investment, always hoping that he could invest a few thousand dollars in a venture financed by Swenson and his wealthy friends, but the prospect for using his two patents was the first to have a good chance of success. The two men had apparently discussed the idea at the Waldorf in New York in October 1901 and Wiley quickly followed their conversation with a letter.

> I am anxious to develop this matter both because I think it is a good thing, from an alcoholic and food side, and also because I want to make some money out of it. I am getting old and I hate awfully to die poor and leave you to live rich. If we could both be poor or both be rich, it wouldn't matter, but as you are bound to be rich I don't see anything I can do but come up to your standard.
>
> Appreciating your great helpfulness in this matter and feeling that you can do what would be quite impossible for me to accomplish, viz, represent this business to men who have money to put into it...[48]

Wiley sent copies of his patents and cautioned Swenson about the critical importance of attaining sterilization temperatures in the drying process so as to prevent fermentation.

Within a year Swenson had built a small plant which he described to Wiley.

> This arrangement that I have made is very similar to that shown in your patent and consists of a box in which I have a number of drags dragging the material back and forth over the perforated plate, and through the entire apparatus I have a current of hot air, which is drawn through by the use of a suction fan.
>
> I have also an arrangement whereby the material, if it should not come out sufficiently dry, can be returned to the inlet and pass through the dryer again. From the first dryer it passes into a second dryer, built exactly in the same manner, but more for the purpose of insuring the materials being heated to the required temperature for complete sterilization. Out of this dryer it passes into a reciprocating press which forms it into bales of 14 inches diameter and of any length that we choose to make them.
>
> I have endeavoured to make the machinery as automatic as possible and also have insured the materials being

> baled while still in a hot condition, so that I do not think we shall have any trouble from fermentation.
>
> I will also skin dry the bales after they are made in another dryer, so as to prevent molding on the outside.49

Wiley was pleased to hear of the progress made and arranged to visit Swenson in Chicago en route back to Washington after an address at Kansas University in October. What happened during that visit we do not know, but the few remaining letters from Wiley never mention that project again! And Swenson seems to have put an end to his idea for preparing animal feed from cotton seed hulls, and sorghum and sugar beet pulp.

But my point is, as I hope their correspondence shows, that Swenson's main interests were as an inventor-entrepreneur and not as an academic. In fact, at the very time that Swenson began these discussions with Wiley, he received a letter from the program chairman of the Western Society of Engineers reminding him of an earlier request for a paper by Swenson on chemical engineering and the need for education in the field. Swenson had previously replied that he might be able to do so but could not name a date. Nor did he this time.50 Every evidence, therefore, suggests that Swenson supported Johnson's efforts to establish a course in chemical engineering but not that Swenson himself had any strong designs on becoming head of it. It is symbolic perhaps of his support that in 1905, the year that Swenson became a Regent of the University of Wisconsin, the Board of Regents voted the chemical engineering course with Burgess as its head.

Eight courses entered the catalogue under the instructional category of Chemical Technology. Several years later, after the Department of Chemical Engineering was formally established, this designation continued under Chemical Engineering, along with Applied Electrochemistry and Electrometallurgy as separate categories, thus indicating that the two constituted different degree programs. The eight courses were as follows:

1. Chemical Machinery and Appliances. General Operations, such as crushing, grinding, separation by the wet and dry methods, filtration, distillation; problems involving the application of thermodynamic principles to evaporation, drying, and refrigeration; various types of apparatus used for carrying on these processes.

2. Technical, Fuel, Gas and Oil Analysis. Laboratory instruction in the technical analysis of fuels, illuminating and fuel gases, products of combustion, etc.

3. Technology of Fuels. Lectures relating to the production, properties, and uses of solid, liquid and gaseous fuels, and a consideration of refractory materials and furnace construction.

4. Chemical Manufacture. Instruction and laboratory practice in the preparation of chemical products on a scale sufficiently large to involve the use of filtering, evaporating, centrifugal, and other chemical machinery, and to obtain data upon which the design of a plant may be based. The design of a plant for the production of one or more materials, with analysis of costs is required of each student.

5. Industrial Chemistry. Class-room work relating to the manufacture of the more important chemical products such as acids, alkalis, cements, fertilizers, glass, paper, etc,; the equipment of plants, and problems connected with the same, chemical resistance of materials.

6. Thermal Efficiencies. A study of heat utilization and losses in chemical and metallurgical reactions involving the use of high temperatures. The purpose of this course is to give instruction and practice in the application of thermal data to chemical transformations.

7. Illumination and Photometry. A laboratory course in the measure of the illuminating power of gas and oils.

8. Manufacture and Distribution of Gas. A class and laboratory study of the development of the gas industry, the chemistry of gas manufacture, and the apparatus and problems involved in the manufacture, distribution and consumption of gas for fuel, power, and illuminating purposes.[51]

As initially proposed the chemical engineering course was scheduled to cover five years, with the two chemical technology courses, Illumination and Photometry and Manufacture and Distribution of Gas, to be required only of those fifth year students specializing in gas engineering.[52] But by the following year this was changed to a four year course leading to the B.S. degree in chemical engineering requiring the first five courses in chemical technology, and an optional fifth year leading to the degree of chemical engineer for those specializing in gas engineering or some other particular line of work.[53]

Burgess seems to have chosen gas engineering as one promising focus for his chemical engineering degree course. The gas industry received top billing as a primary target for the new department in an article that appeared in the alumni magazine in the 1905 commencement issue.

> Perhaps there is no more insistent call for chemical engineers than that which comes from the maker of illuminating and fuel gas. Already several of the largest gas companies in the country have asked for the privilege of having the first men specializing in the new department at the University, and the demand from this one industry alone will, beyond doubt, be greater than can be met by that institution for some time to come.[54]

These new technically trained men, it assured the reader, would find ample opportunity both for their ambitions and for success because the gas industry, despite its long history, still had need to develop its full capacity to improve its service and to find uses for tar, ammonia and other by-products.

In further paragraphs the article went on to single out other industries that would benefit: 1) cement because it "is rapidly replacing stone and steel in all fields of construction, most notable of which is probably that of new reinforced concrete now just being developed;" 2) fertilizer because whether "made of phosphate rocks, from blast furnace slag, from the constituents of the air, or from animal and vegetable material," the technical problems require both the knowledge of chemistry and broad engineering skills; 3) beet sugar because, with United States annual consumption at four billion pounds and with but ten per cent coming from domestic production, the beet sugar industry will rapidly grow in Wisconsin and "the chemical engineer who shall discover new and better processes for extracting all possible sugar from the beet will have no need to ask for kinder fortune;" 4) paper making because, having denuded the forests and being under government pressure to conserve our woodlands from further ruthless waste, the paper maker must seek the aid of chemistry to find new materials so that "the straw of flax, wheat and other grains, the grassy growth of the marshes, all, it is safe to say, can be made to yield the cellulose now obtained from trees, and it will be the special province of the chemical engineer to discover how this may best be done."[55] Finally, and somewhat surprisingly to our age where sanitation has long been the territory of the civil engineer, the article mentions "the scientific solution of sanitary problems, such as the purification of water and air supplies and the disposal of sewage" as a field where the chemical engineer's skills would come into play.

Burgess's efforts to publicize the new course apparently succeeded, for the Daily Cardinal observed in an editorial during the first semester of its life:

> The University has been noted for its unusual
> activity in the matter of new courses and new fields
> of technical learning. Perhaps the latest, and one
> of the most promising fields now offered the student
> in engineering at Wisconsin is the course which fits
> young men to become chemical engineers. Just this
> year an announcement of the branch appeared in the
> annual catalogue, but there are now some twenty men
> in this work.[56]

The university newspaper had previously highlighted the appointment of Judson C. Dickerman as assistant professor of engineering, the man that Burgess chose to strengthen the practical and business dimensions of his envisioned course.

> His training has been such as to peculiarly fit him
> for this work, he being a graduate in the Massachusetts
> Institute of Technology in the class of 1895. Since
> that time he has been engaged in manufacturing chemistry,
> and leaves a position as superintendent at the large
> works of the Merrimac Chemical Company of Boston,
> Mass., to enter the faculty of the University of
> Wisconsin. He has had a wide experience in the
> manufacturing of a large variety of chemicals, the
> construction of plants, installation and operation
> of machinery, and the development of new processes
> and methods...[57]

Dickerman undoubtedly contributed greatly to another development that the article mentioned, the fact that the old chemistry building was to be remodeled in order to accommodate the new department by the first of the coming year.

The account given by McQueen of how Burgess "captured" the old building for chemical engineering seems somewhat apocryphal. As Burgess apparently recounted the story, he had only managed to secure a part of the old Chemistry Building in the bidding war among various departments, though admittedly the most important part.

> Then he unobtrusively put through an order for the
> casting in bronze of some large letters, assorted
> as follows: E-4; I and N-3 each; C and G-2 each; A,
> H, L, M, R - 1 each. The total was nineteen letters;
> the order was "rush." Then one dark night, over a
> certain main entrance, a scaffolding went up.
> Mysterious sounds were heard - heaving and huffing
> and hammering; equally mysterious lights were seen.
> Came the dawn. And there, high above the portal of
> what had been the old Chemistry Building, was a new

name, in letters of imperishable bronze. The nineteen characters, now rearranged, spelled CHEMICAL ENGINEERING.[58]

McQueen's only support for the story is the reminiscence of Otto Kowalke half a century later to the effect that the letters helped resist encroachment by the "Medical crowd."

Fortunately, we have a rather straightforward record of the remodeling from Dickerman's hand. "The old Chemical Building," he said, will "henceforth ... be known officially as the Chemical Engineering Building," the second and third stories, however, were to be reserved "for work in Electrical Testing and Physiological Chemistry."[59] But the Electrical Testing equipment would be for photometry and available to chemical engineering students for determining the lighting values of oils and gases.

Dickerman then described the principal laboratories and their equipment.

> At the north end of the basement will be located the Laboratory of Manufacturing Chemistry. Here the equipment consists of practical working models of the apparatus and machinery used in chemical manufacture. Among some of the apparatus now or soon to be provided, may be mentioned -- centrifugal machine, steam jacketted kettle, vacuum pan, still and condensers, filter presses, gravity and suction filters, tanks with and without agitators, hot air and vacuum dryers, rotating, reverberatory, and gas furnaces, vacuum and compression pumps, hydraulic press, and a multiple effect evaporator.[60]

It was here that students would gain experience in preparing products with real cost accounting and in designing a plant from their data to meet a given productivity per day. There would also be a room with auxiliary equipment in the basement - "crushing, grinding and polishing machinery, including roll crusher, disc grinders, disintegrator, mechanical mixer, ball mill, sand blast apparatus, and polishing wheels" - to prepare the raw materials for the course in chemical manufacture and in applied electrochemistry.

Two rooms on the first floor were to contain the equipment for the course in technical gas, fuel and oil analysis. There would be discussion and practice

> ... in the analysis of chimney, producer, illuminating, metallurgical, and mine gases, and gas analysis for control of chemical processes; also the determination of the heating values of coal, oil and gas, both by

analysis and by direct determination, with several types of combustion calorimeters.[61]

There would also be a room for the microscopic examination of metals and for microphotographic work and a major laboratory devoted to electrochemistry, where the research on the electrolytic production of pure iron, funded by the Carnegie Institute, would be housed.

So the inauguration of the new course seems to have proceeded smoothly, and it gave Charles Burgess his first real chairmanship of a department. The Applied Electrochemistry Course continued to appear as a separate option in the Catalogue, but no students adopted it. Burgess formally asked in 1909 that his title be changed from Professor of Applied Electrochemistry to Professor of Chemical Engineering, and his request was granted.[62] But by 1909 the fate of his department lay in some doubt because two new degree courses had arisen to challenge his domain, one in industrial chemistry and the other in mining and metallurgical engineering.

COMPETITION FROM INDUSTRIAL CHEMISTRY AND MINING ENGINEERING

Louis Kahlenberg became chairman of the Chemistry Department in 1907, the year that also found him serving as chairman both of the newly formed Wisconsin chapter of the American Chemical Society and of the Chemistry Section of the American Association for the Advancement of Science. He sought no less a stature for the department that he now chaired than he himself had attained, and his first venture was to launch a whole complex of degree courses in chemistry.

The *Daily Cardinal* carried the announcement under the headline, "TRAIN EXPERTS IN CHEMISTRY," and indicated that the new four year courses were designed to train a battery of chemists - "analytical chemists, industrial chemists, agricultural and soil chemists, sanitary and food chemists, and physiological chemists" because of the "great demand for well trained men in these special fields of industrial chemistry." The article gave the details for the various degree options, and Kahlenberg shrewdly incorporated Burgess's new courses in chemical technology as part of his plan, especially for the industrial chemists. "The course for the industrial chemist," the *Cardinal* reported, "includes in addition to the elementary work in organic and inorganic chemistry, training in technical fuel and gas analysis, chemical machinery, and chemical manufactures, ..."[63] And when the University Catalogue appeared at the close of the school year with the full curricula for these various new degrees, Burgess and his colleagues in chemical engineering were listed among the Staff of Instruction, and Dickerman's Industrial Chemistry offering had been added as a required course as well.[64]

Moreover, the purpose given for these various chemistry degrees suggests the underlying conflict between industrial chemistry and chemical engineering as training grounds for the needs of American industry.

> And again, since the exploitation of the more readily accessible resources of our country, such as its wealth in timber and minerals, for example, is fairly well accomplished, attention is naturally being directed more and more toward manufacturing articles from materials whose value is not obvious at sight, materials that indeed have often been discarded as worthless. It is in this work that the services of the analytical and industrial chemist are an absolute necessity,... Indeed, in many of our industries it is already realized that it is necessary to have the service of a chemist continually in order to properly inspect the raw materials bought, to control and improve the process of manufacture so as to avoid waste of power and material, and to test the grade of the finished article.[65]

And the final paragraph echoes the essential debate between the two schools, the training of the chemist resting on fundamental principles while the engineering tradition called for practical know-how as well.

> It is not the purpose of this course to prepare men for each individual industry; for that would be well-nigh impossible, since it requires special practical experience to become proficient in any particular line of work. The aim is rather to give the student a training in the fundamental principles of chemistry and cognate sciences... and a proper grounding in mathematics, French and German. With this preparation he will readily be able to adapt himself to any special industry with which he may later on be connected.[66]

At a time when the chemical engineering curriculum still struggled to solve the problem of finding a suitable focus for coping with industrial diversity, Kahlenberg's initiative surely threatened the fledgling program in chemical engineering. It numbered only four graduates in 1907 and five in 1908, and by 1910 the industrial chemistry graduates already matched the chemical engineers with six apiece. Thus, in March of that year when Burgess wrote his wife, he perhaps only feigned his light-hearted words: "I have gotten so I actually enjoy Kahlenberg's solicitation of my students to abandon my department for his - enjoy the knockings of my various alleged friends."[67] The only promising news for the fate of Burgess's program in 1910 was that the other new rival within his own College of Engineering, the new degree course in mining engineering, netted but one graduate.[68]

Word of the new course in mining engineering came in 1908, the same year that the Chemistry Department had announced their new degrees, but the details were not as clearly worked out. In late April the announcement was made that Edwin C. Holden had been appointed as professor of mining engineering at the university.[69] In addition to listing his credentials -- a graduate of the Columbia University School of Mines with extensive experience as "surveyor, assayer, superintendent, general manager and consulting engineer" and with practice that "included the design and operation of mine plants as well as mine examinations" -- the announcement went on to indicate that the variety of mining and geology electives in the College of Engineering would now be integrated into a four year degree course.

> The strong department of geology and chemistry, together with the well developed courses in civil and mechanical engineering make the state university unusually well equipped to give a good course in mining engineering. With the addition of mining engineering (sic) to direct the work and give additional courses, the university will be able to provide a strong course in mining engineering.[70]

The details for the new course, however, apparently had to await Holden's arrival, for the University Catalogue could only report as follows on the Mining Engineering Course.

> Provision has been made for the establishment at once of a department of Mining Engineering in the College of Engineering. A complete course in Mining Engineering will be arranged and instruction offered at the opening of the academic year 1908-09. Details of this course may be had on application.[71]

In the catalogue at the close of the first year under Professor Holden, eight courses and a thesis were listed: Excavation and Quarrying; Tunneling, Boring and Shaft Sinking; Prospecting and Mine Development; Exploitation of Mines; Mine Engineering; Ore Dressing and Coal Washing; Gold and Silver Milling and Cyanidation; and Mining Thesis.[72] Not until the following year, when another professor was appointed and given responsibility for courses in metallurgy, did mining engineering begin to conflict with Burgess's plans for chemical engineering.

The new courses covered the following topics: the assaying of ores; the chemistry that is applied and the mechanical equipment that is used in metallurgical processes; the metallurgy of copper, lead, silver, gold, zinc, tin, antimony, arsenic, mercury, bismuth, and cadmium; and finally the metallurgy of iron and steel.[73]

Burgess must have found these courses in metallurgy competitive with his own plans for chemical engineering because the very same catalogue lists four new chemical technology courses under Chemical Engineering: Technical Pyrometry, Metallurgy of Iron and Steel, Metallography, and Constitution of Alloys. Most of these courses were staffed by James Aston, a new arrival who had been one of Burgess's students in the class of 1898. Burgess invited Aston in 1908 to work under the research grant that Burgess had received from the Carnegie Institute for work on electrolytic iron and its alloys, and Burgess now apparently called upon him to establish metallurgy as a part of chemical technology.[74]

Between 1910 and 1913, the year that Burgess resigned from the university, the numbers of graduates with a B.S. degree from the three competing courses were as follows:

	ChE	Min. E.	Chem.
1910	6	1	6
1911	5	6	4
1912	7	1	11
1913	8	8	15

In 1911 Burgess reported to the Dean of the College of Engineering that the growth of his department had been hampered by the diverting of students to the newly established Mining Department and to the Industrial Chemistry branch of the Chemistry Course. When he submitted his letter of resignation in 1913, however, he singled out the duplication of effort between the two engineering courses as the major problem that demanded a solution. In doing so, he referred to the courses offered under chemical engineering that overlapped with mining and metallurgy offerings.

> Metallography is closely allied to, if not indeed a part of, Metallurgy. Metallurgy of Iron and Steel cannot be taught as a scientific study without at the same time dealing with Metallography. Pyrometry is the fundamental study on which industrial thermal chemistry is based...[75]

So when Burgess resigned the future of chemical engineering at the University of Wisconsin was far from clear. Chemical engineering had emerged from the applied electrochemistry course and yet the chemical engineering degree did not require any of the applied electrochemistry offerings. The catalogue continued to list them under the previous degree program of applied electrochemistry and electrometallurgy. Furthermore, the courses in chemical technology that were added, beginning first with chemical machinery in 1902, and with a good number more in 1905, formed an essential core not only for chemical engineering but for the rival industrial chemistry degree as well. Finally, the latest additions to the chemical technology field – pyrometry, metallography, metallurgy of iron and steel, and the constitution of alloys – were not yet firmly established because the College of Engineering had also authorized a separate degree in mining and metallurgical engineering.

So, although the new chairman of chemical engineering Otto Kowalke, who was one of the first to graduate under Burgess from the Applied Electrochemistry Course in 1903 and who had continued as an instructor and then as assistant professor both in the applied electrochemistry and in the chemical technology fields, recited a lively verse at Burgess's farewell dinner that began, "I've been here eighteen years b'Gosh and started the chemicals well,"[76] Burgess must have had some doubts about the fate of his ambitious efforts over those two decades.

CONSOLIDATION OF THE CHEMICAL ENGINEERING COURSE

Upon Burgess's departure several years of indecision entered the chemical engineering curriculum because it now became necessary to integrate the courses from applied electrochemistry with those that focused on chemical machinery and manufacture, the gas industry, and metallurgy. From 1913 to 1915 only one chemical engineering course was required of all students, Fuel Gas and Oil Analysis. Students were free to elect the remainder of their chemical engineering studies from among the other courses that had been developed. The catalogue indicated that the "arrangement affords the student an opportunity to specialize in Applied Electrochemistry, Metallurgy, Gas and Fuel Engineering, or Chemical Manufacture."[77]

But by the 1916-17 academic year a consolidated curriculum emerged that integrated the diverse offerings. Accommodation with the mining and metallurgy program was apparently reached by dropping the courses on the metallurgy of iron and steel and on the constitution of alloys while retaining the courses on pyrometry and metallography. They were replaced with courses in industrial organic chemistry and in engineering materials. The curriculum now totalled 154 credits with the following requirements in chemical engineering.[78]

> Sophomore Year
> Fuel and Gas Analysis 2 credits
>
> Laboratory instruction in the technical analysis and calorimetry of solid, liquid, and gaseous fuels, products of combustion; photometry of gas, etc..

Engineering Materials 2 credits

Technology of limes, cements, refractories, fuel, and iron and steel.

Junior Year
Technical Pyrometry 3 credits

Laboratory instruction in the calibration and use of pyrometers for tests on refractory materials, slags, and furnaces; thermal analysis of steels, ores, and chemical compounds.

Metallography 3 credits

A laboratory course comprising the microscopic examination of metals and alloys and a study of their structure as affected by heat and mechanical treatment. Special attention given to tool and alloy steels.

Chemical Machinery and Appliances 2 credits

Lectures and recitation relating to the disintegration and sizing of materials; filtration, evaporation, distillation, drying, etc.; materials of construction; transportation and measurement of solids, liquids, and gases.

Summer work

Chemical Manufacture 4 credits

Laboratory practice supplementary to chemical machinery courses; tests of chemical machinery, manufacture and recovery of products, special problems.

Senior Year

Applied Electrochemistry 5 credits

The laws and fundamental principles of electrolysis, with their applications to electro-plating, the refining of metals and electrochemical manufacture.

Industrial Chemistry 3 credits

Lectures and recitations on the technology of the alkali, acid, paint, paper, and other important chemical industries.

Applied Thermal Chemistry 4 credits

The thermo-chemistry and physics of combustion,
chemical and metallurgical processes; drying,
evaporation, furnace calculations and design.

Industrial Organic Chemistry 2 credits

Technology of organic chemical industries.

Ten credits during the senior year were assigned to thesis and electives. The major electives continued the previous emphasis on electrochemistry and gas manufacture.

The Electric Furnace 3 credits

Its construction, operation, and industrial applications.
One lecture and four hours in the furnace room.

Batteries 2 credits

Construction, testing and operation of primary and
storage cells.

Gas Manufacture and Distribution 3 credits

Lectures on the manufacture and distribution of coal,
carburetted water, oil, and producer gas; the uses of
gas for domestic, manufacturing, and power purposes;
recovery of by-products.

Analysis of Bitumens for Highways 2 credits

The chemical properties of oils, residual asphaltic
petroleums, natural asphalts, tars from gas manufacture,
etc., used in the construction of highways.[79]

In addition, advanced courses for 2 credits each were offered in applied electrochemistry, chemical manufacture, and metallography.[80]

This consolidated curriculum stayed in tact with but minor changes; and the chemical engineering course soon outdistanced its rival in engineering, the mining and metallurgy course, and overtook and passed the industrial chemistry course decisively in the 1920's.

	ChE	Chem.	Min. Eng.
1914	7	10	3
1915	19	14	9
1916	17	18	2
1917	19	20	3
1918	14	19	4
1919	5	13	1

UNIVERSITY OF WISCONSIN

	ChE	Chem.	Min. Eng.
1920	15	34	7
1921	27	20	14
1922	29	22	12
1923	46	17	11
1924	31	14	13
1925	20	13	10

Two minor changes might represent turning points in the development of chemical engineering as a discipline. In 1925 the course entitled Chemical Machinery became Manufacturing Operations, reflecting perhaps the appearance of Walker, Lewis and McAdams' <u>Principles of Chemical Engineering</u> in 1923; and the course entitled Applied Thermal Chemistry became Chemical Engineering Calculations.[81]

In addition, a 4 credit course in special problems became required of all students in their final semester, replacing the previous requirement of a thesis.

Thus, the evolution of the chemical engineering curriculum from the innovative but never successful Applied Electrochemistry Course was complete, and a study of the graduate thesis topics over this span reflects that change. The graduate degree for the fifth year was the degree of Chemical Engineer. During Burgess's tenure as chairman, the emphasis was primarily electrochemical and electrometallurgical: "The Influence of Inequalities of Temperature upon the Corrosion of Iron" (1907); "Electrical Fixation of Nitrogen" (1908); "The Magnetic Properties of Electrolytic Iron and Iron Alloys" (1909); "A Further Study of Pyrophoric Alloys," (1909); "The Electrolytic Corrosion of Iron and Steel by Stray Currents" (1911); "A Study of the Variables Entering into the Production of Malleable Iron" (1913); "Notes on the Forging and Other Properties of Alloys of Electrolytic Iron and Other Elements" (1913); "A Microscopic Study of Electrolytic Iron" (1913).

The transition to the more appropriate diversity that characterized chemical engineering seems to begin with Kowalke's thesis in 1909, "A Study of Gas Calorimeters." In the following decade thesis topics dealing with paper and gas manufacture appeared: "Experiments with Jack Pine and Hemlock for Mechanical Pulp" (1912); "The Grinding of Cooked and Uncooked Spruce and Substitutes for Spruce in the Manufacture of Mechanical Pulp" (1915); "The Effect of Varying Humidities on the Physical Properties of Paper" (1916); "Some Fundamentals Affecting the Utilization of Gas Oil in Carburetted Water Gas Manufacturing" (1916); "The Load Carrying Capacities of Magnesia-Silica Mixtures at High Temperature" (1918); "Some Studies of Gaseous Combustion for Heat Treatment of Steel" (1918); "The Disposal of Waste from Ammonia Recovery in By-Product Coke Oven Plants" (1919); "Removal of Sulfur Compounds from Municipal Gas" (1920).[82]

Thus, the research during this period of development in the chemical engineering curriculum reflected the strong bias of the engineering tradition, namely, to contribute to the solution of practical industrial problems. Not until the 1920's will the research begin to focus on the fundamental principles of the emerging discipline of chemical engineering itself through the development of a doctoral program.

Olaf Hougen inaugurated that new research tradition at the University of Wisconsin in 1925 with his doctoral dissertation, "Transfer Coefficients of Ammonia in Absorption Towers." Thus, by the quarter century mark, the early years of struggle for a suitable curriculum in chemical engineering and for a niche in the university had ended, and the new era of research had begun.

ACKNOWLEDGMENTS

I am indebted to Professor Emeritus Olaf A. Hougen for his unpublished study of the history of the Chemical Engineering Department, ON THESE FOUNDATIONS, University of Wisconsin Archives, and to Professor Emeritus Aaron J. Ihde for his many papers on the history of the Chemistry Department. I am further in debt to these good colleagues for their helpful comments and criticisms of earlier drafts of this paper.

NOTES

1. Alexander McQueen, A ROMANCE IN RESEARCH: THE LIFE OF CHARLES F. BURGESS, Pittsburgh, 1951. p. 120.

2. IBID. pp. 120-121.

3. McQueen's work frequently depends upon reminiscences to recount the past. Unfortunately, none of the Burgess papers have survived. (Personal communication from George R. Bell, Chairman, Burgess Vibrocrafters, Inc.)

4. Francis C. Caldwell, "A Comparative Study of the Electrical Engineering Courses Given at Different Institutions," PROCEEDINGS OF THE SOCIETY FOR THE PROMOTION OF ENGINEERING EDUCATION 7(1899) 127-129. (Hereafter designated as PROCEEDINGS SPEE.)

5. Dugald C. Jackson, "The Equipment for Electrical Engineering Laboratories," PROCEEDINGS SPEE 2(1894) 221-235.

6. Storm Bull, "Technical Education at the University of Wisconsin," THE WISCONSIN ENGINEER 3(1899) 1-17, p.12.

7. D.C. Jackson, op.cit.(Note 5) p.222. Jackson went on to indicate the importance for the electrical engineering student to have instruction in physics and experiments in an independent physics laboratory.

8. McQueen, op.cit.(Note 1) pp.30,31.

9. CATALOGUE OF THE UNIVERSITY OF WISCONSIN, 1895-96, p.172.

10. Charles F. Burgess, "Electrolytic Equipment at the University of Wisconsin," WESTERN ELECTRICIAN 20(1897) 157.

11. Charles F. Burgess, "The Relation of Chemistry to Electrical Engineering," THE WISCONSIN ENGINEER 2(1897-98) 135-142, p.137.

12. IBID. 140-141.

13. IBID. 142.

14. C.F. Burgess, op.cit.(Note 10).

15. Charles F. Burgess, "Electro-chemistry at the University of Wisconsin," THE WISCONSIN ENGINEER 3(1899) 279-287, pp.280-281.

16. IBID. p.287.

17. Aaron J. Ihde and H.A. Schuette, "The Early Days of Chemistry at the University of Wisconsin," JOURNAL OF CHEMICAL EDUCATION 29(1952) 65-72, pp.67-68.

18. CATALOGUE OF THE UNIVERSITY OF WISCONSIN, 1898-99, pp.193-195, 121-123, 214-215.

19. THE DAILY CARDINAL, April 21, 1899, p.1.

20. M. Curti and V. Carstensen, THE UNIVERSITY OF WISCONSIN 1848-1925, Volume I, Madison, 1949, pp.445-447.

21. F.E. Turneaure, "John Butler Johnson," THE WISCONSIN ENGINEER 7(1902) 1-4, p.3.

22. THE UNIVERSITY OF WISCONSIN: ITS HISTORY AND ITS ALUMNI, edited by Reuben Gold Thwaites, Madison, 1900.

23. Letter from C.K. Adams to Magnus Swenson, October 6, 1899, MAGNUS SWENSON PAPERS, The Wisconsin State Historical Society, Box 1.

24. John B. Johnson, "Some Unrecognized Functions of Our State Universities," JOURNAL OF THE WESTERN SOCIETY OF ENGINEERING

4(1899) 376-387, p.379.

25. IBID. p.380.

26. IBID.

27. Letter from J.B. Johnson to Magnus Swenson, November 21, 1899, MAGNUS SWENSON PAPERS, Wisconsin State Historical Society, Box 1.

28. IBID.

29. Magnus Swenson, "The Chemical Engineer," BULLETIN OF THE UNIVERSITY OF WISCONSIN, No.39, Engineering Series 2(1900) 195-207.

30. IBID. p.199.

31. IBID. pp.200-201.

32. IBID. pp.201-202.

33. IBID. pp.203-206.

34. IBID. p.207.

35. IBID. p.206.

36. Edward Orton, "The Subdivision of the Field of Chemical Engineering," PROCEEDINGS SPEE 10(1902) 134-161.

37. IBID. p.142.

38. IBID. pp.142-3.

39. IBID. p.148.

40. IBID. p.141.

41. IBID. pp.150-151.

42. Report of Dean J.B. Johnson to President C.K. Adams, March 21, 1900, College of Engineering Archives, Steenbock Library. This important document was discovered by Olaf Hougen in preparing his unpublished history of the Chemical Engineering Department, ON THESE FOUNDATIONS, 1975, University of Wisconsin Archives.

43. THE DAILY CARDINAL, January 16, 1902, p.3.

44. Charles F. Burgess, "Electrochemistry as an Engineering Course," PROCEEDINGS SPEE 10(1902) 124-133, p.133.

45. Olaf A. Hougen, "Magnus Swenson: Inventor and Chemical Engineer" NORWEGIAN-AMERICAN STUDIES AND RECORDS 10(1938) 152-175, p.167.

46. Letter from Magnus Swenson to R.G. Wagner, December 30, 1902, MAGNUS SWENSON PAPERS, Wisconsin State Historical Library.

47. Letter from Magnus Swenson to the Chairman of the 25th Anniversary Dinner for H.W. Wiley, April 6, 1908. MAGNUS SWENSON PAPERS, Wisconsin State Historical Library.

   Wiley does not mention Magnus Swenson in his autobiography and, in his account of his years in Kansas, he claims that he never used whisky. The topic arises in relation to a committee investigation of his use of government funds at a time when there were widespread suspicions of corruption in the Agriculture Department.

   The only record of mismanagement of funds in the report was "that a jug, presumably containing whisky for Doctor Wiley's private use, was sent from Kansas City to Fort Scott, and twenty-five cents of government money were paid for freight thereon." Wiley continues: "As I never used whisky and had no occasion to give it to others I could not refrain from recalling a sentiment from the classics which I happened to remember: _Montes parturiunt, natus ridiculus mus_. (The mountains are in labor and brought forth a mouse.) HARVEY W. WILEY - AN AUTOBIOGRAPHY, Indianapolis, Indiana, 1930, p.184.

48. Letter from H.W. Wiley to Magnus Swenson, October 26, 1901, MAGNUS SWENSON PAPERS, Wisconsin State Historical Library.

49. Letter from Magnus Swenson to H.W. Wiley, September 25, 1902, MAGNUS SWENSON PAPERS, Wisconsin State Historical Library.

50. The two letters were from Andrews Allen. The first dated April 7, 1900 was quite specific.

   We would very much like to have a paper from you on "Chemical Engineering" and the need of a school for training in this line, at a meeting of our Society to be held on June 20th next... The Society will furnish the necessary lantern slides if you will provide us the photographs or original drawing in advance. It would also be well to have four or five parties to prepare discussion on the same ...

   The second letter dated October 9, 1901 simply asked for a

paper "on some subject in connection with Chemical Engineering." MAGNUS SWENSON PAPERS, Wisconsin State Historical Society.

51. CATALOGUE OF THE UNIVERSITY OF WISCONSIN 1905-1906, pp.290-291.

52. IBID. pp.261-263.

53. CATALOGUE 1906-07, pp.268-270.

54. THE WISCONSIN ALUMNI MAGAZINE, Commencement 1905, "Chemical Engineering Department," 303-307, p.304.

55. IBID. pp.304-306.

56. THE DAILY CARDINAL, November 11, 1905.

57. THE DAILY CARDINAL, October 18, 1905.

58. McQueen, op.cit.(Note 1) pp.126-127.

59. Judson C. Dickerman, "The Chemical Engineering Building," THE WISCONSIN ENGINEER 10(1905-06) 102-105.

60. IBID. P.103.

61. IBID. p.104.

62. McQueen, op.cit.(Note 1) pp.131-132.

63. THE DAILY CARDINAL, January 31, 1908.

64. THE UNIVERSITY CATALOGUE, 1907-1908, pp.252-253.

65. IBID. p.254.

66. IBID.

67. McQueen, op.cit.(Note 1) p.135. McQueen further suggests that chemical engineering students suffered from this conflict. "Kahlenberg's physical chemistry course was a required subject for Burgess's chemical engineering students and they had a difficult time with Kahlenberg because he seemed almost antagonistic toward engineers; so as one of them recalls, 'We engineers were an unhappy lot in Kahlenberg's course.'" IBID. p.136.

68. Data for the number of graduates comes from the lists of the names of the graduates under each course that were annually included in the University Catalogue for each year.

69. THE DAILY CARDINAL, April 30, 1908.

70. IBID.

71. THE UNIVERSITY CATALOGUE, 1907-08, p.309.

72. THE UNIVERSITY CATALOGUE, 1908-09, pp.350-352.

73. THE UNIVERSITY CATALOGUE, 1909-10, p.322.

74. The account given by McQueen states that Burgess began offering these courses already in 1908, although they do not appear officially in the catalogue until the 1909-10 academic year when the metallurgical courses under mining engineering also appeared. McQueen wrote:

> It was in that year [1908] that James Aston went to Wisconsin to assist Burgess in his work under the Carnegie grant. Because Aston was skilled in metallography and the metallurgy of iron and steel, Burgess promptly started offering courses in these subjects, taught by Aston. This seemed to fit in with the work done under the Carnegie grant on electrolytic iron, and taken by itself the move was perfectly natural. However, in the same year, 1908, Professor Edwin C. Holden, who had been engaged in the management of mines and as a consulting engineer, was called to Wisconsin to establish a department of mining engineering; there had previously been no such department at Wisconsin. Professor Holden wanted to have a well-rounded-out department and it was natural for him to ask that the teaching of metallurgy of iron and steel be handled by his department, along with the metallurgy of other metals.

McQueen, op.cit.(Note 1), p.145.

75. IBID.

76. IBID. p.149.

77. THE UNIVERSITY CATALOGUE, 1913-1914, p.373.

78. The course descriptions are taken from the UNIVERSITY CATALOGUE, 1915-16, pp.393-396, except for the two new courses, Engineering Materials and Industrial Organic Chemistry, because the descriptions in that issue gave more details than in later catalogues. The requirements and the descriptions for the above named courses derive from the UNIVERSITY CATALOGUE,

1916-17, pp.286-287, 292-293.

79. IBID. 1915-16, pp.394 and 396.

80. IBID. 1916-17, pp.292-293.

81. UNIVERSITY CATALOGUE, 1925-26, p.240.

82. UNIVERSITY CATALOGUES, 1907-1920.

83. Two PhD theses were completed during Charles Burgess's tenure: Oliver P. Watts in 1905, "An Investigation of Borides and Silicides;" Alcan Hirsch in 1911, "The Preparation and Properties of Metallic Cerium." Both, however, appeared before chemical engineering found its own research tradition. They represented the early electrochemical and electro-metallurgical orientation of the department. Olaf Hougen gives a complete list of both Masters Theses and Doctoral Theses from 1905 to 1970 in his ON THESE FOUNDATIONS, pp. A 30 - A 50.

EVOLUTION OF CHEMICAL ENGINEERING FROM

INDUSTRIAL CHEMISTRY AT KYOTO UNIVERSITY

        Fumitake Yoshida*

        Kyoto University
        Kyoto, 606
        JAPAN

    Kyoto University, located in Kyoto, the capital of Japan for nearly eleven centuries until 1868, is today one of the few major centers of learning in Japan. Its predecessor, the Third Higher School, was, despite its name, a college with faculties of law, engineering, and medicine. This became a university in the true sense of the word in 1897, when the national government founded Kyoto Imperial University. It was the second such Imperial University, the first being Tokyo Imperial University founded in 1886.

THE *KOZA* SYSTEM

    It seems appropriate for me first to mention the *koza*, or chair, system which is in a way unique to the major national universities of Japan, many of which are former Imperial Universities. The system was first defined in the Imperial University Act issued in 1886 as an Imperial Ordinance. This was substantiated in the University Standard, which was issued in 1956 as an ordinance of the Ministry of Education and is still in effect today.

    The *koza* (Lehrstuhl in German) is a subdivision of the department, which usually consists of several *koza*. The *koza* has one full professor, one assistant professor, sometimes one lecturer, two instructors, and one technician or secretary. There are no associate professorships in Japanese universities. The professor is responsible for the entire *koza*. He is assisted by all members

---

*Professor Emeritus, Chemical Engineering

of his *koza* in research and teaching. The instructors have usually completed graduate studies and often hold a doctorate.

The size of the budget allocated by the university to a department for expenses other than salaries depends on the number of *koza* it contains. After deducting common departmental expenditures, the department then distributes the remaining funds equally among its *koza*. As a result, the *koza* at major national universities all have practically the same budget. Creating a new *koza* is not a simple matter, because it requires the approval of the national Diet (Congress) as well as the Ministry of Education.

Arguments can be made for and against the *koza* system. Since a *koza* is guaranteed certain funds from the university to be used mostly for research, faculty members can at least pursue research on their chosen subjects without relying on outside sources for research grants. A long-term project, especially one conducted by a team, is easier under the *koza* system. It is also possible, if he so wishes, for a professor to switch into a new field of research, since the research areas specified for a *koza* are not mandatory. On the other hand, teaching staff sometimes find the road to promotion blocked under the *koza* system. For instance, an assistant professor who is only five years younger than the professor of his *koza* must wait until his professor retires to be promoted, however able he may be, unless he moves to another university, which is not necessarily easy in Japan. The same thing can be said about the promotion of an instructor to assistant professor.

THE EARLY YEARS OF INDUSTRIAL CHEMISTRY

Going back to its beginning, Kyoto Imperial University was started officially on June 18, 1897 with only the Faculty of Science and Engineering. In the following year the Faculties of Law and Medicine opened, and later other Faculties followed. The Faculty of Science and Engineering had 21 *koza* in seven Departments: Physics, Chemistry, Mathematics, Civil Engineering, Electrical Engineering, Mechanical Engineering, and Mining and Metallurgy.

The Department of Chemistry had four *koza*, those of organic chemistry, theoretical and inorganic chemistry, inorganic manufacturing chemistry, and organic manufacturing chemistry. The Department was divided into two Majors, Pure Chemistry and Manufacturing Chemistry. Three students majoring in pure chemistry and only one student majoring in manufacturing chemistry were admitted in September, 1989. A new *koza* of applied electrochemistry was added to the Manufacturing Chemistry Major in 1905.

In 1914 the Faculty of Science and Engineering was split into the Faculty of Science and the Faculty of Engineering. At this time the Major of Manufacturing Chemistry became the Department of Industrial Chemistry in the Faculty of Engineering.

As of 1920 the Department of Industrial Chemistry consisted of six *koza* which specialized in the following fields:

Koza I (Prof. T. Yoshioka) Silicate industries, industrial furnaces, and solid fuels.
Koza II (Prof. Y. Nakazawa) Applied electrochemistry, acid and alkali industries, fertilizer industries, and inorganic chemicals.
Koza III (Prof. I. Fukushima) Dye chemistry, textile chemistry, and paper chemistry.
Koza IV (Prof. H. Matsumoto) Fermentation industries, coal and gas industries.
Koza V (Prof. G. Kita) Fat and oil chemistry, hydrocarbon chemistry, chemistry of cellulose derivatives.
Koza VI (Prof. M. Miyata) Photographic chemicals, industrial chemicals.

THE OLD UNIVERSITY SYSTEM

The university system in Japan before World War II was different from those in the U.S. and Europe. After 6 years of primary school and 5 years of secondary school, students who passed a stiff entrance examination were admitted to a *koto gakko*, or 3-year liberal arts college, of which there were about 30 in the whole country. Having passed another entrance examination, graduates of *koto gakko* entered one of the several Imperial Universities. The *koto gakko* had only two majors, literature and science, and taught basic and liberal arts subjects, but not professional or pragmatic subjects. Students majoring in science studied, for instance, mathematics, physics, chemistry, biology, two foreign languages, as well as subjects in the humanities. Although their studies were rather intensive, students had time to read a lot and to think about their future. Thus, they were more mature when they entered university than are today's freshmen. Under the old university system students graduated in three years, which was long enough to teach only professional subjects. Graduates with a Bachelor's degree could proceed to graduate school, where the only requirement was research. On presentation of a suitable dissertation, they would receive a doctorate. There was no residence requirement, and Master's degrees were not granted. Thus, university education under the old system was, in fact, a 6-year two-step process. Some graduates of the old system still seem to harbor nostaligia for that system.

CHEMICAL ENGINEERING IN THE INDUSTRIAL CHEMISTRY DEPARTMENT

With the growth of the Japanese chemical industry after World War I, the professors of the Department of Industrial Chemistry at Kyoto realized the need for a chemical engineering approach to chemical technology. They also heard about the development of chemical engineering in the United States reportedly from executives of Suzuki & Co. in Kobe (the predecessor of Nissho-Iwai Co.), which owned several chemical factories. After negotiation with the Ministry of Education and with the approval of the national Diet, a new *koza* of Chemical Engineering was created in the Department of Industrial Chemistry in 1922. Since no one specialized in this area was available in Japan, it was decided to give engineering training to one of the young graduates of the Industrial Chemistry class of 1920, who was working with Dainippon Celluloid Co. (the predecessor of Daicel Ltd.) He was Prof. Saburo Kamei (1892-1977), who later became the founder of the Chemical Engineering Department at Kyoto. Although Prof. Kamei was appointed Lecturer in 1922 and Assistant Professor the next year, he was obliged to audit all the mechanical engineering courses at Kyoto University for three years from 1922.

In 1926 two courses on unit operations called "Chemical Engineering I" and "Chemical Engineering II", were added to the curriculum of the class graduating in 1928. These courses were started in the spring of 1926 and were taught jointly by professors specialized in various areas of industrial chemistry. The subjects dealt with included:

> Evaporation and drying (Prof.   Matsumoto)
> Crushing and mixing (Prof.   Yoshioka)
> Distillation and condensation (Prof.   Fukushima)
> Filtration and centrifuging (Prof.   Kita)
> Transport of materials (Prof.   Nakazawa)
> Cooling and heating (Assist. Prof.   Nakai)
> Compression of gases (Assist. Prof.   Ohmori)

No textbooks were used for these lectures. One of the graduates who attended these lectures recalls that Prof. Matsumoto, a specialist in fermentation, also gave a lecture on the quantitative aspect of fermentor design, which probably was the first biochemical engineering lecture in Japan. It is also said that Assist. Prof. Kamei taught part of these courses, but it is not clear what subjects he lectured on.

Prof. Kamei was sent abroad on a 2-year study leave beginning in the spring of 1927. He first went to Germany, as most Japanese scholars did in those days. He studied mostly in Berlin, but also managed to visit every technical college (Technische Hochschule) in Germany. He was disappointed, however, to find that chemical engineering in Germany in those days was not well organized and was something different from the kind of chemical engineering he wished to study. After writing to Prof. W.K.Lewis, he was admitted to the Massachusetts Institute of Technology, U.S.A. Although he studied at M.I.T. for only 5 months starting in November, 1928, he audited many courses ardently and was influenced greatly by the teachings of the founders of American chemical engineering. Prof. Kamei, who had difficulty with spoken English, recalled that Prof. W.H. McAdams was very kind and helpful to him. On his return to Kyoto in April, 1929 Prof. Kamei began teaching unit operations more or less along the line of the texbook, "Principles of Chemical Engineering" (1923). He was appointed Professor in 1930.

The curriculum of Industrial Chemistry for the class admitted in 1928 is shown in Table 1. The curriculum did not change much in the period 1926-1940. As I have mentioned, the undergraduate

Table 1. Curriculum of Industrial Chemistry (1928)

1st year
   Physical chemistry (9), Analytical chemistry (6),
   Inorganic industrial chemistry (6), Organic industrial chemistry (6), Solid fuels and industrial furnace (4),
   Liquid fuels and lubricants (4), Gaseous fuels and gas industry (4), Mechanical engineering I (hydraulics) (6)
   Mechanical engineering II (thermal) (6)
   Physical chemistry calculations (3)
   Analytical chemistry laboratory (54), Drawing (9)

2nd year
   Inorganic chemistry (9), Organic chemistry (9)
   Acid and alkali industries (6), <u>Unit operations</u> (6),
   Electrical engineering (6)
   Industrial analysis (20), Industrial chemistry laboratory (34), drawing (6)
   Electives: over 6 units

3rd year
   <u>Unit operations</u> (8), Industrial chemicals (4),
   Cement industry (2), Plant housing design (4),
   Physical chemistry laboratory (3), drawing (6)
   Electives: over 4 units
   Thesis research (82)

Numerals in parentheses indicate (hrs/week) x (number of terms).

One year = 3 terms

program in those days ran for 3 years. The academic year, which
started in April, had three terms, April to July, September to
December, and January to March. Looking at Table 1, one sees a
combination of pure and applied chemistry with a fair amount of
engineering and a great emphasis on laboratory work, including
thesis research. In fact, the third year students spent more than
25 hours a week in the laboratory working on their theses. The
Department of Industrial Chemistry in those days produced approximately 20 to 25 graduates per year. Since there was already a *koza*
of chemical engineering, some students chose topics in chemical
engineering for their Bachelor's theses. Approximately ten graduates of Industrial Chemistry did thesis research in chemical engineering under the guidance of Prof. Kamei during the period 1932-
1941. Prof. Kamei did extensive research in the field of drying
on problems such as the rates of diffusion of water in various
solid materials, the mechanism of drying of solids, and others.

BIRTH OF THE CHEMICAL ENGINEERING DEPARTMENT

The need for chemical engineers was increasingly felt with
the expansion of chemical industries in Japan and the Far East
during the 1930's. After the outbreak of World War II in Europe
in 1939, it became imperative for Japan to build all her chemical
plants domestically, since she could no longer rely in part on
imported plants. This was especially true of plants for liquid
fuels, which Japan, especially her navy, desperately needed for
survival. Under these circumstances, the Ministries of Education
and Finance, and later the national Diet, approved proposals for
the creation of two departments of chemical engineering, one at
Tokyo Institute of Technology and one at Kyoto Imperial University.
It is said that the Ministry of Naval Affairs, which owned oil
refineries, gave support to these proposals. Although a chemical
engineering department had opened one year earlier at Kanazawa
Technical College, this was the start of chemical engineering
education at the university level in Japan. These two departments
of chemical engineering, at Kyoto and the Tokyo Institute, opened
officially on April 1, 1940.

The Department of Chemical Engineering at Kyoto Imperial
University consisted of four *koza* which represented the following
fields:

Koza I   Fluid flow, heat transfer, instrumentation, etc.
Koza II Evaporation, distillation, drying, absorption,
           adsorption, extraction, reaction processes, etc.
Koza III Crushing, grinding, size separation, materials of
           construction, chemical engineering thermodynamics,
           etc.
Koza IV Chemical engineering calculations, design and

fabrication of chemical plant equipment, transport of solids, etc.

Since students of *koto gakko* did not know what chemical engineering was, Prof. Kamei wrote to all the *koto gakko* in Japan to publicize his new department. There were more applicatns to this new department than Prof. Kamei had expected, and seventeen ambitious young men were admitted. For several years after the start of his department, Prof. Kamei continued to write letters to the *koto gakko* publicizing the chemical engineering at Kyoto.

The classes at the new department started in April 1940. Table 2 gives the chemical engineering curriculum for the students

Table 2. Curriculum of Chemical Engineering (1941)

1st year
    Fluid flow (6), Materials of construction (6), Strength of materials (4), Mathematical analysis (6), Physical chemistry (9), Analytical chemistry (4), Outline of inorganic industrial chemistry (6), Outline of organic industrial chemistry (6), Outline of fuel chemistry (6), Physical chemistry calculations (3), Machine drawing (18), Analytical chemistry laboratory (43)

2nd year
    Chem. eng. stoichiometry (6), Unit operations (8), Heat transfer (4), Inorganic chemistry (4), Organic chemistry (6), Mechanical engineering (6), Machine design (6), Electrical engineering (6), Mechanical engineering calculations (3), Chem. eng. drawing I (15), Physical chemistry laboratory (21), Inorganic ind. chem. laboratory (14), Organic ind. chem. laboratory (14), Electives (3-6)

3rd year
    Chemical thermodynamics (2), Chem. eng. design (6), Chem. eng. calculations (6), Chem. eng. drawing II (6), Electrical eng. laboratory (3),
Plant practice (summer),
Thesis research (9)

Numerals in parentheses indicate (hrs/week) x (number of terms).

One year = 3 terms

admitted in 1941. Unlike the curriculum of Industrial Chemistry (Table 1), the Chemical Engineering curriculum included several chemical engineering subjects besides unit operations, such as chemical engineering stoichiometry, heat transfer, thermodynamics, and chemical engineering design and calculations. However, many subjects were shared with students of Industrial Chemistry. Construction of the departmental building, which had to be built of wood due to a shortage of materials, was started in December, 1940 and was completed in January, 1942.

The main difficulty faced by the new department was the recruitment of teaching staff. Since there was not a single chemical engineering graduate in Japan at that time, Prof. Kamei had to look for graduates of industrial chemistry and mechanical engineering who could teach in his department. Prof. Tatsuzo Okada (1895-   ), Assistant Professor of Industrial Chemistry specializing in applied electrochemistry, was appointed Professor of Chemical Engineering in 1940 to teach chemical engineering stoichiometry, thermodynamics, and materials of construction. He returned to Industrial Chemistry in 1943 but held a simultaneous appointment in Chemical Engineering until 1946. Prof. Keiji Hayami (1907-1970), a Mechanical Engineering graduate who had worked as an engineering officer at the naval arsenal, was appointed Assistant Professor in 1942 and Professor the following year. He taught fluid dynamics and strength of materials. He was forced to relinguish his post in 1946 by order of the occupation force on the grounds that he had previously been an officer in the Navy.

The period of Japanese involvement in World War II, December 1941 through August 1945, was a hard time for the Department. Shortages were severe not only in teaching staff but also in laboratory facilities and supplies. Because of the shortage of engineers in industry, the Government ordered all the university engineering schools to shorten the period of education from the regular 3 years to 2 years and 4 months. Thus, the first chemical engineering class of 17 students graduated in July 1942. The shortage of engineers was so serious that the Government introduced a system of allocating graduates in various disciplines to firms and government offices. One problem with this system was that chemical engineers were somehow included in the category of mechanical engineer. Since the quota of mechanical engineers allocated to chemical firms was small, it became difficult for chemical engineering graduates to get jobs in the chemical industry. In collaboration with Prof. Shunichi Uchida, the founder of the chemical engineering department at the Tokyo Institute of Technology, Prof. Kamei negotiated with and finally persuaded government officials to reclassify chemical engineer and put them in the category of "applied chemist" beginning in 1944. In 1945, however, the quota system was abolished with the end of the War.

KYOTO UNIVERSITY

POST-WAR DEVELOPMENTS AND THE NEW UNIVERSITY SYSTEM

The end of the War in 1945 dramatically changed the situation. Faculty members who were in military service or in industry returned to the campus. Shinji Nagata (1913-1974), an Industrial Chemistry graduate of Kyoto who was formerly with Dainippon Boseki Co. (the predecessor of Unitika Ltd.) and had been Assistant Professor since 1941, served in army beginning in 1944. He returned to the campus in 1945, and was promoted to Professor in 1949. Until his death in 1974 he did extensive research in mixing and in reaction engineering. He also did research at Purdue University (1958). Fumitake Yoshida (1913-   ), another Industrial Chemistry graduate of Kyoto who was formerly with Hitachi Ltd. and had been a part-time Lecturer since 1940, was appointed Assistant Professor in 1946 and Professor in 1951. Until his retirement in 1976 he did research in gas-liquid mass transfer operations, biochemical engineering, and biomedical engineering. He also did research at Yale University (1952) and the University of Wisconsin (1959), and taught at the University of California, Berkeley (1963) and the University of Pennsylvania (1970).

Tokuro Mizushina (1920-   ), a 1942 graduate in Chemical Engineering at Kyoto went first into military service, then was appointed Assistant Professor in 1944 and promoted to Professor in 1956. His research areas include simultaneous heat and mass transfer, and heat transfer in turbulent flow. He also did research at the University of Delaware (1953). Prof. Yuzo Nakagawa (1899-   ), a Mechanical Engineering graduate of Kyoto who was Professor in the Department of Fuel Chemistry was transferred to the Department of Chemical Engineering to take over the *koza* professorship which became vacant in 1946 with the resignation of Prof. K. Hayami. He did research in strength of materials, mechanical properties of polymers, comminution, and other areas. He returned to the Department of Fuel Chemistry in 1961, one year before his retirement.

Ryozo Toei (1922-   ), a Chemical Engineering graduate of Kyoto, was appointed Assistant Professor in 1948 and Professor in 1961. His research areas include drying, fluidization, solid-gas reactions, adsorption, and others. Naoya Yoshioka (1920-   ), another Chemical Engineering graduate of Kyoto, was appointed Assistant Professor in 1948 and Professor in 1962. His research areas cover solid-liquid separation, especially with thickeners and liquid cyclones, deep bed filtration, and rheology of viscoelastic fluids.

For several years after the War the shortage of materials and equipment for research and teaching remained severe. However, research work at the Department of Chemical Engineering became increasingly active as Japan and her industries gradually recovered from the damage of the War. The demand for chemical engineering graduates also increased.

In October, 1947, Kyoto Imperial University changed its name to Kyoto University. This accompanied a complete change of the Japanese University system. Influenced by the report of the U.S. Education Mission which visited Japan in 1946 and the conclusions of the national Committee on Education Reform, laws went into effect in 1947 which defined a completely new system of education consisting of 6 years of primary school, 3 years of middle school, 3 years of high school, and 4 years of undergraduate education of university or college.

Kyoto University held an entrance ceremony in July, 1949, for the students admitted under the new system, about 1500 of them. In addition, since at the time there were still graduates of *koto gakko* seeking entrance under the old system, the University also admitted about 1500 of these students in April, 1949, and about 1800 more in April, 1950. These latter were the last students under the old system. Thus, in the spring of 1953 Kyoto University produced graduates of both the old 3-year system and the new 4-year system.

In April, 1953 the graduate school under the new system opened, offering programs for the Master's degree (2 years) and the doctorate (minimum 3 years after the Master's degree). Course work as well as thesis research are required in the new graduate school.

The 4-year undergraduate curriculum under the new system differs from the 3-year curriculum of the old system in including liberal arts and basic subjects. Under the old system these subjects were taught at the 3-year *koto gakko*, before students entered the university. In the new university system the first 2 years are spent mostly in studying liberal arts and basic subjects, and only the last 2 years are devoted to professional subjects including Bachelor's thesis research. Furthermore, the requirement of two foreign languages makes the undergraduate curriculum of Japanese universities heavier than those in other countries. One way to relieve this would be to eliminate thesis research from undergraduate curriula and to encourage students to continue for Master's degree. In fact, most of the engineering graduates of Kyoto wish to proceed to the Master's program. However, only about half of them can be admitted to graduate school after an entrance examination, because of the quota of entering graduate students set by the Ministry of Education. Undergraduates are still required to do thesis research, which is considered very important in Japanese universities. In Japan graduates of major universities seldom move to another university for graduate work.

The exchange of information and personnel between Japan and the United States has increased gradually since around 1950, when Japan was still under occupation by the U.S. army. In 1951 a U.S. mission on engineering education, which included Prof. Barnett F. Dodge, Professor of Chemical Engineering at Yale, visited Japan and submitted a report, in which it pointed out the weakness of Japanese universities in chemical engineering. Prompted by the growth of Japanese industries, the Government decided in 1957 to adopt a policy of increasing the number of science and engineering students at national universities, which resulted in the creation of many new *koza* and departments. Under this policy, the number of *koza* in the Department of Chemical Engineering at Kyoto was increased from four to six in 1962-3, with a corresponding increase in the quota of entering students from 30 to 40.

For six months beginning in October, 1957, Prof. Olaf A. Hougen of the University of Wisconsin visited Japan as a Fulbright visiting professor and lectured at Kyoto and Nagoya Universities. Prof. R. Byron Bird of the same university also lectured at Kyoto and Nagoya on the same program for ten months beginning in October, 1962. The visits of these eminent professors seem to have exerted a good influence on chemical engineering education in Japan.

Koichi Iinoya (1917-   ), a mechanical engineering graduate of Tokyo, who was Professor of Chemical Engineering at Nagoya University, moved to Kyoto in 1964. He specializes in particulate technology, particularly dust collection and aerosol measurement, as well as control of powder processes. He has had research experience at the University of Wisconsin (1959) and Georgia Institute of Technology (1960, 1964)

Takeichiro Takamatsu (1925-   ), a Chemical Engineering graduate of Kyoto, was Professor in the Department of Sanitary Engineering before transferring to Chemical Engineering in 1970 to take up the professorship of a new *koza* created at that time. He specializes in process systems engineering and process control. He did research at M.I.T. (1961).

One section of chemical engineering at the Institute of Atomic Energy, Kyoto University, could be regarded as an extension *koza* of the Department of Chemical Engineering. Wataru Eguchi (1929-   ), a Chemical Engineering graduate of Kyoto, is responsible for this section. He teaches and supervises graduate students in the Department of Chemical Engineering. His research includes mixing in flow reactors, mass transfer with chemical reaction, extraction and other topics.

The *koza* occupied by the late Prof. S. Nagata was taken over in 1975 by Kenji Hashimoto (1935-   ), a Chemical Engineering graduate of Kyoto. He specializes in reaction engineering, adsorption, and other areas. He has done research at the University of Waterloo (1968).

The *koza* professorship which became vacant with the retirement of Prof. F. Yoshida was taken over in 1977 by Eizo Sada (1930-   ), a chemical engineering graduate of Nagoya University who had been teaching at Nagoya. His research includes mass transfer with chemical reaction, mass transfer through membranes, and other themes.

Table 3. Transition of Chem. Eng. Curriculum (1949-1978)

Courses	1949	1958	1967	1978
Applied mathematics	8	4	4	6
Chemistry	20	12	12	6
Mass/energy balance	2	2	2	2
Fluid flow	2	3	4	2
Heat/mass transfer	2	2	4	4
Unit operations	8	6	6	6
Thermodynamics	4	4	6	4
Applied Kinetics	—	4	4	4
Process control	—	4	4	2
Systems engineering	—	—	—	2
Process design	—	3	3	3
Chem. eng. calculations	—	6	2	10
	18	34	35	39
Materials of construction	2	2	3	2
Mechanical engineering	6	6	4	2
Electrical engineering	4	—	4	2
Drawing	15	6	6	2
Analytical chem. laboratory	15	12	12	6
Physical chem. laboratory	6	6	3	3
Ind. chem. laboratory	6	3	3	3
Chem. eng. laboratory	6	3	9	6
Elec. eng. laboratory	3	—	3	—
	36	24	30	18
Electives				
Thesis research				

(1) Subjects given at the Liberal Arts Division are excluded.
(2) Numerals indicate (hr./week) x (number of semesters).

He has done research at the University of Texas (1964) and West Virginia University (1965).

As of 1980 the Department of Chemical Engineering has eight *koza*, including one at a related research institute, with eight Professors (Iinoya, Mizushina, Yoshioka, Toei, Takamatsu, Eguchi, Sada, Hashimoto). The departmental faculty consists of about 30 members above the instructor level.

The Department produces approximately 40 Bachelors, 24 Masters, and several Doctors each year. Very roughly speaking, 70 % them go into process industries, 25 % enter engineering firms, and 5 % work in universities and government institutes.

The gradual change of the chemical engineering curriculum at Kyoto during the 30-year period from the start of the new university system in 1949 through 1978 can be seen from Table 3. During this period the emphasis given to chemistry, chemistry laboratory, mechanical engineering, and drawing has decreased, whereas subjects in chemical engineering, such as applied kinetics, process control, systems engineering, and chemical engineering calculations have been given greater emphasis.

FIVE DEPARTMENTS RELATED TO CHEMICAL TECHNOLOGY

Kyoto University has five departments related to chemical technology in the Faculty of Engineering, namely, Industrial Chemistry, Hydrocarbon Chemistry, Chemical Engineering, Polymer Chemistry, and Synthetic Chemistry. The latter four departments branched off from Industrial Chemistry in 1939, 1940, 1941 and

Table 4. Departments Related to Chemical Technology at Kyoto University

INDUSTRIAL CHEMISTRY (1914)	8 *Koza*
├─ HYDROCARBON CHEMISTRY (1939)	7 *Koza*
├─ CHEMICAL ENGINEERING (1940)	7 *Koza*
├─ POLYMER CHEMISTRY (1941)	7 *Koza*
└─ SYNTHETIC CHEMISTRY (1960)	6 *Koza*
	35 *Koza*

Table 5. Curriculum of Industrial Chemistry (1978)

A Group

    Outline of industrial chemistry (2), Technical foreign language (mainly German) I and II (2+2), Environmental chemistry (2), *Basic chemistry laboratory (5)

B Group

    Physical chemistry I - VI (2 x 6),
    *Physical chemistry laboratory (2)

C Group

    Inorganic chemistry I - IV (2 x 4)
    *Inorganic chemistry laboratory I - II (2 +2)

D Group

    Organic chemistry I - IV (2 x 4),
    Industrial organic chemistry I - II (2 + 2)
    *Organic chemistry laboratory I - II (2 + 2)

E Group

    Analytical chemistry I - IV (2 x 4),
    *Analytical chemistry laboratory (2)

F Group

    Materials for chemical plant equipment (2),
    Applied kinetics (2), Outline of unit operations (4)
    *Chemical engineering laboratory (1), *Drawing (2)

Electives

Thesis Research

---

1) Subjects given in 1st and 2nd year (languages, liberal arts, math, and basic sciences) are excluded.
2) Numerals in parentheses indicate number of units.
3) In addition to 44 units minimum from languages, humanities, etc., students must take at least 96 units, including 30 units minimum from basic sciences and 56 units minimum from subjects in this table, including electives. The 56 units must contain at least 10 from the subjects with * and at least 20 from other subjects, including at least 4 units from each of groups A to F.

1960, respectively (Table 4). Together, these five departments have 35 *koza*, which means they have 35 full professors and an equal number of assistant professors. Together with the related research institutes, these departments have more than 80 faculty members above the assistant professor level and form the largest and strongest group of chemical technology-related departments in Japanese universities.

These five departments at Kyoto University cooperate closely in their teaching programs, providing service courses, including laboratory classes, for each other. Students in one of these departments can take and obtain credits for any course given in one of the other departments providing their schedules so permit.

It is natural that the separation of the four departments from Industrial Chemistry has affected their curricula. The change of the curriculum of Chemical Engineering was shown in Table 3. The change of the curriculum of Industrial Chemistry is more striking, as can be seen from a comparison of the present curriculum (Table 5) with, e.g., the 1928 curriculum (Table 1), which was a sort of combination of applied chemistry and engineering. Today most of the engineering subjects have disappeared from the Industrial Chemistry curriculum, with only a few subjects in elementary chemical engineering remaining. The same is true of the curricula of Hydrocarbon Chemistry, Polymer Chemistry, and Synthetic Chemistry. Such changes seem inevitable, because many of the graduates of Industrial Chemistry and the three other departments which provide research chemists for industry need a strong background of chemistry, even though those industrial chemists who work in plant operation or process development would benefit from a stronger engineering background. On the other hand, some people, especially in the chemical industry, criticize the weakness of chemical engineering graduates in chemistry. It seems to me that, in practice, industrial chemists and chemical engineers make up their respective deficiency by studying on their own or through practical experience, or both, after they have entered industry.

The five departments related to chemical technology at Kyoto University produce roughly 230 Bachelors, 140 Masters, and 50 Doctors each year. The graduates of these departments, totaling nearly 7,000 to date, have made great contributions to the development of Japanese chemical industries. They have also contributed to the advancement of science and technology related to chemical industries.

DOCTORAL THESIS WORK IN CHEMICAL ENGINEERING IN THE UNITED STATES

FROM THE BEGINNING TO 1960

James O. Maloney

Department of Chemical and Petroleum Engineering
University of Kansas
Lawrence, Kansas

SUMMARY

Over 9000 doctoral degrees have been granted by the departments of chemical engineering in the United States, from the first one in 1905 through 1979. This paper is primarily concerned with those 2800 or so awarded up to 1960. A list of them has been assembled, believed to be complete, which contains the name of each recipient, the school awarding the degree, the thesis title and the year of the award. It is this list which constitutes the basis for some observations about doctoral work in this fifty-four year span.

By 1959 thirteen schools had accounted for over sixty percent of the degrees granted, with M.I.T. the leader at eleven percent and Michigan next with nine percent. Michigan at this time was providing fifty-seven teachers with doctoral degrees to departments of chemical engineering, the most of any school. The next three were M.I.T. - forty-one, Wisconsin - thirty-six, and Minnesota - thirty-two. Princeton had the largest percentage of its graduates in teaching - forty percent followed by Minnesota with thirty percent.

The thesis topics have changed with time. Until 1930 work centered around process and product studies. The study of unit operations provided thirty percent of the thesis topics in the decade of the thirties and above fifty percent in the forties. A number of the unit operations dealing with solid particles received scant attention. Furthermore only two thesis had titles that indicated they involved economics and no thesis appeared to consider education.

The utility of this list to the profession is pointed out and some further studies using it, or an amplification of it, are suggested.

INTRODUCTION

The material contained in this paper grew out of a study started in 1960 which had as its purpose a determination of which U.S. schools provided faculty to U.S. chemical engineering departments. The study was limited in that it counted only those who held a doctoral degree in chemical engineering, and who were teaching.

In carry out this study a list was assembled of every person known to have received a U.S. doctoral degree in chemical engineering from 1905 through 1959. In addition to the name of each person, the year the degree was awarded, the school and the thesis title were recorded. Then the booklet "Chemical Engineering Faculties 1959 - 1960", which gives the the faculty membership of every department, was used together with the list to find out where each faculty member received his/her doctoral degree.

The information collected in this study appeared to be of wider utility than originally thought, and it has been used as a basis to develop some additional aspects of doctoral work in chemical engineering in the United States up to 1960. The comments made in this paper will, in some cases, be regarded as unfounded with respect to particular schools, and readers may be moved to prepare specific studies on the doctoral program of a particular school.

In order to give the reader an initial overall view of the number of doctoral degrees granted each year in the United States from the first one in 1905 through 1979, Figure 1 has been prepared. Through 1959 the data from the list referred to above has been used. From 1960 on, the statistical data assembled annually by the American Chemical Society[2] have been used. In all about 9000 degrees have been granted.

In the following pages where the term "degree" is used it means a doctoral degree, either Ph.D. or Sc.D., awarded by a United States university. The term "graduate" is defined to mean a person holding this "degree".

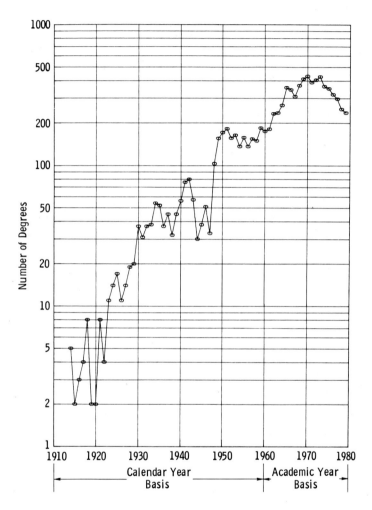

Figure 1. Number of doctoral degrees in chemical engineering - American universities

ASSEMBLY OF THE LIST OF DEGREE HOLDERS

The list was assembled using a series of successive approximations. First a letter was sent to each department awarding the degree asking for a list of their graduates which would include the name, year and thesis topic. Usually the letter was productive. Some chairmen stated that they did not have a list but that one could be prepared using the information presented annually in the January issue of Chemical Engineering Progress[1] or from the list maintained by Dissertation Abstracts. Neither of these statements is totally correct. Chemical Engineering Progress begins its list only in 1951 and Dissertation Abstracts does not include in its list certain important schools. Other chairmen referred the request to their library or their graduate school office. Usually each departmental list was cross checked by the local library or the graduate school.

Next the list was checked against the information found in Chemical Engineering Progress[1] each year from 1952 through 1962 looking for degrees awarded before 1960. Then the information was put on computer cards and listed alphabetically by the name of the degree holder. Several typical entries are shown in Table 1.

Using this list and the names tabulated in "Chemical Engineering Faculties 1959 - 1960", it was possible to locate where a majority of the teachers obtained their degree, but not all. To locate the source of their doctorate the chairman of the appropriate

Table 1.
Typical Entries from the list of Ph.D. Theses
(U.S. Universities 1905-1959)

HIXON, A.N.   1941
    THE SOLUBILITY OF OLEIC ACID, ABIETIC ACID AND THEIR
    MIXTURES IN PROPANE UP TO THE CRITICAL TEMPERATURE
    COLUMBIA UNIVERSITY

HOUGEN, OLAF ANDREAS   1925
    AMMONIA TRANSFER COEFFICIENTS IN ABSORPTION TOWERS
    UNIVERSITY OF WISCONSIN

OTHMER, DONALD FREDERICK 1927
    THE EFFECT OF TEMPERATURE, PURITY AND TEMPERATURE DROP
    ON THE RATE OF CONDENSATION OF STEAM
    UNIVERSITY OF MICHIGAN

DOCTORAL THESIS WORK IN THE U.S.

department was written to and asked to identify the school from which a particular person received his degree. With this information a letter was addressed to the degree-granting school and they were asked to supply the date, full name, thesis title and the department in which the work was done. Then the master list was, as necessary, revised.

In a few cases the issue arose whether a department or a school awarded a doctoral degree in chemical engineering. An arbitrary decision was made that if a school said it awarded a degree in chemical engineering, it did.

To the best of this writer's knowledge this list is the most complete in existence. The author has observed that the data base developed for computer searching to locate thesis work in chemical engineering is not complete. Consequently anyone using a computer searching technique should inspect the range of the data base carefully before accepting it as being complete.

SCHOOLS SUPPLYING THE MAJORITY OF THE TEACHERS

On the staffs of the chemical engineering departments in the United States in 1959, there were 573 persons holding doctoral degrees. This number included persons holding degrees from foreign schools and in fields other than chemical engineering. Table 2 presents a list of the schools which were supplying the majority of the doctoral-holding staff members to the chemical engineering departments.

The thirteen schools listed in Table 2 account for 1737 of the total 2824 awarded through 1959 and 37 of the 573 degree-holders teaching in the U.S. The number of contributions of the schools to the teaching profession is not entirely accurate since some individuals could have been in teaching and died before 1959. They would not have been located in this study.

Several comments may be made about the information in this table. Michigan is the leader in providing faculty to other schools in the nation (49). Wisconsin (28), Minnesota (28) and M.I.T. (26) are distant seconds. The percentage of the graduates from a given school who enter teaching is also of interest. Here Princeton is the leader with about 40 percent followed by Minnesota with 30 percent and then Wisconsin and Carnegie with 27 percent.

The conditions which motivated graduates to enter teaching would be an interesting matter to explore. A sociological study of the effects of staff behavior, of the desire for prestige, of the wish to serve other students, etc., on a person's decision to teach might be useful today as we attempt to increase the number of teachers.

Table 2. Schools Having Ten or more Doctoral Recipients on American Chemical Engineering Faculties 1959-60

SCHOOL	TOTAL DEGREES AWARDED THROUGH 1959	TOTAL ON CH.E. FACULTIES	TEACHING IN SCHOOL FROM WHICH DEGREE WAS RECEIVED	TEACHING AT OTHER SCHOOLS
MASSACHUSETTS INSTITUTE OF TECHNOLOGY	321	41	15	26
MICHIGAN	253	57	8	49
OHIO STATE	187	24	3	21
COLUMBIA	175	20	2	18
ILLINOIS	140	16	0	16
WISCONSIN	133	36	8	28
PURDUE	106	15	1	14
MINNESOTA	107	32	4	28
IOWA STATE	95	18	7	11
YALE	82	18	4	14
CARNEGIE INST. OF TECHNOLOGY	52	14	2	12
PENN. STATE UNIV.	45	10	4	6
PRINCETON	41	16	3	13

Today so many schools offer the doctoral degree that the effects which Michigan or the other large suppliers of staff might have had on chemical engineering education before 1959 probably will not occur again.

PRODUCTION OF DOCTORAL DEGREE HOLDERS

Once the master list had been established a number of arrangements of the information were possible. The number of doctoral degrees granted from the first one through 1959 is found to be close to 2824 and the way in which the number granted grew with time has been calculated and is shown in Table 3. Also included at the end of the table are numbers for the 60-69 and the 70-79 decade. The decade of 50-59 produced more doctoral degrees than the entire 1905-1950 period and the 60-69 decade in turn more than the entire 1905-1959 period. Figure 1 shows the overall picture of degree production by year and indicates that the peak was reached in the early seventies and has been dropping rapidly since then. Up through the fifties only a few graduates seemed to have come from foreign countries and of those the majority seemed to be from China, or at least had Chinese names. There are indications that about half of the degree recipients now (1980) are foreign nationals.

Table 3. Doctoral Degrees granted by American Schools From the Beginning through 1959

YEAR	NUMBER	YEAR	NUMBER	YEAR	NUMBER
1905	1				
DECADE TOTAL	1				
1910	1	1930	37	1950	171
11	1	31	31	51	182
12	1	32	37	52	156
13	-	33	38	53	164
14	5	34	54	54	136
1915	2	1935	52	1955	157
16	3	36	37	56	136
17	4	37	45	57	154
18	8	38	32	58	149
19	2	39	45	59	184
DECADE TOTAL	27		408		1589
1920	2	1940	56		
21	8	41	76		
22	4	42	80		
23	11	43	57	TOTAL ALL DEGREES	
24	14	44	30	THROUGH 1959	2824
1925	17	1945	38	DEGREES	
26	11	46	51	1960 - 1969	2881
27	14	47	33	DEGREES	
28	19	48	103	1970 - 1979	3472
29	20	49	155	GRAND TOTAL	9177
DECADE TOTAL	120		679		

The next table, Table 4, shows the number of graduates by school. Additional tables, not included, were also prepared which show the number of degrees awarded by each school each year.

The author of this paper was able to identify only two degrees awarded to women.

THESIS TOPICS

The list may also be used to find out in what fields of chemical engineering the thesis work was done. Before beginning two preliminary remarks are in order. No thesis seems to be concerned with any educational aspect of chemical engineering despite the fact that one graduate in six became a chemical engineering educator. Only two thesis titles indicated any consideration of economics. The concept that the economics of chemical processes or of the chemical industry were of considerable importance and were worth university research work obviously never took hold.

Table 4.
Doctoral Degrees--American Universities from the Beginning
Through 1959 by School

School	Count	School	Count
BROOKLYN, POLYTECHNIC INSTITUTE OF	73	MISSOURI, UNIVERSITY OF	5
CALIFORNIA INSTITUTE OF TECHNOLOGY	24	MONTANA STATE COLLEGE	8
CALIFORNIA, UNIVERSITY OF (BERKELEY)	23	NEW YORK UNIVERSITY	23
CALIFORNIA, UNIVERSITY OF (LOS ANGELES)	13	NORTH CAROLINA STATE COLLEGE	9
CARNEGIE INSTITUTE OF TECHNOLOGY	52	NORTHWESTERN UNIVERSITY	43
CASE INSTITUTE OF TECHNOLOGY	41	OHIO STATE UNIVERSITY	187
CINCINNATI, UNIVERSITY OF	33	OKLAHOMA STATE UNIVERSITY	7
COLORADO SCHOOL OF MINES	1	OKLAHOMA, UNIVERSITY OF	7
COLORADO, UNIVERSITY OF	13	OREGON STATE COLLEGE	8
COLUMBIA UNIVERSITY	175	PENNSYLVANIA STATE UNIVERSITY	45
CORNELL UNIVERSITY	61	PENNSYLVANIA, UNIVERSITY OF	32
DELAWARE, UNIVERSITY OF	57	PITTSBURGH, UNIVERSITY OF	50
FLORIDA, UNIVERSITY OF	10	PRINCETON UNIVERSITY	41
GEORGIA INSTITUTE OF TECHNOLOGY	28	PURDUE UNIVERSITY	106
HOUSTON, UNIVERSITY OF	6	RENSSELAER POLYTECHNIC INSTITUTE	26
ILLINOIS INSTITUTE OF TECHNOLOGY	38	RICE UNIVERSITY	5
ILLINOIS, UNIVERSITY OF	140	ROCHESTER, UNIVERSITY OF	3
INSTITUTE OF PAPER CHEMISTRY	8	SOUTHERN CALIFORNIA, UNIVERSITY OF	2
IOWA STATE UNIVERSITY OF SCIENCE AND TECHNOLOGY	95	SYRACUSE UNIVERSITY	20
IOWA, STATE UNIVERSITY OF	48	TENNESSEE, UNIVERSITY OF	17
JOHNS HOPKINS UNIVERSITY	22	TEXAS, AGRICULTURAL AND MECHANICAL COLLEGE OF	9
KANSAS, UNIVERSITY OF	8	TEXAS, UNIVERSITY OF	95
LEHIGH UNIVERSITY	12	UTAH, UNIVERSITY OF	6
LOUISIANA STATE UNIVERSITY	25	VIRGINIA POLYTECHNICAL INSTITUTE	30
LOUISVILLE, UNIVERSITY OF	1	WASHINGTON UNIVERSITY	31
MARYLAND, UNIVERSITY OF	23	WASHINGTON, UNIVERSITY OF	53
MASSACHUSETTS INSTITUTE OF TECHNOLOGY	321	WEST VIRGINIA UNIVERSITY	22
MICHIGAN STATE UNIVERSITY	8	WISCONSIN, UNIVERSITY OF	133
MICHIGAN, UNIVERSITY OF	253	YALE UNIVERSITY	82
MINNESOTA, UNIVERSITY OF	107	TOTAL	2824

Table 5 shows the research topics during the first fifteen years period 1905-1919, and Table 6 for the period 1920-1929. In these two periods only about twenty thesis had titles which indicated work in the unit operations. The concept of the unit operations is generally regarded as one of the major unifying ideas in chemical engineering, and it was explicitly developed in 1915 by A.D. Little. Yet, studies in them developed slowly. In part this may be explained by the fact that many of the thesis advisors in those early days were originally industrial chemists and consequently they tended to focus their attention and that of their students on process studies. Another possibility might be that there was a close connection between faculty members and industry in those days. One could also observe that educational systems have sufficient inertia so that change is slow. In addition changes which tend to depreciate anyone's store of intellectual capital will be resisted. For whatever reasons process studies predominate up to the thirties. Then affairs change.

In the thirties the writer estimates that about 30 percent of the thesis work was in the unit operations and in the forties this jumps to over 50 percent and thus the unit operations research was dominating the field.

In Table 7 is assembled some information on certain topics in the unit operations that were not heavily studied. This table shows that studies in the areas of agitation/mixing, and solids creation and processing have not been popular. They total less than three percent of all the thesis. One possible reason for lack of interest in thesis topics is that findings on the behavior of a particular system are not readily generalizable to other systems or even to the same system at another time. In addition these operations do not lend themselves to precise mathamatical formulation. These studies frequently employ equipment whose performance is highly dependent upon a particular mechanical design. Even though these areas have been excluded from any intensive study in this period, one should

Table 5.
Thesis Topics during the Period 1905-1919

COAL AND COAL PRODUCTS	8
METALS AND ALLOYS	6
GAS PRODUCERS	4
ORGANIC AND INORGANIC TECHNOLOGY	5
WOOD UTILIZATION	1
LEATHER TANNING	1
BRICK MANUFACTURE	1
GAS MASKS	1

NOTE: NO RESEARCH IN THE UNIT OPERATIONS

Table 6.
Thesis Topics during the Period 1920-1929 (120 Total)

CHEMICAL REACTIONS	18
COMBUSTION AND OXIDATION	12
PHYSICAL PROPERTIES	8
COAL AND COAL PRODUCTS	7
OIL SHALE	6
METALLURGY AND CORROSION	6
UNIT OPERATIONS	
HEAT TRANSFER	5
DISTILLATION	5
ABSORPTION	4
CRYSTALLIZATION	2
DIFFUSION	1
GRINDING	1
DRYING	1
RATE OF SOLUTION	1
SPECIFIC INDUSTRY STUDIES	
SUGAR	4
ELECTROCHEMICAL	4
PAPER MAKING	3
CEMENT AND LIME	3
STEEL	2
TANNING	2
SULFURIC ACID	1
WOOD UTILIZATION	1
RUBBER	1

Table 7.
Number of Thesis in Selected Categories
(1905-1959)

AGITATION AND MIXING	14
CENTRIFUGATION	6
COMPUTERS	7
CONTROL	4
CRUSHING AND GRINDING	11
CRYSTALLIZATION	27
ECONOMICS	2
EDUCATION	0
FILTRATION	8
FLOTATION	5
SETTLING	16
SOLUTION OF SOLIDS	4

DOCTORAL THESIS WORK IN THE U.S.

not draw the conclusion that these operations are not widely employed in industry or that the engineering problems associated with them disappeared because they were ignored by the universities.

A signification number of studies were made in mass transfer and Table 8 shows a breakdown of them. The vapor-liquid equilibrium studies far out number equilibrium studies for other types of systems. The number of studies on the performance of distillation, absorption and solvent extraction columns are somewhat equivalent.

Some suggestions may be advanced to explain why these fields have been so popular. In part there was a need for the information in order to design equipment, but other influences may have been at work. Most departments carry out research in areas for which funds can be provided. In general many equilibrium studies can be carried out in relatively inexpensive apparatus. A few thousand dollars would suffice for most of the apparatus used through the fifties. The investigations were labor intensive and data gathering and reduction was done manually. Futhermore an original thesis was almost certain to result because of the infinite number of possible systems that could be studied.

Somewhat the same situation existed with respect to column studies in mass transfer. As long as one operated at relatively low pressures equipment costs were low, and a wide range of choice existed with respect to the systems to be separated and the type of tower internals. Here again a thesis was almost certain to be produced.

Many other sortings of the thesis titles could be made and the waxing and waning of topics shown.

The author of this paper has personally read several hundred of these thesis or their abstracts and would observe that it is a rare thesis that explicitly discloses how the work reported in the

Table 8.
Thesis Topics in Mass Transfer 1905-1959 (Approximate)

EQUILIBRIUM STUDIES	
VAPOR-LIQUID	123
GAS-LIQUID	4
LIQUID-LIQUID	9
SOLID-LIQUID	4
COLUMN PERFORMANCE STUDIES	
DISTILLATION	100
ABSORPTION	70
SOLVENT EXTRACTION	89

thesis advances the subject of chemical engineering. In some cases the significance can be surmised, in others it is fairly clear that the significance of the work has escaped both the student and his advisor and in still others any significance would be most difficult to find.

In developing this list and working with it each thesis title had to be read. A comment or so about thesis titles may be appropriate. It is this writer's opinion that a thesis title should provide a statement of the contents of the thesis in sufficient detail that a potential reader could tell if it were of interest. Any thesis title which is so short that a reader could easily imagine a score of different subjects it might be covering probably is an inadequate title.

Some examples, which in this writer's opinion are inadequate, are shown in Table 9. Another type of title is one which calls up puzzling connotations. Table 10 lists a few of these titles. The idea of stirred tanks moving around is not what the student meant to convey. One wonders if unsanitary surveys of Mamaroneck Harbor have been done. Particle size in any industry is an idea that is amusing to comtemplate.

UTILITY OF THIS LIST

This list has a number of potential uses. It can be read item-by-item in less than a day and from it can be produced a short-list of all doctoral degrees granted to students working on a given topic. This list together with the ones in Chemical Engineering Progress[3] after 1959 can be totally scanned in a week. Such a study

Table 9.
Thesis Titles which Need Amplification

A STUDY OF LIQUID EXTRACTION
LIQUID-LIQUID EXTRACTION
LIQUID-LIQUID EXTRACTION
STUDIES IN DISTILLATION
ENTRAINMENT STUDIES

PERFORMANCE OF A PACKED ABSORPTION TOWER
A PROBLEM IN ADSORPTION
THERMAL PROPERTIES ON THE N-PETANE - BENZENE SYSTEM
SPECTROPHOTOMETRIC STUDIES
A STUDY OF POLAR LIQUIDS

SETTLING

Table 10.
Thesis Titles Having Misleading Connotations

PARTICLE SIZE IN THE CEMENT INDUSTRY
THE DYNAMICAL BEHAVIOR OF STIRRED TANKS
TEMPERATURE STUDIES IN A PACKED LIQUID-LIQUID EXTRACTION COLUMN
A SANITARY SURVEY OF MAMARONECK HARBOR

can give a student an excellent historical outline of the university work done in a field.

Cases exist in which the thesis material has never resulted in a journal article even though the work done had a certain importance. A study of this list would locate such works. In general a thesis will contain many more details about the work than will be found in the published paper, and these details can be useful in planning additional work. Frequently the thesis will contain extensive reviews of the literature which can be helpful. The reading of theses done by students at other schools can be a useful experience in that they give an indication of the level of work done at other universities. These theses also may be helpful when a student comes to write up his own work.

A copy of the list may be borrowed from the author.

REFERENCES

1. "Chemical Engineering Faculties, 1959-1960," Amer. Inst. Chem. Engrs. New York 1959
2. "Committee on Professional Training", Annual feature, March or April, Chemical and Engineering News.
3. "Research Roundup for 19XX", Annual feature, January issue Chem. Engr. Prog. First one, 1952

ACKNOWLEDGEMENTS

The author wishes to thank Maung Maung Win who worked so diligently on the early assembly task and Gregory T. Eagleman who helped in the final verification of the entries in the list.

HISTORICAL HIGHLIGHTS IN THE CHEMICAL ENGINEERING LITERATURE

Herman Skolnik

Hercules Incorporated
Wilmington, DE 19899

ABSTRACT

Historical highlights in the chemical engineering literature are discussed from antiquity through the Middle Ages, the Renaissance, the 18th and 19th centuries, when chemical engineering became a profession and an educational curriculum, and into the 20th century.

INTRODUCTION

The literature of chemical engineering is a fascinating story that has yet to be told. It is an extensive story whose beginnings are associated with human basic needs of food, shelter, and clothing and with tools and systems for satisfying these needs. The story does not begin with chemical engineering nor with chemistry, not with science, but with technology.

Mankind's conquest of fire, that is the ability to make and use fire, was undoubtedly one of the first and major technological breakthroughs. It provided warmth, changed eating habits, and led to the achievement of operations such as heating and drying for making containers, kitchen utensils, and tools by the working of clay, ores, and metals. Fire was the means by which civilizations progressed from the Stone Age to the Bronze and Iron Ages and into the industrial revolution.

The next primary thrust was the introduction of agriculture, the technology that promoted the need for human beings to live with and to understand nature, to live with other human beings in a cooperative society, and to harness the energy of water, the wind, and animals.

Agriculture, mining, metallurgy, medicine, and textiles were the primary impetus to the development of industrial arts many centuries before knowledge could be accumulated and categorized as physics, chemistry, botany, or biology. Until relatively recently in the time scale of civilizations, technology was a matter of art, handed down from generation to generation and transferred from one civilization to another by word of mouth and example. With the advent of writing, the amount of information and knowledge that could be fixed and transferred was many times that of the spoken word. Written records gave society a storehouse of knowledge considerably greater than any one mind could possess, and allowed the preservation and transmittal of knowledge to future generations. The ascendency of writers and philosophers during the Golden Age of Greece, their books (scrolls) and libraries, and the establishment of Greek as the common language throughout the Mediterranean world for nearly a thousand years created the first universal basic literature of science and technology. Through this communication channel, the work of the Egyptian, Persian, Indian, and Chinese alchemists entered into the European culture, first by the Greek and then later by the Latin universal language.

Antiquity is without records, except those which archeologists and anthropologists have been able to piece together from discoveries of fossil skeletons and of products from early civilizations. Analyses of articles of metal, pottery, glass, mortars, and dyed textiles from excavations tell us that copper as a product of mining and metallurgy is at least 5000 years old. Bronzes of copper and tin were used at least 3000 years ago. Iron and products from iron were known as early as 3000 B.C. in Egypt and in Babylon and mercury was known to the Egyptians in 1500 B.C. Much of what we know about the technology of antiquity is from the writings of the Greek philosophers and classicists, such as Democritus, Plato, Aristotle, Archimedes, and others.

Among the more important early writers was Pliny the Elder (23-79 A.D.), whose "Historia Naturalis", although hardly good science, was an encyclopedic attempt of summarizing all that was known. He covered 20,000 topics, with references to about 500 authors, in astronomy, geography, mining, metallurgy, medicine, botany, biology, and alchemy.

Another excellent source of information on technology during the Roman Empire is the book by Marcus Vitruvius Pollio (25 B.C.) "De architectura" (On Architecture) which, for example, describes the presence of pozzalana sand. This sand when mixed with limestone formed a hydraulic cement, capable of hardening under water, that made possible the construction of magnificent buildings and bridges. With the downfall of Rome, hydraulic cement disappeared until the 19th century when Portland cement was invented.

The Middle Ages (5th to mid-15th Century)

The Greco-Roman period was followed with a cultural, scientific, and technological decline that lasted for about ten centuries throughout Europe. The world, however, was not without technology and technologists, for much of the Greco-Roman civilization continued and expanded in Byzantium, particularly in the university centers of Alexandria, Beirut, Antioch, and Nisibis, where Greek science and philosophy were preserved with translations into Arabic. It is through these translations that much of the works from the Golden Age of Greece and the engineering advances of the Romans are known today, from Greece and Rome to Byzantium, to Eastern and Spanish Islam, and then throughout Europe.

Advances, nevertheless, were made during this period in Europe, such as the plow, windmills, horse harness, horseshoes, canal lock chambers, and ship rudders and ship designs that made tacking by sails possible.

By the introduction of the plow with wheels, vertical blade (coulter), horizontal plowshare, and mouldboard to turn over the sod, a new system of agriculture evolved from the 6th century on. As this new plow required eight oxen, its adoption led to communal farming. The horse harness appeared in the 9th century and by the 12th century horses were widely used for plowing, replacing oxen, with double the productivity. The invention of nailed horseshoes just before the 10th century made transportation of foods possible throughout Northern Europe.

The Renaissance (mid-14th to 17th Century)

Although the Renaissance brought forth an intellectual reawakening in novels, poetry, art, and scholarship, science was slow in making significant advances, even though the explorations of Columbus, Magellan, and others presaged an expanding world. There were at least two reasons for the low position of science, especially in Italy, where the Renaissance began: patrons favored the writers, artists, and scholars; and the church did not look with favor on those whose work might be in conflict with its views on the nature of things.

Technology, on the other hand, experienced a new momentum during the Renaissance, primarily because of three factors: the dispersion of scholars through Europe from the Byzantium empire with the fall of Constantinople in 1453; the development of gunpowder and the consequent elimination of the medieval castle as a fortress and center of power and feudalism; and, most important, Gutenberg's introduction of movable type in the 1450's and the democratization of communication by the printed word. With the development of printing, a flood of pamphlets and books became readily available to the general public.

Iron and steel were greatly in demand for weapons, nails, anchors, tools, etc. With the greatly rising demand for iron and the increasing need for bronze (copper and tin alloys) for guns, statues, and home appliances, and for brass (copper and zinc alloys) for instruments, metallurgy became of prime importance. Vannuccio Biringuccio (1480-1539) of Siena, Italy, acquired an extensive knowledge of mining and metallurgy as practiced in Italy and in Germany and was well read in the existing literature. His single publication, "De la Pirotechnia", a ten-volume work printed in Venice in 1540, was the first systematic text on mining and metallurgy. Books I and II describe the recovery of silver, gold, copper, lead, tin, iron, mercury, sulfur, antimony, aluminum, arsenic, etc., from their ores. Book III covers testing, roasting, and cupellation. Book IV describes the preparation of aqua fortis and its use in parting. Book V covers gold, silver, copper, lead, and tin. Books VI, VII, and VIII are concerned with casting, molds, and furnaces. Books IX and X describe the arts of distillation, sublimation, extraction, amalgamation, and the manufacture of gunpowder, saltpeter, and projectiles. The first English edition of this landmark work, translated by Cyril S. Smith and Martha T. Grudi, was published in 1942 by the American Institute of Mining and Metallurgical Engineers.

"De Re Metallica", Georgius Agricola's (1494-1555) major work, published posthumously in 1556 in Basel, was of greater influence than that of Biringuccia's which was written in Italian and later translated as a French edition. Agricola's folio of about 600 pages was written in Latin, which was the universal language of the educated, and then translated as a German version in 1557 and the Italian version in 1563. It was translated first in English from the original Latin by Herbert C. Hoover and his wife (L.H.) in 1912 with biographical notes on the development of mining, metallurgy, geology, and mineralogy from antiquity to the 16th century. "De Re Metallica" consisted of 12 books, reviewing the existing literature and discussing mines and mining, assaying, smelting, the obtaining of metals from ores, the preparation of soluble salts, and the manufacture of alum, vitriol, sulfur, and glass, among others. This was the standard reference work for over a century and by which Agricola was named the father of mineralogy.

Lazarus Ercker (1530-1593), like Agricola, was a resident of Central Europe mining areas, wrote a treatise of 288 folio pages on mineral ores, mining, and smelting. Ercker's book, "Beschreibung Allerfurnemstem Mineralischen Ertzt und Berckwerksarten" (Description of the most noble mineral ores and types of mines), published in 1556, enjoyed the popularity of Agricola's treatise.

The works of Biringucio, Agricola, and Ercker were among the first technological treatises to be produced by printing.

Although there was little progress in science well into the 17th century, there were many significant advances associated with and motivated by the expanding industries and commerce. Among the advances were the invention of the logarithm, calculus, compound microscope, telescope, barometer, and scientific instruments.

Jan Baptist Van Helmont (1577-1644), educated as a physician, devoted his talents to chemistry. He coined the word "chemistry", deriving it from the Greek "chaos", and was the first to use the word "gas", the first to use quantitative measurements in experimental work, and to emphasize the value of using a balance. His work in chemistry was collected and published in 1648, the English edition appearing in 1662 under the title "Oriatricke or Physik Refined".

Robert Boyle (1627-1691), primarily a physicist, has been called the father of modern chemistry through his work on gases and analytical chemistry. He lives in our literature with the Boyle's law and through the 20 books he wrote, the most famous being "The Sceptical Chymist", published in 1661.

One of the most outstanding works of chemistry of the 17th century was "Furni novi philosophici" (New Philosophies), published in 1648 in Amsterdam by Johann Rudolf Glauber (1604-1670), best known throughout his chemical career as an iatrochemist. This book extensively covers plant equipment, such as furnaces, distillation apparatus, and sources, processing, and uses of oils and spirits. Some historians of science consider Glauber as one of the first chemical engineers as he developed processes for the manufacture of sulfuric, nitric, acetic, and hydrochloric acids. His ideas on chemical engineering were expounded in his book "Teutschlands Wohlfahrt" (Germany's Prosperity), published in 1656. He was the first to describe the preparation of sodium sulfate, which he called sal mirible, but which became known as Glauber's salt and is still used as an aperient.

The 18th Century

Arnold Toynbee (1815-1883), uncle of the famous 20th century historian of the same name, popularized the term "Industrial Revolution" to describe England's economic and manufacturing developments from 1760-1840. As we know now, many elements of the industrial revolution date back far in history with the major evolutionary period occurring from 1600 to 1760. It was initially a slow shift from an agricultural basis to a bilateral economic and manufacturing undertaking, from the rural area to the city, and from the small workshop to the factory. During this period scientists learned from technology and scientists hardly affected at all the many technological advances that materialized into the industrial revolution. Science, nevertheless, was a creative element in this technological climate, resulting in the formation of the Royal Society of England in 1662, the Academia Royale des Sciences of Paris in 1666, and the Society of Arts of England in 1754. The utility of scientific knowledge dominated the objectives of these societies.

The word engineer entered our language about 200 A.D. when the term "ingenium" was applied to military machines and "ingeniator" to those who designed them (the word ingenious has the same root). Throughout most of history engineering had a military context, and even in the young United States, the first engineering education was established at the U.S. Military Academy of West Point, New York, when it was founded in 1802. It was not until the industrial revolution that engineers acquired status as designer and builder of nonmilitary works. To distinguish these endeavors from military engineering, the term civil engineering was introduced. Systematic training and education for engineers was first given at the École des Ponts et Chausées (school of Bridges and roads) which was founded in Paris in 1747. The École Polytechnique, founded in 1794 in Paris, was the next major school for engineers who were given two years of education in basic sciences in a three-year curriculum. Among the graduates of the École Polytechnique were Sadi Carnot, best known for his work in thermodynamics, and Gustave Eiffel, best known for the tower in Paris that bears his name.

LITERATURE HIGHLIGHTS

Mechanics schools were established in England rather than engineering colleges. Engineering as a curriculum was first introduced in 1840 at the University of London and at the University of Glasgow.

During the 18th century the phlogiston theory dominated the thinking of chemists such as Georg Ernst Stahl (1660-1734), Guyton de Morveau (1737-1816), Joseph Black (1728-1799), Henry Cavendish (1731-1810), and Joseph Priestley (1728-1804).

A monumental four-volume "Dictionnaire de la Chymie", the first great encyclopedia of chemistry, was published in 1778 by Pierre Joseph Macquer in France.

It was the work and publications of Antoine Laurent Lavoisier (1743-1794) that invalidated the phlogistion theory and initiated the foundations of a new chemistry. An important publication of this period was "Methode d'une Nomenclature Chimique", written by Lavoisier, Guyton de Morveau, and Antoine de Fourcroy and issued in 1787. This publication eliminated many arbitrary names that were then currently used and introduced a new systematic nomenclature for chemicals.

From the engineering viewpoint, the most notable achievement of the 18th century was the steam engine. Prior to the 18th century, there were four basic sources of power: human, animal, wind, and water. A new source of power was needed particularly by the mining industry for pumping water from deep mines. This need was provided by the steam engine, first by Thomas Savery (1650-1715) who invented a full-scale steam engine constructed in 1699, then the low-pressure steam engine of Thomas Newcomen (1663-1729) in 1712 and that of James Watt (1736-1819) in 1775, and finally the high pressure steam engines of Richard Trevithick (1771-1833) of England and Oliver Evans (1755-1819) of the United States which were in operation by the turn of the century. Newcomen was an ironmonger, Watt an instrument maker and mechanic, Trevithick a mechanic, and Evans a millwright. The literature of these inventors was patents.

The 19th Century

Whereas the 18th century saw the emergence of mechanics and mechanical engineering, chemistry became a discipline of science in the 19th century and by the end of the century the industrial chemist and then the chemical engineer were products of the educational system and a factor in the expanding chemical industry. At the beginning of the 19th century, engineers in the United States came from three sources: self-education, Europe, and the U.S. Military Academy. In 1824, engineering became a curriculum at Rensselaer Polytectnic Institute, then at Harvard in 1847, Yale in 1850, state universities in the 1850s and 1860s, and many others through the remaining years. The first curriculum in chemical engineering was established in 1888 at MIT and the second at the University of Michigan by the turn of the century.

Through most of the 19th century, chemists and scientists in general were a relatively rare breed, especially in the United States. Chemistry as a professional way of life came of age over the past 150 years because chemistry evolved into both a science and an industry. Paralleling this evolutionary period of the profession and industry, the literature went through an initially slow growing period into its current rate of doubling every ten years. Visits, lectures, correspondence, books, and a few journals constituted the literature through the 19th century. By the end of the 19th century, the literature could be classified into analytical, inorganic, organic, physical, and electrochemistry and a budding new area of industrial chemistry.

Chemistry and the chemical industry were first tied together in Germany during the second half of the 19th century, making Germany the dominant chemical industrial nation well into the 20th century. Industrial laboratories were thus introduced in Germany in close cooperation with the government and academic institutions.

Although industrialization was occurring rapidly in the United States, the coordination of chemistry and the chemical industry was slow in developing. In 1872, 16 manufacturers of sulfuric acid founded the Manufacturing Chemists' Association to work against unjust freight regulations and to lobby for protective tariffs. Petroleum refining, a new industry in the 1850s, hired George Saybolt as its first chemist in the 1880s. The first full time chemist was hired by Andrew Carnegie to perform research on steelmaking problems. The output of chemicals and allied products in the United States in 1880 was valued at $140 million. In 1876 and 1878, Josial Willard Gibbs, Jr. published in two parts his paper "On the Equilibrium of Heterogeneous Substances" in the Transactions of the Connecticut Academy of Science. Although these publications created the new science of chemical thermodynamics, they were ignored in the United States, but fortunately not in Europe.

Thermodynamics arose from the need to calculate the efficiency of steam engines, and is an outstanding example of the

LITERATURE HIGHLIGHTS

debt science has to technology. Sadi Carnot's "Reflections on the Motive Force of Fire", published in 1824, was the first systematic attempt toward the theory of heat engines. James Prescott Joule, an English businessman with an interest in science, provided his experimental proof in the 1840s of the first law of thermodynamics, i.e., that energy cannot be created nor destroyed and worked out the mechanical equivalent of heat. The German physicist Rudolph Clausius introduced in 1865 the definition of entropy, the basis for the second law of thermodynamics. William Rankin, professor of engineering at the University of Glasgow established the practical application of thermodynamics to the design of steam engines in his "Manual of the Steam Engine and Other Prime Movers", published in 1859.

The literature started to become large by the end of the 19th century with the issuance of patents, publication of many books, and the introduction of numerous journals. Products of the chemical industry were beginning to become numerous, among which were: sulfur, gypsum, quicklime, alkalies, sulfuric and hydrochloric acids, soda, potash, aluminum (Hall's electrochemical process was the basis for the founding of Alcoa in 1888), rubber (first vulcanized by Goodyear in 1839), nitrocellulose (celluloid was introduced by John W. Hyatt in 1868), refined petroleum products (the Drake well was the first to produce petroleum in 1859), and dyes (the W. H. Perkin's patent on mauve in 1856 initiated a fast growing dye industry which produced 29 synthetic dyes by 1862).

By the end of the 19th century, there were about 500 journals of some interest to industrial chemists. In addition to the general references listed at the end of this paper, a fairly good picture of industrial chemistry is given by the following books, listed by author in a chronological order, which were published in the 19th and very early 20th centuries:

1. Samuel Parkes, "Chemical Essays', 5-vols., 1815.

2. S. F. Gray, 'The Operative Chemist", 1828.

3. Andrew Ure, "Dictionary of Arts, Manufactures, and Mines", 1839.

4. Michael Faraday, "Experimental Researches in Electrochemistry", 3 vols., 1839-1855.

5. F. L. Knapp, "Lebrbuch der chemische Technologie", 2-vols., 1847.

6. Charles Goodyear, "Vulcanized Gum-Elastic," 1853, and "Gum-Elastic," 1855.

7. Sheridan Muspratt, "Chemistry", 2-vols., 1860.

8. A. P. Bolley, "Handbuch der chemischen Technologie", 7-vols., 1862-1870.

9. H. E. Dussance, monographs on "Dyes from Coal Tar", 1863; "Manufacture of Explosives", 1864; "Preparation of Cosmetics", 1868; "Manufacture of Vinegar and Acetic Acid", 1871. Editor of Industrial Chemist, 1862-1864, and Journal of Applied Chemistry, 1864-1866.

10. Henri Erni, "Coal, Oil, and Petroleum", 1865, Editor of Journal of Applied Chemistry, 1866-1868.

11. Albert Landenburg, "History of Chemistry" and "Lectures on the Development of Chemistry During the Last 100 Years", 1869.

12. Edward Bancroft, "Experimental Researches Concerning the Philosophy of Permanent Colours and the Best Methods of Producing them by Dyeing, Calico Printing, etc", 1874.

13. A. W. Hofmann, Editor, "Berichte über die Entwicklung der Chemischen Industrie während des letzten Jahrzehnts", 2-vols., 1875.

14. J. W. Richards, "Aluminum", 1890.

15. Ernest Sorel, "Le rectification de l'alcool", 1893.

16. V. B. Lewes, "Acetylene', 1900.

17. F. M. Thorp, "Outlines of Industrial Chemistry", 1901.

18. G. E. Davis, 'Handbook of Chemical Engineering", 1901.

19. Eugen Housbrand, "Evaporating, Condensing, and Cooling Apparatus", translated by A. C. Wright, 1903.

20. Sidney Young, "Fractional Distillation", 1903.

21. E. C. Worden, "Nitrocellulose Industry", 2-vols., 1911.

22. Wilhelm Ostwald, "L'Evolution de l'electrochimie", 1912, translated from German.

The 20th Century

Chemical engineering emerged in the early part of this century as an amorphous mixture of chemistry, physics, and general engineering, and much of the conceptional and theoretical directions chemical engineering took were toward the development of links with these separate sciences and technologies. The industrial chemist and chemical engineer were thought of as synonymous or related terms, as revealed in the name "Division of Industrial Chemists and Chemical Engineers", formed in 1908 as the first division of the American Chemical Society - the name was changed in 1918 to Division of Industrial and Engineering Chemistry. Other early divisions of the American Chemical Society closely allied to what chemists and chemical engineers do in the industrial world were: Agricultural and Food Chemistry, formed in 1908; Water, Sewage, & Sanitation Chemistry (now Environmental Chemistry), formed in 1913; and Fertilizer Chemistry (now Fertilizer and Soil Chemistry), formed in 1908. Eleven other divisions (such as, Cellulose, Paper and Textile; Chemical Marketing and Economics; Fluorine Chemistry; Fuel Chemistry; Nuclear Chemistry and Technology; Petroleum Chemistry; etc.) took root first as subdivisions of the Division of Industrial and Engineering Chemistry.

Specialization of scientists and technologists was the impetus not only for the formation of ACS divisions but also for the advent of new groups, such as the New York Section of the Society of Chemical Industry in 1906, the American Electrochemical Society (now the Electrochemical Society) in 1902, and the American Institute of Chemical Engineers (AIChE) in 1908. Many charter members of AIChE and all of its first officers were also ACS members. Whereas ACS membership was open to "any person interested in the promotion of chemistry" AIChE membership was restricted to those "not less than 30 years old...proficient in chemistry and in some branch of engineering as applied to chemical problems, and...engaged actively in work involving the application of chemical principles".

During its first 25 years or so, the American Chemical Society was dominated heavily by academe, despite the fact that by the turn of the century industrial chemists constituted nearly 50% of the membership. The academic oriented members considered themselves as "pure scientists" and those in industry "impure scientists". Although industrially oriented papers were published in the Journal of the American Chemical Society (JACS), ACS members engaged in some area of technical chemistry felt that papers of interest to them were given short shrift in JACS and they argued for their own journal. They also pressed for an abstract journal that covered the whole of chemistry and chemical technology, not just theoretical chemistry as preferred by the academic members. These desires were realized with the introduction of Chemical Abstracts in 1907 to cover both theoretical and applied chemistry and a new journal in 1909, Industrial and Engineering Chemistry. Industrial and Engineering Chemistry spawned Chemical and Engineering News in 1923, Analytical Chemistry in 1929, and Journal of Chemical and Engineering Data in 1956, and was replaced in 1963 with three quarterlies: IEC Product Research & Development, IEC Fundamentals, and IEC Process Design. In addition, the ACS also introduced the following related journals: Journal of Agricultural and Food Chemistry in 1953, Journal of Chemical Information and Computer Sciences in 1961, Environmental Science and Technology in 1967, Macromolecules in 1968, Chemtech in 1971, and Journal of Physical and Chemical Reference Data in 1972.

Arthur D. Little's categorization of chemical engineering processes into unit operations in 1915 found ready acceptance and adoption in the MIT curriculum. The distinction between unit operations and unit processes was clarified in 1923 by W. H. Walker, W. K. Lewis, and W. H. McAdams in their outstanding book, "Principles of Chemical Engineering", in which unit operations referred to primarily physical treatments or steps that are inherent in a chemical manufacturing process and unit processes referred to a distinct chemical change. Another classical book was P. H. Groggin's "Unit Processes in Organic Synthesis", published in 1935, which covered processes such as nitration, halogenation, alkylation, etc.

The British World War I blockade barred Germany's shipment of chemicals to the United States, forcing American companies to expand into the manufacturing of dyestuffs, pharmaceuticals, and other organic chemicals and stimulated expansion of the steel, petroleum, and other industries. The rising demand for explosives required sharply increasing supplies of toluene, phenol, nitric acid, and other chemicals. These expanding industries as well as staffing for the Chemical Warfare Service and other governmental agencies were dependent on an increasing recruitment of chemists and chemical engineers to the degree that the membership of the ACS doubled over the period from 1915 to 1920, from 7170 to 15582.

After a short postwar slump, the WWI industrial expansion accelerated from 1922 to 1929, marked by democratization by the automobile, the introduction of nitrocellulose lacquer, and the phenomenal growth of the petroleum, rubber, coal tar, plastics, and rayon industries. From 1920 to 1929, the number of papers and patents abstracted in Chemical Abstracts expanded some 250%. During the depression years, the annual dollar value of the chemical industry's output rose some 40% as all industry declined about 10%. Industrial research and development resulted in new and improved products, such as alkyd resins, synthetic methanol, fluorocarbon refrigerants, polystyrene, neoprene, and nylon.

World War II was a major impetus to the chemical industry and initiated a close cooperation between the federal government and those who were involved in science and technology. Thus arose the National Defense Research Committee (NDRC) to channel funds into weapons research, then the Office of Scientific Research and Development (OSRD) for funding of research for the armed services, including initiation of the atomic bomb project. Many products resulted from this funding as well as from the independent research activities in the chemical industry, such as butyl, neoprene, N-type, and general purpose GR-S synthetic rubbers; high test gasoline; pharmaceuticals, such as antimalarials, new sulfa drugs, and penicillin; DDT; napalm; and the atomic bomb. Big science under the umbrella of federal funding continued through the 1950s, 1960s, and into the 1970s with AEC, NIH, NASA, NSF, etc., and most recently with energy programs. Some 237,000 scientists and engineers were employed in R&D in the U.S. in 1954; 495,000 in 1965; and 550,000 in 1970 (of these in 1970, 52,000 were chemical engineers). The number of abstracts published annually by Chemical Abstracts reached 100,000 in 1957; passed 200,000 in 1966, and 300,000 in 1971. In 1980, it is estimated the number of abstracts may exceed 500,000.

The 20th century has been a period of rapid growth in the chemical industry and in governmental laboratories, R&D funding by both the chemical industry and the federal government, and the numbers of chemists and chemical engineers, all resulting in a corresponding rapid growth of the literature in a multitude of fragmented disciplines and subdisciplines of science and technology. Since 1940, the individual chemist and chemical engineer no longer could hope to maintain a current working knowledge of the literature without aid. In response to this need, a new specialist, the chemical information scientist, and new information services arose, to direct relevant patents, books, and journal articles to the attention of chemists and chemical engineers.

There are five types of organizations engaged in indexing/abstracting services:

Governmental Agencies, e.g., NTIS of the Dept of Commerce

Scientific and Technical Societies, e.g.,
    ACS, Chemical Abstracts Services
    American Institute of Physics, Physics Abstracts
    Engineering Index, Inc.
    API Information Services

Commercial, e.g.,
    H. W. Wilson, Co., Applied Science and Technology Index
    University Microfilms, Dissertation Abstracts
    Institute for Scientific Information, Citation Index
    and Current Abstracts

    Derwent Publications Ltd., World Patents Index

    IFI/Plenum Service, Uniterm Index

University, e.g.,
    University of Akron, Bibliography of Rubber Literature
    Institute of Paper Chemistry Abstract Bulletin

Industrial - for internal use only

Over the past 20 years, there has been a rapid expansion of computerized data bases, and today online data bases are a major tool in facilitating the utilization of information. There are three relatively large data base vendors, viz., Bibliographic Retrieval Services, Lockheed Information Systems, and Systems Development Corp., that supply online services for data base producers, such as

- Chemical Abstracts Service called CA Search
- University Microfilms and its 550,000 references to doctoral dissertations
- IFI/Plenum citations to U.S. Patents
- Engineering Index, Inc. 672,000 citations
- National Technical Information Service's 600,000 references
- American Society for Metals 300,000 citations from its "Metals Abstracts Index" and "Alloys Index"
- Predicasts, Inc., citations and statistics on industries and products
- Derwent Publications, Ltd., to the world patent literature
- Institute for Scientific Information's "Citation Index"
- American Institute of Physics citations to physics journals

Journals relevant to chemical engineering, and their starting year are:

LITERATURE HIGHLIGHTS

Angewandte Chemie (Germany) 1888
American Society for Testing Materials
                                     Bulletin, 1921
                                     Proceedings, 1898
British Chemical Engineering - 1956
Canadian Chemical Processing - 1917
Chemical Engineering - 1902
Chemical Engineering Progress - 1947
Chemical Engineering Science (England) - 1951
Chemical Processing - 1938
Chemical and Process Engineering (England) - 1920
Chemie - Ingenieur - Technik (Germany) - 1928
ChemTech - 1970
IEC, Fundamentals - 1962
IEC, Process Design - 1962
IEC, Product Research and Development - 1962
International Chemical Engineering - 1962
Journal of Applied Chemistry (England) - 1951
Journal of Chemical and Engineering Data - 1956
Journal of the Society of Chemical Industry (England) - 1882
Transactions of the Institution of Chemical Engineers (England) - 1919

In addition to these journals, a large number is published to cover specific areas of industrial chemistry, such as agricultural chemistry; cellulose, pulp, and paper; ceramics; cosmetics; environmental chemistry; explosives; food industries; petroleum; plastics, polymer chemistry and technology; protective coatings and inks; textiles; and toxicology.

The more important sources of data compilations on properties of chemicals are:

Landolt-Bornstein's Tables
International Critical Tables
National Standard reference Data
API Research Project 44 and TRC Data
CINDAS (Center for Information and Numerical Data)
ASTM Data Banks
Single volume handbooks, such as
      CRC Handbook of Chemistry and Physics
      Perry's Chemical Engineering Handbook
      Lange's Handbook of Chemistry

There are about 70,000 scientific and technical books currently in print which are produced by some 2000 publishers. An excellent source for knowing the existence of specific books is the "Cumulative Book Index" published by H. W. Wilson Co., New York. A current awareness of new books is possible by examining Chemical Abstracts issues and indexes and the publication of book reviews in many journals.

Compendia of relevance to chemical engineering and to chemical technology are:

- Kirk-Othmer "Encyclopedia of Chemical Technology", Interscience-Wiley
- Ulmann's "Encyclopedia der technischem Chemie", Urban and Schwarzenberg
- "Encyclopedia of Science and Technology" - McGraw-Hill
- Thorpe's "Dictionary of Applied Chemistry" - Longman-Green
- "Encyclopedia of Chemical Process Equipment" - Reinhold
- Houben-Weyl's "Methoden der organischen Chemie" - E. Muller, ed., George Thiemie Verlag
- "Techniques of Organic Chemistry" - Wiley
- Seidell's "Solubilities of Inorganic, Organic, and Metal Organic Compounds" - Van Nostrand
- ACS Advances in Chemistry Series and Symposium Series
- AIChE Monograph Series and Symposium Series

The Future

I think the future will look back on the past century as a period during which a substantial information and knowledge base in chemical engineering was formed. Yet information science was not an object of interest to the vast majority of chemical engineers.

We have inherited a rich literature from the achievements of the past. This literature comprises books, journals, theses, patents, government publications, trade catalogs, papers presented at technical meetings, and company research reports, and this literature is growing larger at an accelerating pace - doubling every ten to twelve years at the present growth.

In the face of this growth of the literature, how does one keep up with it? The answer of course is that no individual can. The solution is a matter of selectivity and the enlistment of every aid possible, some of which were mentioned in this paper, and the development of new methods, procedures, and organizations to cope with the problems.

The education of chemical engineers has metamorphosed from applied chemistry to unit operations and unit processes to an emphasis on mass transfer, mathematical analysis, and computer science at the expense of the chemical content of chemical engineering. I think there are cogent reasons for placing some emphasis on information science in the academic chemical engineering curriculum. If technical obsolescence is to be avoided, no chemical engineer can rest on the oars of learning gained in a four-year sojourn in college and additional years of graduate work. Academe is but the first stage in the guiding and shaping of a career.

I would hope that sometime in the future an information science course will be a part of the curriculum for all chemical engineering majors, a course that covers their literature, history, philosophy, and ethics, and that information scientists be a member of the faculty to conduct the R&D this phase of chemical engineering needs so desperately.

Suggested References for Further Reading

1. American Chemical Society, "Chemistry - Key to Better Living", ACS Diamond Jubilee Volume, Washington, D.C., (1951).
2. D. J. Boorstin, "The Americans. The Democratic Experience", Random House, (1973).
3. C. A. Browne, Editor, "A Half-Century of Chemistry in America, 1876-1926", ACS Golden Jubilee Number, Washington, D.C., (1926).
4. C. A. Browne and M. E. Weeks, "A History of the American Chemical Society", ACS, Washington, D.C., (1952).
5. M. P. Crosland, "Historical Studies in the Language of Chemistry", Harvard University Press, Cambridge, Mass., (1962).
6. G. H. Daniels, "Science in American Society", A. A. Knopf, N.Y., (1971).
7. W. Durant, "The Story of Civilization", Vol. I-X, Simon and Schuster, N.Y., (1954-1957).
8. M. Kranzberg and C. W. Pursell, Jr., Editors, "Technology in Western Civilization", Vol. I, Oxford University Press, N.Y., (1967).
9. H. M. Leicester and H. S. Klickstein, "A Source Book in Chemistry, 1400-1900", McGraw-Hill, N.Y., (1952).
10. "Literature of Chemical Technology", Advances in Chemistry Series No. 78, ACS, Washington, D.C., (1968).
11. "Manufacturing Chemists' Association, MCA 1872-1972, A Centennial History", MCA, Washington, D.C., (1972).
12. J. A. Michener, "The Source", Random House, N.Y., (1965).
13. W. D. Miles, Editor, "American Chemists and Chemical Engineers", ACS, Washington, D.C., (1976).

14. F. J. Moore, "A History of Chemistry", McGraw-Hill, N.Y., (1939).
15. R. L. Pigford, "Chemical Technology: The Past 100 Years", Chem. & Eng. News, ACS Centennial Issue, April 6, pp. 190-203 (1976).
16. E. L. Piret, Editor, "Chemical Engineering Around the World", Am. Inst. of Chem. Eng., N.Y., (1958).
17. H. Skolnik and R. M. Reese, "A Century of Chemistry", ACS, Washington, D.C., (1976).
18. H. Skolnik, "Milestones in Chemical Information Science", J. Chem. Info. Comput. Sci. 16:187-193 (1976).
19. J. M. Stillman, "The Story of Alchemy and Early Chemistry", Dover Publications, N.Y., (1960).

PIONEERS IN THE FIELD OF DIFFUSION

Farhang Mohtadi

Department of Chemical Engineering
The University of Calgary - Canada

INTRODUCTION

History reveals that man is the explorer of nature and has a strong urge to discover and to invent. Each discovery or invention by man opens in turn new avenues of knowledge and experience. In this way the sum total of human knowledge increases with time and every new discovery becomes more subtle and penetrating. It therefore becomes tempting to assume that the most important discoveries or inventions are also the most recent. But this is not always true. The wheel, invented by Persians at least two millennia before Christ is, undoubtedly, one of man's greatest inventions of all time.

There is much truth in Sir Isaac Newton's contention that, *"To explain all nature is too difficult a task for any one man, or even for any one age."* So, to admire only the contemporary achievements and to ignore the achievements of the past would, in the words of Bronowski,[1] *"make a caricature of knowledge."*

This paper is a brief historical review of the life of Joseph Priestley (1733 - 1804), Thomas Graham (1805 - 1869) and Adolf Fick (1829 - 1901) and a short account of the unmatched contributions of these great men of science to our understanding of the phenomenon of diffusion. Priestley, whose unorthodox religious and political views are as fascinating as his numerous scientific books and papers, discovered that when various gases are mixed, they remain *"diffused"*. Graham's extensive and ingeniously designed experiments on diffusion in gases, liquids and solids led him to the formulation of the first quantitative law for diffusion and effusion. Fick complemented the work of Graham by developing a fundamental mathematical model of diffusion, generally referred to as *Fick's Law*.

JOSEPH PRIESTLEY (1733 - 1804)

Joseph Priestley was born in the small village of Birstal Fieldhead, near Leeds, England, on March 13, 1733. His father, Jonas Priestley, was a cloth dresser and a staunch Calvinist who wanted his son to become a clergy. Priestley attended local parish schools and supplemented his formal schooling with private lessons and self-directed studies in mathematics and natural philosophy as well as in such languages as Latin, Greek, Hebrew, French, German, and Arabic. At the age of eighteen, Priestley rejected the stern Calvinism of his father and at nineteen he entered the dissenting academy at Daventry which was the successor to the famous Northampton Academy of Philip Doddridge.

During his three years at Daventry academy Priestley was introduced to a wide variety of subjects including logic and metaphysics and he developed a strong leaning towards the Newtonian view of natural philosophy and the Arian position in theology. After completing his studies at Daventry in 1755, Priestley went to preach first at Needham Market and then at Nantwich. Neither of these undertakings was a success, partly because of Priestley's unorthodox views on Christianity and partly because of an inherited speech impediment. In 1761 Priestley moved to a recently founded academy at Warrington and he stayed there for the following six years. At Warrington Priestley taught a variety of subjects and he wrote a number of important books and essays including *The Rudiments of English Grammar* (1761), *The Theory of Languages* (1762), and *Liberal Education for Civil and Active Life* (1765). In recognition of these literary contributions, Priestley was awarded a doctorate degree (LL.D.) by the University of Edinburgh in 1764. He was also ordained in Warrington that year.

In 1762 Priestley married Mary Wilkinson whose father, Isaac Wilkinson, was the greatest ironmaster of eighteenth-century England. After five years of marriage, because of growing family responsibilities and financial problems, Priestley resigned his teaching position at Warrington to become minister of Mill Hill Chapel, a large Presbyterian church in Leeds. The period between September 1767, when Priestley moved from Warrington to Leeds, and the summer of 1773, when he left Leeds to become "resident intellectual" and librarian to Lord Shelburne, is probably the most important period of Priestley's life. In this relatively short time Priestley published twenty eight non-scientific works, four books on science (including *The History and Present State of Electricity* for which he received much encouragement from Benjamin Franklin), and several essays in theology. The latter made

Priestley the leading publicist for the Unitarian cause in England. He also read to the Royal Society seven papers describing scientific experiments. These were the years of Priestley's most interesting work on electricity and also the period during which he began his work on chemistry and on the properties of gases. In recognition of these scientific accomplishments, Priestley was elected a Fellow of Royal Society in 1766.

Priestley's *Memoirs*[2] contain brief but lucid accounts of these years, including the following passage:

". . . But nothing of a nature foreign to the duties of my profession [as a dissenting clergy] *engaged my attention while I was at Leeds so much as the prosecution of my experiments relating to electricity, and especially the doctorine of air. The last I was led into in consequence of inhabiting a house adjoining to a public brewery, where I first amused myself with making experiments on fixed air* [carbon dioxide] *which I found ready made in the process of fermentation. When I removed from that house, I was under the necessity of making the fixed air for myself; and one experiment leading to another, as I have distinctly and faithfully noted in my various publications on the subject, I by degrees contrived a convenient apparatus for the purpose, but of the cheapest kind. . . . When I began these experiments I knew very little of Chemistry, and had in a manner no idea on the subject before I attended a course of chemical lectures delivered in the academy at Warrington by Dr. Turner of Liverpool. But I have often thought that upon the whole, this circumstance was no disadvantage to me; as in this situation I was led to devise an apparatus, and processes of my own, adapted to my peculiar views. Whereas, if I had been previously accustomed to the usual chemical processes, I should not have so easily thought of any other; and without new modes of operation I should hardly have discovered anything materially new."*

Lord Shelburne's patronage solved Priestley's financial worries, at the time when he had reached his peak as a scientist. During 1774 Priestley and Shelburne toured the continent of Europe and there Priestley met many of his fellow scientists including Antoine Lavoisier. Priestley was to remain with Lord Shelburne altogether for seven years (1773 - 1780). During this period he published seventeen books and essays on religion, metaphysics and politics - including his most important philosophical study, *Disquisitions Relating to Matter and Spirit* (1777). But these seven years were also the most productive period of Priestley's scientific career and, perhaps, the only period of his life that was dominated by science. The published record of these years include the major part of Priestley's publications on chemistry and properties of gases. For example, all three volumes of *Experiments and Observations on Different Kinds of Air* and two of the three volumes of *Experiments and Observations Relating to Various Branches of Natural Philosophy* were published between 1774 and 1779.

There are some interesting passages in Priestley's *Memoirs*[2] relating to this period of his life:

> ". . . When I went to his Lordship, I had materials for one volume of Experiments on Air, which I soon after published, and inscribed to him; and before I left him I published three volumes more, and had materials for a fourth, which I published immediately on my settling in Birmingham. He encouraged me in the prosecution of my philosophical enquiries, and allowed me 40 £ per annum for my expenses of that kind, and was pleased to see me make experiments to entertain his guests, and especially foreigners . . ."

During 1779 the relationship between Priestley and Lord Shelburne became less friendly and in 1780 Priestley chose to leave Shelburne's service and, with his family, he moved to Birmingham. There he became a preacher at the New Meeting House and met a number of friends who induced him to join the Lunar Society. The Lunar Society, an informal collection of liberal intellectuals, scientists and enlightened industrialists, included amongst its members Mathew Boulton, Erasmus Darwin, James Watt and Josiah Wedgwood. The Society met on Monday nights nearest the full moon and discussed the latest ideas in science and philosophy. Members of the Society supported Priestley's researches and ideas - even his doctorine of phlogiston - to which Priestley adhered, in opposition to Lavoisier, all his life.

Priestley's major preoccupations at this time were theological and he became the chief protagonist for the Unitarian views as well as the chief critique of the established Christian beliefs. In his *History of the Corruptions of Christianity* published in 1782, Priestley rejected miracles, the Atonement and the Trinity and he considered all these doctorines to be *"corruptions"*, not at all essential to Christian faith. In his view, the essential thing in Christianity was the belief in God, based on the evidence of divine truth. Priestley also firmly believed in the simple humanitarian principle that:

> "The good and happiness of the majority of any state is the great standard by which everything relating to that state must finally be determined."

He defined a just government as one that aimed at the happiness of its citizens and he advocated that any unjust government should be overthrown by the people[3].

These views, combined with Priestley's unabashed support of the cause of the American Colonies and of French Revolution, made him exceedingly unpopular with both the church and the establishment. He was denounced in Parliament as a heretic by Edmund Burke and on 14 July, 1791, the second anniversary of the fall of Bastille, a *Church-and-King* mob in Birmingham gathered round Priestley's house and joyously burned it down. Priestley managed to escape to London from where he addressed a sternly worded letter to the people of Birmingham, admonishing them for their misdemeanour. He joined the dissenting academy at Hackney and gave some lectures at New College. However, continued political and religious persecution made life difficult. On 8 April, 1794, at the age of sixty-one, Priestley sailed for America. He refused various offers of teaching and preaching in New York and Philadelphia and built himself a house and a laboratory in the small town of Northumberland. He died in Northumberland on 6 February, 1804.

Experiments on Diffusion of Gases

A lucid account of Priestley's experiments on diffusion of different kinds of gases appears in Section V, Book VII of *Observations and Experiments Relating to Fixed Air*[4]. This section is concerned with gases that have *"no mutual action"* and starts with the following interesting passage:

"Considering the very different specific gravities, and other remarkably different properties of different kinds of air, it might naturally enough be taken for granted, that those which differ very much in specific gravity at least, would separate from each other after they were mixed, the heavier occupying the lower place, and the lighter the upper; and that, by this means, the heavier kind of air might be able to expel the lighter, if there should be an opportunity for its escape from the upper part of the vessel. As different kinds of air will often be unavoidably mixed in a variety of experiments, I thought it a matter of some consequence to ascertain precisely how the fact was in this respect; that I might not, upon any occasion, deceive myself, by imagining that I had one kind of air only, when, in reality, I might have a much greater proportion of some other kind than I suspected. The result of my trials has been this general conclusion: that when two kinds of air have been mixed, it is not possible to separate them again, by any method of decanting, or pouring them off, though the greatest

possible care be taken in doing it. They may not properly incorporate, so as to form a third species of air, possessed of new properties; but they will remain equally diffused through the mass of each other; and whether it be the upper or the lower part of the air that is taken out of the vessel, without disturbing the rest, it will contain an equal mixture of them both."

Priestley conducted a fairly large number of experiments involving such binary gas mixtures as *fixed air and common air, inflammable and nitrous air, fixed air and nitrous air, vitriolic acid air and fluor acid air,* . . . etc. His results showed, consistently, that after mixing, the gases always remained *"equally diffused"*. Priestley's experiments on diffusion of gases were ingeniously simple and his description of them is extremely short and sweet, as the following quotation shows[4]:

"Having mixed equal quantities of inflammable and nitrous air in a phial which had a perforation at the bottom, and letting it out in five different parts, I observed that all burned with a lambent flame, without any sensible difference between them."

Although Priestley's contributions to science are monumental - he discovered oxygen and several other gases, demonstrated the phenomenon of diffusion, wrote books on electricity, optics, etc. - he spent most of his life as a teacher or preacher. Priestley's science was an activity of his leisure. Theology was his main pursuit and he prized his scientific reputation because, as he wrote in his *Memoirs*[2], it gave weight to his efforts *"to defend Christianity"*. A remarkable supposition!

THOMAS GRAHAM (1805 - 1864)

Thomas Graham was born in Glasgow, on December 21, 1805. His father, a wealthy business man, wanted him to become a minister of the Church of Scotland but Graham was not interested in a clerical career. He entered the University of Glasgow in 1819 and there he acquired a strong liking for scientific experiments and research. He graduated from the University of Glasgow with a degree of Master of Arts in 1826, worked for two years at Edinburgh University and then returned to Glasgow to teach chemistry at the Mechanic's Institute. In 1830, he succeeded Alexandre Ure as professor of chemistry at the Andersonian University (later known as the Royal College of Science and Technology). Graham moved to London in 1837, upon his appointment as professor of chemistry at University College. In 1841, he became the first President of the Chemical Society of London. After the death of Dalton in 1844, Graham became the acknowledged Dean of English Chemists and a worthy successor to Black, Priestley, Cavendish, Davy, and Dalton. He resigned his professorship at University College in 1854 to succeed Sir John Hirschel as Master of the Mint.

The most fruitful period of Graham's long and brilliant scientific career which lasted over forty years was, probably, between 1861 and 1869. During this time he communicated four elaborate papers to the Royal Society. For one of these papers, *On Liquid Diffusion Applied to Analysis*, and for his Bakerian lectures, *On the Diffusion of Liquids* and *On Osmotic Force*, Graham was awarded the Copeley Medal of the Royal Society in 1862. In the same year he received the Jecker prize of the French Academy of Science. His last major paper, *On the Absorption and Dialytic Separation of Gases by Colloid Septa* was presented to the Royal Society in 1866. This was followed by four supplementary notes, giving account of further discoveries on liquid diffusion and dialysis, the last of which was communicated to the Royal Society just a few months before his death on September 13, 1869.

Graham is universally acknowledged as the founder of colloid chemistry. Based on his work on osmosis, he developed what he called a *"dializer"*, which he used to separate colloids from crystalloids. In this way he prepared colloids of silicic acid, alumina, ferric oxide and several other hydrous metal oxides and he distinguished between sols and gels.

Graham was a chemical philosopher with an unusual zeal for accurate quantitative work. His experiments were original in conception and simple in execution. His mode of thought was inductive and he developed many of his ideas directly from his own experimental work which he recorded in the most precise and unbiased manner. His celebrated book *Elements of Chemistry, Including the Application of Science in the Arts* which was first published in 1842 was widely used, not only in Britain but also on the Continent. He was elected a Fellow of the Royal Society in 1837, a corresponding member of the French Academy in 1847 and a Doctor of Civil Law at Oxford in 1855.

Graham's Work on Diffusion in Gases

In 1803, Dalton, in a paper, *On the Tendency of Elastic Fluids to Diffusion through Each Other*, had described the remarkable action of inter-mixture that takes place, even in opposition to the influence of gravity, when any two gases are allowed to communicate with each other. The subject was investigated a few years later by Berthelot who, after extensive and careful experiments, corroborated Dalton's results but opposed his theoretical conclusions. This was the state-of-the-art when Graham entered the debate.

The first of Graham's several outstanding papers relating directly to the subject of gas diffusion appeared in 1829, entitled *A Short Account of Experimental Researches on the Diffusion of Gases through Each Other, and Their Separation by Mechanical Means*[5]. The experimental procedure used by Graham was basic and simple. Each gas investigated was allowed to diffuse from a horizontally placed bottle through a narrow tube, upwards or downwards into the air, in such a way that the diffusion had to take place in opposition to the influence of gravity. The rate of diffusion was found to vary inversely with the relative density of the diffusing gas. Thus, hydrogen was found to escape four or five times more quickly than the much heavier carbon dioxide. Also, with a mixture of two gases, the lightest of the two was found to leave the bottle in largest proportion, so that a sort of mechanical separation of gases could be effected by this unequal diffusibility - a process first used at Oak Ridge, Tennessee, during the second World War to separate the fissionable isotope U235 from the non-fissionable isotope U238.

Graham complemented the simple method of measuring diffusion by free communication of gases with the study of diffusion through porous media. He devised a simple instrument which he called *"a diffusion-tube"*, consisting of a glass cylinder open at both ends, half an inch in diameter, and from six to fourteen inches in length. A wooden stick, slightly less in diameter, was then inserted into the glass tube so as to occupy all but one-fifth of an inch of its length. This empty space was filled with a paste of Plaster-of-Paris. After the plaster had set, the wooden stick was withdrawn and the glass tube thus left with an immovable plug of porous plaster. The tube was now filled with hydrogen over mercury, and hydrogen was allowed to diffuse into the air through the pores of the plug. Within a few minutes, much of the hydrogen had diffused out of the tube and had been replaced by air.

On December 19, 1831, Graham read before the Royal Society of Edinburgh a paper describing results of his diffusion experiments for which he received the Keith prize of the Society. This work published later in the *Edinburgh Philosophical Transactions* is entitled *On the Law of the Diffusion of Gases* and it starts with the following statement of the objective of the work[6]:

"It is the object of this paper to establish with numerical exactness the following law of the diffusion of gases: The diffusion or spontaneous intermixture of two gases in contact, is effected by an interchange in position of indefinitely minute volumes of the gas, which volumes are not necessarily of equal magnitude, being, in the case of each gas, inversely proportional to the square root of the density of that gas."

These replacing volumes of the gases termed *equivalent volumes of diffusion* by Graham had the following numerical values:

hydrogen	3.7947
water vapor	1.2649
nitrogen	1.0140
oxygen	0.9487
chlorine	0.6325

The numbers were found to be reciprocals of the square roots of the relative densities of these gases - the denisty of air being assumed as unity.

Since the unequal diffusion volumes of different gases were regarded as consequences of their unequal diffusion velocities, it was concluded that the relative velocities at which different gases diffused into one another were identical with the rates at which they diffused under pressure into a vacuum - a result which is in accordance with Dalton's maxim. But, although the relative rates of diffusion and effusion were found to be alike, Graham stated subsequently[7] that:

"the phenomena of effusion and diffusion are distinct and essentially different in their nature. The effusion movement affects masses of gases, the diffusion affects molecules; and a gas is usually carried by the former kind of impulse with a velocity many thousand times as great as is demonstrated by the latter."

Graham's address to the Royal Society of Edinburgh ends as follows:

"The law at which we have arrived (which is merely a description of appearances, and involves, I believe, nothing hypothetic), is certainly not provided for in the corpuscular philosophy of the day, and is altogether so extraordinary, that I may be excused for not speculating further upon its cause, till its various bearings, and certain collateral subjects, be fully investigated."

Some thirty years after presentation of the above paper, Graham again subjected the phenomena of gas diffusion to elaborate experimental investigations. His results were communicated to the Royal Society of London in a paper *On the Molecular Mobility of Gases* and published in *Philosophical Transactions* in 1863. These later experiments were made principally using a porous plug of compressed graphite. Referring to such media Graham observed[8]:

"The pores of artificial graphite appear to be so minute, that a gas in mass cannot penetrate the plate at all. It seems that molecules only can pass; and they may be

> *supposed to pass wholly unimpeded by friction, for the smallest pores that can be imagined to exist in the graphite must be tunnels in magnitude to the ultimate atoms of a gaseous body. The sole motive agency appears to be that intestine movement of molecules which is now generally recognized as an essential property of the gaseous condition of matter. According to the physical hypothesis now generally received, a gas is represented as consisting of solid and perfectly elastic spherical particles of atoms, which move in all directions . . . If the containing vessel be porous, like a diffusiometer, then gas is projected through the open channels, by the atomic motion described, and escapes. Simultaneously, the external air is carried inwards in the same manner, and takes the place of gas which leaves the vessel . . . The molecular movement is accelerated by heat and retarded by cold . . ."*

Diffusion in Liquids and Occlusion of Gases by Metals

In some ways, the most interesting experiments of Graham are those concerned with diffusion in liquids and with the *"occlusion"* of gases by metals.

Details of some of the former work appears in a substantial paper entitled *Liquid Diffusion Applied to Analysis* which was read before the Royal Society on June 13, 1861. It is in this paper[9], that Graham introduced the word *colloidal condition of matter*. He also coined the word *dialysis* in the following passage:

> *"It may perhaps be allowed to me to apply the convenient term dialysis to the method of separation by diffusion through a septum of gelatinous matter. The most suitable of all substances for the dialytic septum appears to be the commercial material known as vegetable parchment or parchment-paper, which was produced by M. Gaine, and is now successfully manufactured by Messrs. De La Rue."*

In this paper Graham describes in detail, and with great accuracy, results of his experiments on diffusion of various solutions (salt, sugar, gum, tannin, albumen, iodine, caramel, etc.). His results indicated that:

> *"Diffusion is promoted by heat, and separations may be effected in a shorter time at high than at low temperatures."*

He also showed that the purification of many colloid substances could be achieved with great advantage by dialysis whereby:

> "*Accompanying crystalloid is eliminated and the colloid is left behind in a state of purity.*"

Graham recognized that colloids and crystalloids constituted *"different worlds of matter"* but he concluded, nevertheless, that:

> "*in nature there are no abrupt transitions, and the distinctions of class are never absolute.*"

Graham's experiments on occlusion of gases included several metals, but the most dramatic results were those obtained with silver, iron, platinum, and above all, palladium. Palladium was found to have the property of occluding many hundred times its volume of hydrogen. In one experiment, for example, a specimen of electrolytically deposited palladium, heated to 100°C and then slowly cooled in an atmosphere of hydrogen, was found to occlude 982.14 times its volume of the gas. Some idea of such enormous capacity of palladium to absorb hydrogen may be formed by remembering that water at room temperature absorbs 782.7 times its volume of ammonia gas, one of the most soluble gases known. Graham took advantage of the selectivity of metals for various gases to separate gaseous mixtures. For example, he used a silver leaf, heated and cooled in ordinary air and subsequently heated in a vacuum, to produce a highly enriched air containing 85 per cent oxygen.

Graham will be remembered as a great pioneer and philosopher of science with exceptional gifts for very accurate experimental work.

ADOLF FICK (1829 - 1901)

Adolf Fick was born in Kassel, Germany, on September 3, 1829. His father, Friedrich Fick, was a senior municipal architect in Kassel and had nine children. Adolf was the youngest in the family and from early childhood he showed great aptitude for mathematics and physics. However, when in 1847 Fick entered the University at Marburg, he was persuaded by his elder brother who was professor of anatomy, to become a medical student. At Marburg Fick developed a particularly close relationship with Carl Ludwig and soon became one of the main proponents of the new orientation of physiology towards physics, to which Helmholtz, Brücke, and Emil du Bois-Reymond also subscribed.

Fick wrote his first scientific paper in 1849, whilst he was still a student at Marburg. In this paper which dealt with the torque exerted by the motor muscles of the femur in the hip joint, he clearly demonstrated his unusual talent for the mathematical analysis of physiological processes. In the Fall of 1849 Fick went to Berlin to continue his studies and soon he became a friend of

Helmholtz. Two years later, upon receiving his doctorate, he accepted the position of a demonstrator in anatomy at Marburg. He stayed in this position for a very short time before moving to Zurich, where he became an associate of Ludwig and began a brilliantly diversified scientific career. Fick remained in Zurich from 1852 until 1868, first as a demonstrator in anatomy and then as a professor of physiology. In 1868 he was appointed professor of physiology in the faculty of medicine at Würtzburg. Ten years later he became Rector of the University of Würtzburg. He retired in 1899, at the age of seventy and he died on August 21, 1901.

Most of Fick's work is related primarily with problems that interface medicine, physiology and physics. His monograph, *Medizinische Physik,* published in 1856, for example, deals with a great variety of subjects ranging from molecular diffusion, endosmosis, mechanics of solids, statics and dynamics of muscles, motion of fluids in elastic tubes (blood vessels), recording of pulse variations, conservation of energy in human body, color perception and the origin and measurement of bioelectric phenomena. The physics of vision was of particular interest to Fick who wrote his doctoral thesis *De Errore Optico* on the subject. He built, in 1888, the first practical instrument for measurement of intraocular pressure (ophthalmotonometer) and he developed a principle that permits calculation of cardiac output from the measurement of the minute volume of oxygen consumption and arteriovenous oxygen difference in the living organism. Fick was deeply interested in the philosophy and methodology of scientific investigation and he considered mathematics to be the only adequate language of science. He was a crystal-clear thinker and yet a bacchanalian who enjoyed good wine and carefree life, particularly in his younger days. When he died, he had three children, one of whom was an anatomist in Berlin.

Fick's Law of Diffusion

Fick's celebrated paper, *On Liquid Diffusion* was communicated by the author, and published in *Philosophical Magazine* in July, 1855[10]. The paper starts with the following paragraph:

> *"A few years ago Graham published an extensive investigation on the diffusion of salts in water, in which he more especially compared the diffusibility of different salts. It appears to me a matter of great regret, however, that in such an exceedingly valuable and extensive investigation, the development of a fundamental law, for the operation of diffusion in a single element of space, was neglected, and I have therefore endeavored to supply this omission."*

Fick thought it quite natural to suppose that the diffusion of a salt in a solvent must be identical with the diffusion of heat in a conducting body, for which Fourier had already formulated his celebrated theory of heat conduction. He therefore postulated that:

> *"the transfer of salt and water occurring in a unit of time, between two elements of space filled with differently concentrated solutions of the same salt must be caeteris paribus, directly proportional to the difference of concentration, and inversely proportional to the distance of the elements from one another."*

He then went on to say:

> *"Exactly according to the model of Fourier's Mathematical development for a current of heat, we can obtain from this fundamental law for the diffusion-current, the differential equation*

$$\frac{\partial y}{\partial t} = -k \left(\frac{\partial^2 y}{\partial x^2} + \frac{1}{Q} \frac{dQ}{dx} \cdot \frac{\partial y}{\partial x} \right) \quad (1)$$

> *when the section Q of the vessel in which the current takes place is a function of its height above the bottom. If the section be constant (i.e., the vessel cylindrical or prismatic), the differential equation becomes simplified to*

$$\frac{\partial y}{\partial t} = -k \frac{\partial^2 y}{\partial x^2} \quad (2)$$

Fick proposed several methods for the solution and experimental verification of this differential equation. For example, for determining the decrease of concentration of a solute held in a cylindrical vessel, he lowered into the vessel a glass bulb suspended from the beam of a balance, and determined the density gradient within the solution. In this and similar experiments involving funnel-shaped vessels (variable cross-section) attempts were made to ensure that:

> *"the diffusion-current has become stationary, and a so-called dynamic equilibrium has been produced."*

These attempts were only partially successful.

The constant k in equation (1) was referred to as *diffusibility* by Fick and was defined as:

> "the quantity of salt which, during a unit time, passes through the sectional-unit, out of one stream into the next adjacent one, when the rapidity of the determination of concentration $(\frac{dy}{dx})$ is equal to unity."

He determined and tabulated the value of k for various salts and he pointed out that:

> "as might be expected from Graham's experiments, the value of k increases with increase of temperature; probably, however, this dependence upon temperature is not a simple one."

On the relationship between diffusibility and other fundamental properties of the salts tested, as, for instance, their molecular mass, Fick suggested that:

> "nothing can be said until extensive series of experiments with different substances have been made."

Fick was brilliant intellectual but he had no intellectual arrogance. His modesty clearly manifests itself in the last paragraph of his famous discourse on liquid diffusion with which the present paper is concluded:

> "The comparison of the experiments adduced above with the hypothesis developed on the foundation of the diffusion law, shows, though not absolutely, that the truth of this hypothesis may be determined; and it is in fact highly probable that, with or without modification, such an hypothesis may serve as the foundation of a subsequent theory of these very dark phenomena."

SUMMARY

Joseph Priestley, Thomas Graham, and Adolf Fick were the great pioneers of diffusion. Priestley was a dissenting English clergyman. His vocation was chemistry. His extensive experiments on mixtures of different gases led him to postulate that *"when two kinds of air have been mixed, it is not possible to separate them again . . . but they remain equally diffused through the mass of each other."* Graham, a Scotsman, was a brilliant academic. In 1831, he proposed his celebrated law of diffusion of gases which states that *"the diffusion, or spontaneous intermixture of two gases in contact is effected by an interchange in position of infinitely minute volumes of the gases . . . being, in the case of each gas, inversely proportional to the square root of the density of that gas."* Fick, a German, was a physiologist with strong interest in mathematical analysis. In an historical paper published in 1855, he proposed a fundamental law for diffusion which states that *"the diffusion flux is directly proportional to the difference in concentration and inversely proportional to the distance of elements from one another"*. The paper has briefly reviewed the life and work of these great men.

REFERENCES

1. J. Bronowski, "The Ascent of Man", Little, Brown & Co. Boston, (1973).
2. J. Priestley, "Memoirs of Dr. Joseph Priestley, to the Year 1795, Written by Himself, with a Continuation to the Time of His Decease, by His Son", London, (1806).
3. J.A. Passmore, "Priestley's Writings on Philosophy, Science and Politics", New York, (1965).
4. J. Priestley, "Observations and Experiments Relating to Fixed Air", Book VII, Section V - Birmingham, (1790).
5. T. Graham, "A Short Account of Experimental Researches on the Diffusion of Gases through Each Other, and Their Separation by Mechanical Means", Quarterly Journal of Science and the Arts, 27, 74, (1829).
6. T. Graham, "On the Law of Diffusion of Gases", Edinburgh Phil. Trans., 2, 175, 269, 351, (1833).
7. T. Graham, "On the Law of Diffusion of Gases", Edinburgh Roy. Soc. Trans., 12, 222, (1834).
8. T. Graham, "On the Molecular Mobility of Gases", Phil. Trans. Roy. Soc., 153, 385, (1863).
9. T. Graham, "Liquid Diffusion Applied to Analysis", Phil. Trans. Roy. Soc., 151, 183, (1861).
10. A. Fick, "On Liquid Diffusion", Phil. Mag., 10 (63), 30, (1855).

Other References

1. W. Durrant, and A. Durrant, "The History of Civilization", Vol. 9, "The Age of Voltaire" - Simon and Shuster, New York, (1965).
2. C.C. Gillespie, Editor-in-Chief; "Dictionary of Scientific Biography", Charles Scribners, New York, Volumes 4, 5, and 9.
3. R.E. Schofield, Editor; "A Scientific Autobiography of Joseph Priestley, 1733-1804", M.I.T. Press, (1966).
4. W. Olding, "On Professor Graham's Scientific Work", The Royal Institution Library of Science (Physical Sciences) Vol. 2, Elsevir, (1970) (Editors, Bragg W.L. and Porter G.) Text of an Address to the Royal Institution on January 28, 1870.
5. F. Schenk, "Adolf Fick Gesammelte Abhandlungen", 4 Volumes - Würtzburg, (1903 - 1905).

DISTILLATION – SOME STEPS IN ITS DEVELOPMENT

Donald F. Othmer

Polytechnic Institute of New York
333 Jay Street
Brooklyn, NY 11201

THE BEGINNINGS

Distillation as an art or science is a chemical engineering operation which even the man in the street may be willing to try to define, so general is the use of the word and of the operation so known. The broadest usage of the word covers what is probably the chemical engineer's most widely used separating operation. The old alembic or retort for centuries was the symbol of the alchemist and now is the logo of the chemist; and the worm condenser is the symbol of the entire spirits industry.

On one hand, distillation must be differentiated by the chemical engineer, if not by the man in the street, from the simpler concept of evaporation, and also from the technical concept of rectification. On the other hand, the extractive metallurgist has used distillation as a word, as a concept, and as a tool along with many other terms which he uses in place of or in close parallel thereto. He has been the greatest user of distillation associated with chemical reaction, e.g., in his winning of zinc and other metals, including possible processes for iron, aluminum and others.

The first definition in English of the term rectification was by A. Cooper (1), "Rectifying is freeing the spirit from its essential oil and phlegm," these latter being the higher boiling, more viscous liquids. This was by various chemical or other treatments rather than by dephlegmation as our modern terminology of distillation would indicate, through partial condensation and reflux-wash. In general usage, the word rectification means – "improving" or "correcting." In the spirits industry, a rectifier may be one who blends or "improves" spirits into cordials, etc., either with or without distillation.

Everyone concerned with distillation has his own concepts of the many facets of the operation which may come under this title. The first attempts to understand and utilize the basic art were long ago. The ancient Greeks knew that the evaporation of sea water allowed the condensation of a potable water without salt. This differentiates evaporation, meaning the vaporization of a pure liquid away from its solution with a non-volatile solid. Distillation, however, usually means the vaporization of a more or less purified liquid away from its solution with another volatile liquid - also from any solids contained in the original liquid.

Some interpretations of the writings of these Greek natural philosophers suggest that they recognized, without actually separating, an inflammable ingredient in wine. According to their idea of the four elements which made up all substances - earth, water, air, and fire - wine would thus be composed of some fraction of water and some fraction of fire - while brandy, if indeed they did know it, would have had a larger percentage of "fire." And none would gainsay!

By the same token, gasoline would be even more "fire," with less water; and crude oil would have a bit of earth worked in, somehow, at one end of what we now call a molecule!

All knowledge slept, with little growth, during the Dark Ages. However, some believe the Arabs made some developments in distillation, possibly before Mohammed, who prohibited even wine. There were some stirrings in middle Europe about the 11th century.

THE RENAISSANCE

Then were heard the incantations and the incessant mumbo-jumbo of the alchemists. Almost all of their "experimental work" used distillation. Indeed, a cleric in the early 1200's noted that distillation was a most important part of all alchemical practice. The alchemists did attempt to make, over hundreds of years, the Elixir of Life and the Philosopher's Stone. Many have discussed the "operations" used by the alchemists. These may be compared to a list of our modern unit operations. Distillation and other processes were well described by Paracelsus (2) who regarded himself as an adversary of the alchemists. He, Theophrastus Bombastus Paracelsus von Honenhein, 1493-1541, however, was half chemist and half alchemist, half physician and half quack, half scientist and half magician. These add up to more than unity - indeed they should; he was indeed quite a person! He had (2) his own seven (magical 7) unit operations; 1. Calcination; 2. Sublimation;

3. Dissolving; 4. Putrefaction or Fermentation; 5. Distillation with three types : Ascension, Washing, and Imbibition which might indicate both a column and its reflux; 6. Coagulation, of the Heat, and of the Cold; 7. Coloration.

Paracelsus openly despised the alchemists, and might be regarded as at least a generation advanced toward the ultimate species of the scientist. In many ways he did show a much more logical and quantitative approach to natural philosophy. The alchemists talked - rather than taught - in elegant riddles; each chemical engineer can decide how the operations philosophy and exhortations of Paracelsus fit into modern thinking.

George Agricola (Bauer) 1494-1555, was also one of the great scientific writers of all time. He was much more objective, understandable, and less prolific than his contemporary, Paracelsus. He also described (3) distillation in his De Re Metallica, particularly of Quicksilver and Nitric Acid, also evaporation, also crystallization. There were many editions of his major work in Latin, the accepted language of the time for all learned discussions - continuing to the present in names in chemistry, botany, zoology, etc.*
There were even more editions in the "vulgar tongues," the common languages, and these - the German, Italian, and Spanish editions - like old cookbooks - have been worn out from constant use in the field and in the workshops.

The English had insufficient interest in metallurgy in the 16th century to justify the expensive publication of this treatise (over 600 folio pages and hundreds of fine woodcuts). Not until 1912 was an English translation privately made and published by an American mining engineer and his wife. A copy cost this writer 25 years later more than twice what he had earlier paid for a copy of the first complete Latin edition, because the translator had meanwhile become famous as the President of the United States, Herbert C. Hoover.

Agricola's great work was for 180 years the accepted text on metallurgy and some aspects of inorganic chemical technology. He was a very practical man, reporting chemical and metallurgical processing operations as they existed and were practiced. He had his feet planted firmly on - or in - the ground. Nevertheless, his other major work - often bound with De Re Metallica - discussed supernatural elves, goblins, etc. which lived under the ground in mines!

A contemporary of Agricola was Birringguccio (4) who regarded

---

*Today, probably a hundred copies are known, world-wide of the Latin editions. They have been unused and preserved mainly in the libraries of the monasteries.

distillation as a working form of his ideas of practical thermodynamics, which he called "Pirotechnie," the technology of heat. He likewise was concerned with the practical production of inorganic materials as metals, salts, etc. by thermal physical and chemical effects, although his last chapter is devoted to "that fire which burns in the heart of a youth for a maid."

Save, then, for a few pragmatists, as Agricola and Biringuccio- distillation for half a millenium was almost a mystic art of the alchemists and their contemporary physicians who also were practicing magicians. The latter used distillation not so much for beverages but for medicines made from weird concoctions of many herbs, seeds, flowers, fruits, and miscellaneous parts of fauna as well as flora. Often this involved, among other aspects, a partial steam distillation of essential oils which flavor aromatic plant material. In 1512 in Strassburg was printed in German the great Book on the Art of Distillation by Brunschwygk (5) full folio size of hundreds of pages and many woodcuts, following his Small Book of Distillation, 1500. These gave many recipes for medicines and cosmetics. Much lesser attention was given to beverages.

Thereafter, in the 16th century, spirits for beverages slowly came into production, always scarce and precious - knowledge of their making was quite guarded. Brandy-wine (from branden - to burn + wine) came in France and middle Europe. Usqubaugh - "water of life," later corrupted to whiskey, in Ireland, then Scotland, came sometimes, it was said, after a thousand distillations, one of "much highness of glorification." (6)

In several of these books of the 1500's a wealth of distilling equipment has been pictured in woodcuts. The stills were made of copper, glass, stoneware, and other materials in an imaginative and fascinating array of shapes of flasks, columns, condensers, receivers, furnaces, and their accessories. Problems arose through lack of knowledge and difficulties in working the materials of construction. They were solved by ingenuity instead of by calculations and deduction. Sometimes these problems were quite the same as those of today. Thus one philosopher, Porta (1569), needed taller columns (7) and complained that this need could not be fit into his home, so he moved to another house with higher ceilings!

Intriguing - often amusing - is any study of both the equipment and the practices of distillation of the alchemists. Clumsy as these are, looking back four centuries, they developed slowly by many steps from the rituals for supernatural help and the practice of magic of the alchemists. From this chicanery the art of modern distillation and the science which underlies its practice were formulated later. There were so many "witches' brews" in the products of those who called themselves distillers, and these were to make potions, lotions, and similar fluids - all in an amazing variety of methods

DISTILLATION DEVELOPMENT

and equipment. These were usually purported to be of magical, often even they were assumed to be of <u>divine</u> design.

THE CONTRIBUTION OF ALCOHOLIC SPIRITS

The progress of distillation, after the alchemists, was largely motivated or stimulated by the thirst for alcoholic beverages which contained more alcohol than the organisms which fermented it could tolerate. These were made in many countries from the beers or wines which could be fermented from the available carbohydrate material; grapes and other fruits having sugars into brandies, molasses from cane and sugar beets into rums, grains into whiskey, and potatoes into vodka. Roughly, the alcoholic strength approximated, with every different spirit so distilled, the alcoholic content of the first condensate shown on the diagram of the composition of the vapor from the first vaporization of the fermented liquid charged.

By the early 1800's many of the techniques of modern practice were anticipated if not developed. Preheating of the liquid to be fed to the still was accomplished first by indirect heat transfer, then by direct bubbling of the vapors through feed liquid charged into chambers counter-currently. This led to the use of steam-heated pots and columns with reflux. These contributions to the art were mainly from the French. All of these and more were included, however, in 1830 in a British patent to Coffey (8) of Dublin. His continuous still - which had plates somewhat like the modern valve tray - was very suggestive of what might be called 20th century practice.

Thus when distillers from England set up shop to make English gin in the United States after the repeal of Prohibition in the mid-30's, they used Coffey stills, which were little more than enlargements from the patent drawings of a hundred years before. Such is the traditional conservatism of the distillers of spirits that these stills had to be made in their own workshops in Britain and hammered out by the same British coppersmiths who had made the ones already in use there.

Tit for tat, American distillers - and their chemical engineers-invaded Scotland at almost the same time. They built much improved, American designed units to make neutral spirits to blend the Scotch whiskey made for the enlarged American market. In this case, the Americans took advantage of a hundred years of developments - largely during the previous 20 years in processing petroleum and chemicals therefrom. And American chemical engineers went to Scotland to build and start up the units, more stingy with steam than the Scotch thought possible. The new processing included azeotropic and extractive distillation to make purer spirits, and vapor reuse distillation (9) to reduce steam consumption, by a combination of the principles of multi-effect evaporation and usual rectification.

The production of neutral spirits - i.e., 95% alcohol practically chemically pure for blending in beverages, resulted in several developments in economy of steam use. For example, the Wentworth system (10,11,12,13) used in large plants since the 30's, more recently greatly improved (14), gives pure absolute alcohol at less than half the steam requirements of common current processes for making impure 95% alcohol. This considerable saving in energy is important in the design and operation of present plants for the production of alcohol of the very highest quality, for use in perfumes - or quite anhydrous for fuels for automobiles.

The years of the '30's saw the downfall of Prohibition and of the world's economy in the Great Depression. They were also the years of the first large-scale use of alcohol in motor cars, either as Gasohol with 10 to 20% of anhydrous-absolute-alcohol, or of alcohol as the complete fuel. Now - 50 years later - there is a great surge in building of plants for alcohol for motor fuels, mainly to give out the benefits of farm relief subsidies - direct or in the elimination of taxes. This has resulted in many primitive plants and much production - also many failures - by do-it-yourself farmers. However, these same artificial benefits will bring forth also the world's largest-ever alcohol plants.

THE PETROLEUM ERA

Oil directly from coal came in the two decades before the beginning of the century-long flood of petroleum - 1870-1970. This will probably be resumed and continue for many more decades, while this flood slowly recedes. "Coal oil" was made by dry distilling coal, and was fractionated from heavier cuts in huge pot stills heated by fires beneath. These same plants in Pennsylvania were converted to become simple petroleum refineries after Drake's discovery well. Using nearby crude and the same stills, fractions were cut out by specific gravity, as the condensate from the classic coil in a tank of water ran through a "look box" in which was a hydrometer. First came gasoline, next was the important fraction - kerosene - for a burning oil, and finally lube stocks were diverted by appropriate valves to their respective tanks as the operator watched the density of the condensate increase. Cuts were made by density instead of temperature because the thermometer, if any, was of mercury in glass on top of the pot - outside and some distance away. "Cracking" may have been discovered accidently when the operator of an overfired still took a too-long "coffee break." On his return, he found light fractions running again, from the decomposition of what had been intended to be drawn from the pot as "bottoms" or residual oil.

DISTILLATION DEVELOPMENT

Because of the size of the horizontal stills, 12-18 ft. in diameter and 30-45 ft. long, they were not enclosed in buildings. When the necessarily large bubble towers to be superimposed came to the industry, they also were not enclosed by a building. Usually they were free-standing in the open. The operator and his controls continued to be in a small control house; and this practice persists in petroleum distillation. The many, and always more, remote sensing and control devices which have been developed for modern petroleum processing are always so located and housed. These devices allow the operator to understand fully the variables and to exercise his controls to optimize the separating functions which are taking place over a large expanse of ground area and height. Such sensors and controls are centralized for a multiplicity of distillation operations interspersed with chemical reactors, e.g., cracking units, which results in a maze of complex piping, heat interchangers, vessels for various purposes for the flow and processing of the many fluid streams involved.

Contrariwise, for many other reasons including the maintenance of security of its valuable and enticing product, the alcohol industry has almost invariably housed its stills in enclosed buildings. Very large units have in some cases, however, been partially in the open, particularly the beer stills. Generally the chemical industry encloses its smaller units and has its larger ones in the open.

PHYSICAL MECHANISM OF DISTILLATION

Basically the separations possible by distillation are due to the greater tendency to vaporize, i.e., the greater volatility, of one liquid of a mixture compared to that of another - or others - in the mixture. Thus, throughout the liquid range of all but the very highest concentrations of alcohol in water, the vapor distilling away from the liquid is significantly higher in its content of alcohol. All quantitative studies and designs of the processes and equipment to be used for separations by distillation are thus based on data demonstrating this equilibrium relation between liquid and vapor. Fortunately, these data are determined simply and quickly by equilibrium stills in the laboratory. Vapor from the mixture of liquids is boiled off in substantial equilibrium with the liquid. These vapors are condensed and the condensate is recycled to the still until its composition does not change and represents the true equilibrium vapor composition. Samples of the condensate and of the still pot are then analyzed. These data, so determined, allow the design of plants for the separation of systems of mixtures of liquids. Many modifications of the original system (15) have been published.

All forms of distillation which are multistage, usually those with a column, well exemplify that type of separation operation which is controlled by the approach to a thermodynamic equilibrium, here that between the vapor leaving and the liquid being boiled. In a plate column, heat input successively boils-up vapors, more concentrated in the more volatile component than the liquid on the plate. These vapors contact the liquid on the next higher plate. There, some of the vapors are condensed which have a higher composition of the less volatile component. The heat of condensation of these vapors in the liquid on the higher plate then reboils-up vapors which are even more concentrated in the more volatile component. The vapors leaving the top plate are most concentrated in the lower boiling component, pass to a condenser, and are condensed. Some part of this condensate is refluxed back to the top of the column to wash the rising vapors. As the condensate descends the column, this reflux condenses out successive amounts of the higher boiling component as it goes down through the successive stages. The less volatile component is discharged in its highest concentration in the liquid at the base.

The physical mechanism of the interchange between molecules of the two components at the interface of the liquid phase with its vapor phase depends on molecular diffusion and heat tranfer. It is too involved for design calculations of the chemical engineer for any particular column. Some aspects of these molecular movements at the interface of a vapor being condensed into its liquid have been discussed (16) and these become much more complicated when two or more species of molecules are considered in their relations in a distillation column.

Necessary for plant design, however, is an understanding of this mechanism of separation of the components and the heat balance in such a column still. Such understanding is aided, however, on a macroscopic basis, if: - a) compositions of the vapor and of the liquid are considered in terms of fractions of the total number of moles present in each phase, b) the difference of the boiling points of the components is not great, hence the heats of vaporization of moles of the different components may be considered to be equal, c) sensible heats, heats of mixing and radiation losses from the column are neglected as being small compared to the latent heats involved.

These assumptions then determine that there is a substantially equal heat flow in and out and from bottom to the top of the column, and that balances of heat substantially represent balances of materials, i.e., as indicated by the compositions in mol fractions of the two phases. Therefore the flow of molecules in the liquid from plate to plate is constant in number, as is also the flow of vapor up from plate to plate. These allow the mathematical

DISTILLATION DEVELOPMENT

modeling and calculations for column design by the McCabe-Thiele (17) method. Also they allow an understanding or feeling of what is taking place in a distillation column.

DISTILLATION COLUMNS

Volatile liquids have been quite difficult to separate in the laboratory. For example, the German and British chemical journals of only a hundred years ago tell of the infinite patience of chemists in distilling a newly discovered compound to separate out the impurities and to identify its exact boiling point. A flask would be charged 500 to 600 times in succession with a small fraction of a liter of liquid which was then batch distilled. Condensate fractions were cut out each time over small boiling ranges and systematically reassembled, then redistilled. There were then no worthwhile laboratory distillation columns to reduce the tremendous cost in time and thermal energy, by dividing the time and effort - and the heat requirement - through the use simultaneously of as many as fifty theoretical distillations or boil-ups on column plates having a reflux. Each plate would act as both a still and a condenser. Now, however, columns and accessories are available in the process development laboratory (18-21) which are as efficient as those in the plant. They may be assembled and operated together as will be units in the large plant units, and thus show first the successful operation of the proposed process. Such columns have been used in pressures from high vacuum to 10 atmospheres and at temperatures up to over 200°C.

The most illustrious scientist this writer ever had the privilege of working with closely was 50 years ago. A renowned organic chemist, he had developed, to simplify his separations of a liter or less of liquids, an ingenious laboratory plate distillation column. In it he separated many materials with never a thought of energy or material balances. He wanted only the product, but he derided the separation out of part of his hard-won overhead product and refluxing it back to the top of his own rather efficient glass column. He would say - "What a good chemist has separated let it not be put back together by any damn fool." - meaning, of course, the chemical engineer. But for a century, industry had already been refluxing a part of the same liquid as came over the top as product!

In the alcohol industry up until 1930 there have been used hundreds of designs of plate type distilling columns which depend, for the essential vapor-liquid contact, of vapor bubbles rising through thin layers of liquid. By contrast, packed columns depend on the thin films of liquid trickling down the surface of the packing being contacted by the rising vapors. Some of these have been compared (22).

In the alcohol industry the alcohol must be separated from numerous materials in the "mash" liquid resulting from the fermentation, along with the water. This has always been done by distillation (23). The beer still separates non-volatiles with most of the water at the base. Alcohol comes over the top with many more volatile materials and a small part of the water. The beer still may have superimposed upon it the rectifying column to give a One Column Unit. If these two columns are separated, a Two Column Unit results, usually with a side liquid stream taken from the beer still to remove mainly higher alcohols in a mixture called fusel oil. (Fusel oil steam distills at about 90°C, thus is kept by the distilling action from going to either the head of the column at about 78° or to the bottom at 100°C plus - the higher than atmospheric boiling point is due to the pressure necessary at the base of the column. The oil is drawn off in an emulsion with water containing some alcohol, decanted, and the water is returned to a lower plate). A Three Column Unit separates a liquid stream taken from the top or near the top of the rectifying column to improve the quality of the product, and a Four Column or Six Column Unit will do this better, while allowing a reduction in steam requirements.

Before the use of continuous instead of batch columns in the petroleum industry, in about 1920, one of the most sophisticated distillation systems ever was developed, the so-called Barbet unit with 5 continuous columns purified methanol: - wood alcohol coming from wood carbonization. There were two, sometimes three, chemical treating sections in these columns, and one or two side stream draws of liquid for removal of "wood oils." These oils, which steam distill with an intermediary boiling point collect in the middle of a column. They neither could go overhead with the more volatile methanol or out of the bottom with the water, as with fusel oil in the alcohol system. From the 180 other volatile products which had by then been identified in the pyroligneous acid obtained as condensate from the pyrolysis of hardwoods, pure methanol was separated in these five columns. Also separated were several other commercially acceptable fractions by simultaneous physical processing and chemical treating steps within the columns themselves.

Columns which allow these multiple successive distillations of many, many natural and synthetic mixtures of liquids have been used in diameters from 6 mm. (1/4") in the laboratory to some 13.5 meters (45'). The heights have been up to 65 meters (200') and as many as 1000 plates have been superimposed. Pressures at the top have varied from a few millimeters of mercury up to 30 or 40 atmospheres. Rectification columns separate the cryogenic liquids at almost the universe's lowest possible temperature. Also they separate boiling metals some two thousand degrees higher. Ratios of the condensate

DISTILLATION DEVELOPMENT

to be refluxed to that taken off the top of the column as product vary from a low fraction for some easily separable liquids to hundreds of thousands of times for heavy water from water as suggested almost 50 years ago by Urey (24), and long practiced since (25).

DISTILLATION PROCESSES

Numerous processes have been developed which separate liquids by distillation with some modifications, or using an additional physical or chemical mechanism.

Azeotropic Distillation

In its simplest form, azeotropic distillation depends on the use of an added liquid - the entrainer. This is usually infinitely miscible with one of the components to be separated. The entrainer is quite insoluble in the other component, with which it forms an azeotropic mixture. This has a total pressure equal to the sum of the partial pressures of the two, and a boiling point - the azeotropic boiling point - which is much lower than that of either of the original liquids of the mixture. Thus the entrainer brings over this azeotropic mixture as the vapors at the column head in what is essentially a steam distillation. These vapors condense in the condenser, the condensate separates into two layers because of the mutual insolubility, the insoluble entrainer layer is returned to the column as reflux, and the layer of the one component of the original mixture is removed as product (26, 27).

Azeotropic distillation was developed by Young (28) at the turn of this century using benzene as the entrainer to separate the last of the water from its hydroxy-homologue, ethanol, which has very similar properties. The literature since is full of related data for the azeotropes of other liquids with alcohol which were found in attempts to find something other than benzene. All of these, however, like benzene, increase the volatility or lower the boiling point of the ethanol as well as that of the water. Many people searched for one which would lower the boiling point only of the water. Wentworth (10-13) did indeed find the liquid to simplify the process of removal of the last of the water by lowering the effective boiling point of only the water. This system is now available in a considerably improved form (14), which will give at the lowest cost the large amounts of absolute alcohol being planned for use as motor fuel in this country as well as in other countries of the world.

Azeotropic distillation was developed industrially for separating water from another hydroxy-relative, acetic acid, in another series, again by effectively making the water more volatile (27, 29). This was before the days of stainless steel. Copper and bronze were then the accepted materials of construction, with special parts and condensers being made of silver, often many thousands of ounces in a single unit (30). In a few years, to accomplish this separation at one plant, probably the largest mass of copper ever to be used in one area was in huge distillation units of massive construction to withstand the inevitable corrosion. Numerous other plants in this and other countries use this process for the separation of water from the spent acetic acid from synthesis processing where acetic may be not only an acid in the reaction but also a solvent for reactants or products.

Now again a large grass roots project, larger even than the first, is under way using the same entrainer - butyl acetate - for the water.

Extractive Distillation

As just mentioned, a selected third liquid is added to a mixture of two other liquids for an azeotropic distillation to increase the vapor pressure - or effectively to lower the boiling point of the one of these two in which it is largely insoluble. Contrariwise, for an extractive distillation a selected third liquid, of a higher boiling point, is added to the top of a column in which is being separated a mixture of two other liquids. With one of these, the added liquid has large mutual solubility, i.e., for this one it acts as a solvent. The high boiling solvent carries the liquid down and out the bottom, while the other liquid goes, almost, pure in the vapors off the top.

Over 50 years ago this method was used in several installations in central Europe, two in the United States, to separate that difficult pair of liquids - acetic acid and water (27), It has also been used more recently in separating hydrocarbons by class, i.e., aromatics from paraffins, rather than as in ordinary distillation by volatilities. Also, liquid mixtures, as acetone and methanol, which cannot otherwise be separated by distillation may be separated by extractive distillation, as may some of the impurities obtained in fermentation of carbohydrates to alcohol.

Another variety of extractive distillation involves the use of a salt in solution as the solvent. This effectively reduces the vapor pressure of one component of a mixture - such as water or a hydroxy relative, i.g., a lower alcohol. Professor W. F. Furter has been a protagonist for such separations both as to the determination of the underlying physical data with his co-workers, and then their use in designing separation processes (31, 32, 33).

DISTILLATION DEVELOPMENT

Other Distillations

The separation of oxygen from air is the oldest, largest, and most widely practiced industrial cryogenic process. More recently the greatest tonnage product may be liquified Natural Gas. Both depend on carefully designed and operated distillation towers which are so specialized that this mechanical design is considered almost as integral with the process used. These designs represent very important advances in distillation practice. Such towers, improved continually during the last 75 years along with their equally fine-tuned heat exchangers, supply a vast amount of oxygen for many and diverse industries. In a single synfuels plant now on the drawing boards, oxygen for partial oxidation of coal to produce synthetic gas will be rectified away from air in high purity in several towers at the rate of 25 to 50 tons per minute (34).

At the far other end of the temperature scale are the quite different stills which vaporize zinc and other metals - sometimes their salts - from ores or mixtures with other metals or salts at temperatures up to 1500-2000°C. In these examples of extractive metallurgy - some of which go back to the ancients - frequently there is involved a chemical reaction, e.g., the reduction of an oxide of the metal with carbon, with another metal, or with chlorine. The metal so released or a salt, e.g., a chloride (35) is distilled off, sometimes with rectification.

Numerous other chemical reactions have been practiced for many years in more ordinary distillation columns (27). One common one in industry is the esterification in a continuous column of an organic acid fed in as one stream with an alcohol entering in another stream (36,37,38). The column may have a greater than usual amount of liquid holdup on each plate to allow increased residence and reaction time. The ester formed, or some other added liquid, may act as an entrainer to withdraw the water at a lower temperature at the top of the column. The simultaneous combination of the volatilization and of the chemical reaction functions have been analyzed. Two or three columns may be arranged to insure a complete reaction and a separation of the pure product.

Other chemical reactions, wherein products may be simultaneously separated in a distillation tower are the nitration of benzene (39) and of toluene (40) without the use of sulfuric acid, and the formation of acetamide from acetic acid and ammonia. The water of reaction is removed overhead in a constant boiling mixture and the higher boiling product is separated in a distilling column from both the original reactants and other products of the reactions (41). One physical separation, demulsification of various crude oils - especially those from oil sands, may use a form of azeotropic distillation as a means of separating the oil from the water in the emulsion also with the almost microscopic particles of silica. This uses a form of azeotropic distillation (42) and saves the wear and tear on centrifuges caused by the fine silica particles.

CONCLUSION

In the mid-20's the late Professor G. G. Brown taught this writer at the University of Michigan somewhat of the thermodynamics of separations by distillation. In the mid-'40s, noting an improvement in an alcohol distilling process (13) he said, "I can remember that 25 years ago - before columns were introduced into the petroleum industry, it was the alcohol industry that was teaching the petroleum industry how to use columns. I think the last years show that the petroleum industry is perhaps in a position to teach the alcohol industry something!" (43). That reciprocal help between these two industries and, more recently, from a third one, the chemical industry, in the use of distillation as a separating process is well shown in the current SYNFUELS program of the government and may well improve greatly this separating process when used alone or when used in combination with numerous other separations, often chemical in nature, as well as for an adjunct to many as yet unused chemical syntheses.

REFERENCES

1. A. Cooper, "The Complete Distiller," London. 1760.
2. Paracelsus, "The Writings of Paracelsus," Ed. A.E. Waite, London, 1894.
3. George Agricola (Bauer), "De Re Metallica," Basel, 1556. Hoover translation, London 1912.
4. V. Biringguccio, "De La Pyrotechnia," Napoli, 1550.
5. Hier Brunschuygk, "Liber de Arte Distillandi de Compositis," Strassburg, 1512.
6. John Reed, Alchemy under James IV of Scotland, Ambii Vol, II, 1938, pp. 60-67
7. G.B. Della Porta, "Magia Naturalis," Napoli, 1569.
8. Aeneas Coffey, British Patent No. 5974, August 5, 1830.
9. D. F. Othmer, "Vapor Re-Use Process," Ind. Eng. Chem. 28, 1435, 1936.
10. T. O. Wentworth, "Distillation Process," U.S. Pat. 2,152,164, March 28, 1939.
11. T. O. Wentworth and D.F. Othmer, "Absolute Alcohol, an Economical Method for its Manufacture," Trans. A.I.Ch.E. 36 785, 1940.
12. T. O. Wentworth and D.F. Othmer, "Absolute Alcohol," Ind. Eng. Chem., 32 1588, 1940.
13. T. O. Wentworth, D. F. Othmer and G. M. Pohler, "Absolute Alcohol II Plant Data," Trans. A.I.Ch.E., 39,565, 1943.
14. T.O. Wentworth, US and Foreign Patent Applications, to be issued.

15. D. F. Othmer, "Composition of Vapors from Boiling Binary Solutions," Ind. Eng. Chem., 20, 743, 1928 and Anal. Chem. 20, 763, 1948.
16. D. F. Othmer, "Condensation Coefficient of Heat Transfer," Chem. & Process Eng. 49, 6, 109 (1968).
17. W. L. McCabe and E. W. Thiele, "Graphical Design of Fractionating Columns," Ind. Eng. Chem. 17 605 (1925)
18. D. F. Othmer, "Large Capacity Laboratory Condensers," I.&E.C. (Anal. Ed.) 1, 153 (1929).
19. Ibid. "Large Glass Distillation Equipment," I.& E.C. 22, 322 (1930).
20. Ibid. "Glass Temperature and Float Regulators," I.&.E.C. (Anal. Ed.) 3, 139 (1931).
21. Ibid. "Glass for Pilot Plant Construction," Chem. Met. Eng. 41, 547 (1934)
22. D. F. Othmer, "Distillation Practices and Methods of Calculation," Chem. Met. Eng. 49. 84, (1942).
23. Kirk-Othmer, Encyl. for Chem. Tech. "Industrial Alcohol - Distillation," 1st Ed. Vol. 1, p. 264, (1947).
24. H. C. Urey, Private Communication to D. F. Othmer, December, (1935).
25. Kirk-Othmer, Encycl. of Chem. Tech. "Deuterium," 3rd Ed. Vol. 7 p. 539 (1979).
26. D. F. Othmer, "Azeotropic Separation" Chem. Eng. Prog. 59, 67, (1963)
27. Kirk-Othmer, Encycl. of Chem. Tech. "Azeotropic & Extractive Distillation" 3rd Ed. Vol. 3 p. 352 (1978)
28. S. Young, "Absolute Alcohol" Trans. Chem. Soc. 81, 707 (1902)
29. D. F. Othmer, "Azeotropic Distillation for Dehydrating Acetic Acid" Chem. Met. Eng. 48, No. 6, 91 (1941)
30. D. F. Othmer, "Corrosion of Acetic Acid Equipment" Chem. Met. Eng. 48, No. 6, 91 (1941)
31. W. F. Furter and R.A. Cook, "Salt Effect in Distillation: A Literature Review" Int. J. Heat &Mass Transfer 10 (1), 23 (1967)
32. W. F. Furter, "Salt Effect in Distillation: a Literature Review II" Can. Jour. of Chem. Eng. 55, (3) 229 (1977)
33. W. F. Furter, "Extractive Distillation by Salt Effect" Advances in Chemistry Series, Ser. 115, Am. Chem. Soc., Washington, DC (1972)
34. D. F. Othmer, "Methanol - A versatile Fuel and Chemical Feed Stock." Paper at Am. Soc. Mech. Eng., Energy Conference, Houston, Texas, Jan. 22, 1981
35. D. F. Othmer and Rudolf Nowak, "Halogen Affinities - A New Ordering of Metals," A.I.Ch.E., Jour. 18, 1, 219, 1972
36. C. E. Leyes and D. F. Othmer, "Continuous Esterification of Butanol and Acetic Acid," Trans. A.I.Ch.E., 41, 157 (1945)
37. C. E. Leyes and D. F. Othmer, "Esterification of Butanol & Acetic Acid," Ind. Eng. Chem. 37, 968, (1945)

38. Saul Berman, H. Isbenjian, A. Sedoff, and D. F. Othmer, "Esterification Continuously of Dibutyl Phthalate," Ind. Eng. Chem. 40, 2139 (1948)
39. D. F. Othmer, J. J. Jacobs, J. F. Levy, "Nitration of Benzene," Ind. Eng. Chem. 34, 286, 42
40. D. F. Othmer and H. L. Kleinhaus, "Nitration of Toluene," Ind. Eng. Chem. 36, 447, (1944)
41. D. F. Othmer, "Partial Pressure Processes," Ind. Eng. Chem. 33, 1106, (1941)
42. L. L. Balassa and D. F. Othmer in "Extractive and Azeotropic Distillation," Advances in Chemistry Series, No. 115, pp. 108-121, American Chemical Society, Washington, DC, 1972
43. Brown, G. G., Comments on Paper on Absolute Alcohol, Trans. Am. Inst. Ch. Eng., 39, 577 (1943)

CHEMICAL ENGINEERING AND THE CHEMICAL INDUSTRY IN SOUTH AFRICA

O.B. Volckman*, and F. Hawke**

*Department of Chemical Engineering, University of the
 Witwatersrand, Johannesburg, South Africa, 2001
**Formerly Research Manager, AECI Ltd. Present address
 P.O. Box 782324, Sandton, South Africa, 2146

This paper traces the development of the chemical industry and the concurrent development of South African chemical engineering from the end of the nineteenth century to the present time.

The chemical industry in South Africa may be said to have its roots in the period 1866 - 1886. It was in 1866 that the first diamond in South Africa was found near Hope Town in the Cape Province, not far from the present city of Kimberley, (see map at end of text), and it was in 1886 that gold was discovered in the Transvaal and that Harrison and Walker found the so-called Main Reef on the Witwatersrand and that Johannesburg, now the major commercial centre in South Africa, was founded.

Diamond mining, and gold mining in particular, needed blasting explosives and it was this need that led to the setting up of a flourishing blasting explosives industry which became the nucleus of the chemical industry in South Africa. Mining also needed engineers. Provisions were made for training mining engineers at Kimberley which led to developments in engineering education at Cape Town and Johannesburg which were later extended to the field of chemical engineering. Indeed, the South African chemical industry and chemical engineering in South Africa have always had strong connections with the mining industry, and more lately in the areas of ore benefication and extraction metallurgy.

It is unlikely that South Africa would be in its present state of industrial advancement had it not been for the discovery of diamonds and gold, and for its huge mineral wealth which includes such metals as platinum, uranium, chromium, nickel, copper, manganese and iron, phosphates and vast deposits of readily mineable coal, a source not only of energy but of a wide range of chemicals.

South African industry, and not least the chemical industry, has had to face many constraints which are often due to geographic or climatic origin and these constraints have had their effect on determining the nature of the education and training of chemical engineers, and of chemical engineering practice.

The siting of Johannesburg, and the development of the industrial complex of the Witwatersrand and of the adjacent area is a case in point. The discovery of gold on the Witwatersrand is undoubtedly the only reason why the Witwatersrand and adjacent areas became the centre of commerce and industry, and Johannesburg the major South African city. Geographically and topographically the area is indeed unpromising for such development. The nearest South African port is Durban, some six hundred kilometres away, to which it was linked in the early days by a rather rudimentary railroad system. This was not a very satisfactory feature for gold mining which was a very import-dependent industry at the time. Moreover, Johannesburg is approximately 1750 metres above sea level - about the same altitude as Denver, Colorado - and much of the terrain between it and the coast is difficult for railroad working. The only major reliable source of water was the Vaal River, a small river by most standards. Even though water supplies were adequate for those early days they were later to form an obstacle to industrial development, and in fact the water supplies have to be augmented now by pumping water from the Tugela river in Natal over the Drakensberg mountain range.

Much of the country, the so-called "highveld", is at an average altitude of about fourteen hundred metres, and over much of this area very low atmospheric humidities, considerable diurnal temperature changes and a high incidence of severe electrical storms are representative of the physical conditions for which allowance must be made. By contrast, some of the coastal areas have a semi-tropical climate and suffer high humidities. There are no South African rivers that are navigable in the commercial sense.

Lack of adequate water supplies has led industrialists to adopt many novel procedures and process modifications to reduce dependence upon water. South Africa to-day has probably one of the most stringent sets of laws controlling effluent disposal and the use and re-use of water. This has naturally had its effect on chemical engineering practice, many chemical processes being prodigious users of water as well as producers of water-borne, aerial and solid wastes.

In the early days there was an almost complete dependence on imports, both of materials and of technical skills. The main

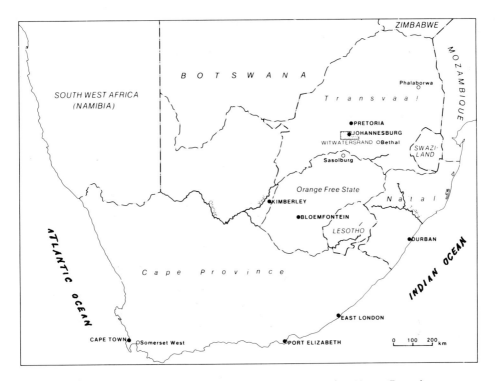

Figure 1. Map of South Africa Showing the Four Provinces Before the Formation of the Independent State of Transkei (Between Natal and East London) and Features Referred To in the Text.

source was Britain. Distances from supplies were great. Ocean voyages of eleven thousand kilometres or more were common, often taking several weeks. This led to problems of maintaining adequate stocks and to delays in communications.

These and other factors such as the long distances between major towns and cities, together with the limitations imposed by a 1,066 metre (3'6") gauge railway system, often over difficult terrain, and the general scarcity of adequate and reliable water supplies were serious obstacles to the economic viability of a number of enterprises.

Prior to the Second World War much of the country's industrial activity was aimed at meeting the needs of the gold mining industry. This was largely on a 'fill-in' basis, as the industry relied on imported materials and technical skills for many of its requirements. Since the early nineteen sixties, however, very considerable economic expansion has taken place, arising partly from the desire to rely less on imported goods and raw materials, and to become more self-sufficient in these and in technical skills.

These factors, and many others outside the scope of this paper, have bred a tradition of independence and ingenuity amongst those connected with the chemical industry, and not least amongst chemical engineers themselves.

Let us then go back to the early years on the Witwatersrand. In 1980 the gold mines struck pyritic ore from which it was not possible to recover a payable percentage of gold using the mercury amalgamation process, which had proved a satisfactory method of extracting the gold from the ores mined so far. For a short period the future of the whole region was in doubt as a more efficient means of extraction and recovery was needed, but fortunately the problem was solved by the application of the McArthur-Forrest cyanidation process which had been developed in Scotland by a Scotsman and two Canadians and patented only three years previously. This technical breakthrough put both the gold mining industry, and the Witwatersrand, on their feet, and established a base for the sustained economic and industrial growth that has continued to this day. Few people in those days would have guessed that this would have been the case, and that the gold mining industry would still be flourishing ninety years later. Early estimates for the life of the gold mines were about ten years.

Mining requires explosives, so what more natural than that the first major chemical industry should be one concerned with the production of blasting explosives. The first dynamite factory began production in 1896. This factory was operated by De Zuid Afrikaansche Fabrieken voor Ontplofbare Stoffen Beperkt, the Government of the South African Republic in the Transvaal having commissioned the Nobel-Dynamit Trust of Hamburg, Germany to establish it at Modderfontein some twenty kilometres north-east of Johannesburg.

Relations between the Kruger Government of the South African Republic in the Transvaal and the diamond industry, controlled by the British company of De Beers, were far from cordial. De Beers therefore decided to build their own explosives factory outside the Transvaal and so established a plant in Somerset West, in the Cape Province, in 1901. The project was carried out under the direction of W.R. Quinan of the California Powder Works, who became the first General Manager when production started two years later.

A third explosives factory began production in 1909. This was a factory established by the British firm of Kynoch. It was situated at Umbogintwini on the coast about twenty kilometres south west of Durban. This factory, together with those of Modderfontein and Somerset West, constituted the heavy chemicals industry of the time. All three factories manufactured their own nitrating acids, but the raw materials, nitrates and sulphur, were imported.

Explosives manufacture for the mining industry was the start of the chemical industry in South Africa, and the major growth of the industry can be traced, until comparatively recent times, through the expansion of the commercial explosives industry to include the manufacture of heavy inorganic chemicals, fertilizers and related products, and more recently, organic chemicals, and a range of chemicals derived from coal, many of them through the acetylene chemistry route.

Chemical engineering has been intimately linked with these developments. With the increasing importance of extraction metallurgy, especially hydrometallurgy, this discipline has again become an important attribute to the mining and metallurgical, as well as the chemical industries.

For the first quarter of the twentieth century development in the chemical industry was slow and was essentially along the lines resulting in the production of consumer goods such as sugar, soaps, fats and paints. Important developments in the field of agricultural products took place during this period. These were the local manufacture of superphosphate, crop protection products and cattle and sheep dips.

1930 was a memorable year for the chemical industry in South Africa, as it was the year when one of the first decisions to make the country less reliant on imported materials was taken, culminating in the building of a synthetic ammonia plant at Modderfontein, together with ammonia oxidation plants both there and in the Cape.

Also about this time a Tentelew contact sulphuric acid plant was built (also at Modderfontein), using Herreschoff furnaces to burn pyrites from the gold mines of the Witwatersrand.

In the early days of the history of South Africa's mining and industrial development much reliance was placed on technical and scientific manpower from overseas. However, the need for locally trained engineers for the diamond industry led to the establishment in 1896 of the South African School of Mines. This school was centered at Kimberley. Students had to complete a two year theoretical course at the South African College at Cape Town, (established 1829) followed by two years of practical training at Kimberley. Cape Town was at that time the major educational centre, although by the turn of the century Johannesburg had become a contender, at least as far as technical education was concerned.

By 1903, at the end of the Boer War, it was clear that with the decline in the diamond mining industry and the growth of gold mining on the Witwatersrand the focal point of the mining industry was moving to the Transvaal. The School of Mines in Kimberley was closed down and in August 1903 the Transvaal Technical Institute came into being in Johannesburg. This Institute did not have university status, but it did have a number of highly qualified staff who provided professional training in engineering (civil engineering, mechanical and electrical engineering, mining engineering, hydraulics and sanitary engineering, metallurgy and chemical engineering, and architecture). It is interesting to note the term 'chemical engineering' appearing at this stage, and to note the close association of this discipline with metallurgy - a result of the predominant influence of the mining industry. In 1906 the Institute was upgraded in response to the demand for higher education and was renamed The Transvaal University College.

It was not until the Union of South Africa was established in 1910 that any further significant advances took place in higher technical education. In that year the Transvaal University College was split into two centres, one being at Pretoria, the administrative capital, and the other remaining at Johannesburg, the centre of commerce and industry. The Pretoria College catered for non-engineering subjects and that at Johannesburg for technical and professional subjects. The Johannesburg centre once again had its name changed, this time to The South African School of Mines and Technology, while that at Pretoria retained the title of The Transvaal University College.

In the same year the South African School of Mines and Technology at Johannesburg first offered courses leading to its Diploma in Chemical Engineering. Seven years later, in 1917, the Diploma was dropped in anticipation of authority being obtained for the courses to be recognised for the Bachelor of Science in

Engineering degree of the University of South Africa, a federal degree-awarding institution established by the Government in 1918 and on whose Council the constituent colleges were represented.

The first degrees were awarded in 1918 to 108 candidates to whom retrospective recognition of the Diploma was given. No specialisation in the branches of engineering was mentioned.

In 1920 the South African School of Mines and Technology was renamed The University College, Johannesburg, in keeping with its status as a constituent college of the University of South Africa, and on March 1st, 1922 it became the University of the Witwatersrand, Johannesburg, awarding its own degrees. One of these was the degree of Bachelor of Science (Chemical Technology). This was a four-year degree, the first courses for which were offered in 1920. Eight years later the degree was changed to that of Bachelor of Science in Engineering (Chemical). The course remained much the same as before except for normal updating. It did, however, now offer two options. One option was known as the "Industrial Option", the final year of which comprised essentially industrial chemistry, chemical engineering and a course in physical chemistry. In the "Metallurgical Option" emphasis in the final year was on metallurgy and metallurgically oriented subjects. The first three years of both options included courses in chemistry, physics, mathematics, and a selection of mechanical engineering courses, electrical engineering, fuels and elementary metallurgy.

At Cape Town developments had been somewhat similar to those at Johannesburg. In April 1918 the establishment of a course in Chemical Engineering was proposed. The curriculum was published in the Engineering Prospectus of the University in 1920 and the first degree of Bachelor of Science (Chemical Engineering) awarded on the 19th December, 1922. The technical work of the course was given by staff of the engineering and chemistry departments, and for the first time there is mention in the prospectus of what is general practice in the engineering curricula at South African Universities, namely, the requirement that students spend prescribed periods during their summer vacations at approved works to gain practical engineering experience. Much of the chemical engineering syllabus at that time seems to have been descriptive, with detailed descriptions being given of plants used in the manufacture of acids, explosives, fertilizers, cement, soaps, candles, greases, sugar and so on. This is vastly different from today's highly sophisticated chemical engineering syllabuses, but reflects the state of the chemical industry in South Africa at that time. Amongst the reference books listed are "Chemical Engineering" by Allen; "Chemical Works" by Dyson and Clarkson; and the Proceedings of the Institution of Chemical Engineers.

Due to a clash of personalities at the University of Cape

Town the degree of Bachelor of Science (Chemical Engineering) was abolished in 1924 and degrees with that title were not awarded again until 1957. In its place a new course was introduced by the chemistry department leading to the degree of Bachelor of Science in Applied and Industrial Chemistry. In spite of the change in title the curriculum was not greatly different from what had gone before and was fairly similar to the curriculum at Johannesburg.

Engineering education had developed more slowly in Natal. It was not until 1927 that any mention was made of chemical engineering or the like. In that year the college calendar mentions a course in "Metallurgy and Chemical Technology". Apart from this reference to Chemical Technology there appear to have been no degree courses in either Applied Chemistry, Chemical Technology or Chemical Engineering until well after the end of World War II.

In the mid-nineteen-thirties the chemical industry in South Africa was thriving although the organic chemicals sector was probably the weakest. There was a fermentation industry which, in addition to the manufacture of potable liquors and wines, produced industrial ethanol, acetone/butanol and acetic acid. Diethyl ether, in both industrial and anaesthetic grades was also made. Initially the primary raw material for industrial fermentations was maize (corn), but for economic reasons this was replaced by molasses, available from the extensive South African cane sugar industry in Natal.

A chlor-alkali industry was established in association with paper pulp manufacture, itself based on straw and local varieties of wood as sources of cellulose. Calcium carbide was also manufactured and in turn used for the production of acetylene.

As early as 1935 processes were being sought for the manufacture of liquid fuels from indigenous carbonaceous materials. The first of these to be exploited commercially was Torbanite shale, the distillate from which was refined and blended with ethanol, and benzol (obtained as a by-product from the coking ovens at the major steel works). The resulting automotive fuel was known as SATMAR. A few years later the Anglo Transvaal Consolidated Investment Company sought to license the Fischer-Tropsch process from Ruhrchemie in Germany for the manufacture of oil from coal, but this venture fell through with the outbreak of war.

Even before the Second World War attention had been paid to the greater use of indigenous raw materials and this became of increasing importance under war time conditions. Prior to the war South Africa had been essentially an agriculture and mining based economy. The chemical industry provided fertilizers and certain crop protection and veterinary products as well as explosives, accessories and certain heavy chemicals for the mining industry.

With the imposition of war time conditions there was a need for greater self sufficiency.

Although this period was one of "keeping going" with whatever human and other resources were available several important developments were initiated. An organic chloro-chemicals plant was established near Johannesburg in 1941. Subsequently this plant formed the basis for a more sophisticated insecticide industry which first produced DDT and later BHC, both of which were used extensively on the Southern sub-continent, particularly for combating malaria.

The Cyanamid Company of America established an electrothermal plant in Witbank on a site adjacent to a large calcium carbide factory. The major product of Cyanamid's factory was calcium cyanide used in the extraction of gold. Some years later a cyanide plant was erected adjacent to a large sewage treatment works. Here calcium cyanide solution was manufactured and was transported to customers in bulk road tankers.

A third important development during this period was that of a process for the extraction of uranium from uraniferous gold bearing ores. The research and development work was carried out by teams of American, British, Canadian and South African scientists and engineers working both in South Africa and abroad.

Each of these three developments illustrates a trend that has become a permanent feature of South African industry. First and foremost there has been the concentration on the exploitation of local raw materials even if these are not "fashionable" at the time. Of these the most important is the use of coal as the source of carbon. Secondly, and largely as a result of the success of uranium extraction processes, there has been considerable development in the metallurgical industry (and in extraction metallurgy in particular), and with it the application of chemical engineering techniques and the use of sophisticated hydrometallurgical processes. This has brought the chemical engineer into very close association with the powerful mining industry as a whole.

Immediately after the war the Government reviewed the needs and capabilities of the country with respect to scientists and engineers. Immigration and contract labour from overseas had up to this time provided most of the additional skilled manpower needed in the chemical industry. In particular, ways of improving the local supply of scientists and engineers were considered. The need was also stressed for the formulation of programmes to accelerate and extend the exploitation of local resources.

From these deliberations came the founding of the South African Council for Scientific and Industrial Research (CSIR).

This semi-government organisation was established in the latter part of 1945. The CSIR was centered in Pretoria where the first laboratories were set up in 1946.

The CSIR was made responsible for the encouragement and support of research in universities and similar institutions; the organisation of industrial research associations; and to advise on future developments where needed and to set up National Laboratories. The first five National Laboratories to be established were those of Physical, Chemical, and Building Research, Personnel Research, and Telecommunications Research. Industrial Research Associations or Institutes included those for Leather, Fishing, Paint, Sugar Milling and Wool. Later a much wider range of activities was incorporated including the co-ordination of Medical Research and the responsibility for the allocation of Government funds for university research, and the setting up of a Water Treatment Research Institute, later to become the National Institute for Water Research.

The only area of importance outside the ambit of the CSIR was the exploitation of South African minerals. This was already being handled by the Government Metallurgical Laboratory, later to become the National Institute for Metallurgy, (NIM).

As far as chemical engineering was concerned the immediate post war period was characterised by the realisation that, in developing the country's natural resources, the chemical engineer was a most important link in the whole techno-economic chain needed if the requisite degree of industrial development and self sufficiency was to be attained. This was recognised at the CSIR by the establishment of a Process Development Division which was wholly chemical engineering based. In the universities, which were almost entirely financed by the Government, changes in chemical engineering education had to wait until the late fifties when pressure from certain sectors of the chemical industry and from the profession itself led to the establishment of autonomous chairs of chemical engineering at Cape Town, Pretoria, the Witwatersrand and Natal.

With the war over the country began to put into effect some of the earlier plans aimed at achieving greater self sufficiency. In the chemical field the CSIR concentrated its efforts mainly on investigations into the active principles of natural products, in wool fibre technology, pelagic fishing and agricultural by-products, in the use of indigenous timbers, in water purification and re-use, and in the study of local clays and other silicate minerals. These were but a few of the fields in which the chemical engineer found himself actively engaged.

A factory was established on the Natal coast to manufacture rayon grade cellulose pulp from local hardwood. Problems had to be overcome associated with an unusual raw material (the wood of blue gum or 'eucalyptus saligna') and with effluent disposal. This venture is typical of the successful adaption of overseas processes to suit local raw material and conditions, often, as in this case, resulting in an improved product or process.

Intensive research by Government laboratories and a major fertilizer manufacturer enabled the vast phosphate rock deposits at Phalaborwa in the North Eastern Transvaal to be exploited commercially. This venture has eliminated, almost completely, dependence upon imported phosphate rock for fertilizer manufacture, and has also led to the profitable sale of by-products from the beneficiation of the ore.

In the period 1950 to 1958 intensive research and development was undertaken by the CSIR, supported by a group of South African gold mining companies, and in close association with the US Office of Saline Water, the TNO in the Netherlands, and various commercial interests in the United Kingdom, into the desalination of water by electrodialysis on a multi-million litre per annum scale. The source of the saline water was from the deep level gold mines in the Orange Free State. A secondary objective of the programme was the treatment of other naturally occurring brack waters to render them suitable for animal and human consumption. Although technically successful the results of this work did not provide an economically viable solution to the problem as a whole.

South Africa possesses great mineral wealth but one of the few minerals it lacks is mineral oil. With the exception of the small Torbanite shale deposits, liquid hydrocarbons are absent from the country's mineral wealth. Even before the war consideration had been given to the establishment of an oil-from-coal industry. Under the conditions then obtaining such an industry would have proved uneconomic. However, the Anglo Transvaal Consolidated Investment Company still retained its licence from Ruhrchemie to erect a Fischer-Tropsch plant. In 1950 these rights were transferred to the South African Government and SASOL (derived from the Afrikaans name of the South African Coal, Oil and Gas Corporation - now a public Company) was conceived. The major plant items, including a Kellogg fluidised bed converter, were commissioned five years later and SASOL came into being, sited in the new town of Sasolburg in the Orange Free State, a few miles south of the Vaal River, with the Sigma colliery as its coal source on an adjacent site. By 1955 South Africa was producing its own range of hydrocarbon products from coal, and the foundations were laid for a large petrochemicals complex at Sasolburg.

Even more recently South Africa's two largest chemical companies (AECI and Sentrachem) collaborated in building a complex for the manufacture of coal-based chemicals. This has become known as 'Coalplex' and comprises four major sections:

The acetylene section which includes calcium carbide manufacture and its conversion to acetylene; a chlor-alkali plant using diaphragm cells to minimise pollution problems; a plant for the production of vinyl chloride monomer (VCM) via the acetylene route; the PVC plant in which VCM is polymerised and PVC compounds manufactured. Output of PVC is at the rate of 100 000 tonnes per year.

In the twenty-five years that SASOL has now been operating, a body of knowledge and technology has been built up such that to-day South Africa leads the world in oil-from-coal manufacturing technology. SASOL II has reached the commissioning stage and SASOL III is under construction: both are situated on a coal field near Bethal in the Eastern Transvaal at a site named Secunda. Yet again coal is South Africa's source of carbon.

The original synthetic ammonia plant built at Modderfontein almost fifty years ago was coal based: so is the 1000 tonne per day anhydrous ammonia plant recently commissioned there. The original plant used semi-water gas generators, the new plant, the largest coal based ammonia plant in the world, employs Koppers-Totzek gasifiers.

Today the chemical industry accounts for more than ten percent of the country's gross domestic product and in 1978 (the latest year for which official figures are available at the time of writing), gross sales were of a value of more than two-and-a-half billion rands. Statistics based on the value of production can be misleading, especially in the light of present day inflation and possible changes in the basis of collating data. Prior to 1922 the chemical industry was so small that no separate statistics for it are available. Even the value of manufacturing as a whole was a small contribution to the gross domestic product, as can be seen from Table 1. From 1922 better data are available and the following growth of the chemical industry can be seen from Table 2. In 1972 the manufacture of chemicals contributed 12,1 percent of the gross value of sales in the manufacturing area. Food (15,5%) was the largest and metal products (8,2%) the third largest.

Chemical engineering and the employment of chemical engineers, with whom are included the applied chemists and chemical technologists of earlier days, has followed a similar growth trend to that of the chemical industry.

In the early mining days and in the period when the heavy inorganic chemicals industry was being established the demand for

Table 1. Growth of the Manufacturing Industry 1911 - 1928

Year	Gross Domestic Product (At Current Prices) (Rm)	Gross Output for Manufacturing Sector (At Current Prices) (Rm)	As a % of GDP
1911	300	12	4,0
1914	303	16	5,3
1920	558	41	7,4
1922	439	35	8,0
1930	551	52	9,4
1938	884	99	11,2

Source : Department of Statistics R.S.A.

Table 2. Growth of the Chemical Industry 1922 - 1978

	Value (At Current Prices) (Rm)	As a % of the Manufacturing Sector	As a % of the Gross Domestic Product
1922	4,0	11,4	0,9
1930	6,1	11,7	1,1
1940	18,2	14,8	1,8
1950	40,2	8,6	1,6
1960	306,2	11,2	6,2
1970	791,1	10,7	6,8
1972	1089,6	12,1	7,6
1976*	2659,7	13,9	10,3
1977*	3864,2	17,9	12,3
1978*	4461,3	18,2	- (+)

* Provisional, (+) Figures for GDP for 1978 not available
Source : Department of Statistics - R.S.A.

"chemical engineers" was met predominantly by recruitment from overseas as mentioned earlier, supplemented by the small number of diplomates, and, later, graduates from Cape Town and the Witwatersrand. This situation continued into the fourth decade of the present century although there was a proportionately greater contribution from South African graduates.

The immense development of chemical engineering in the United States, and to a lesser extent in Britain, during and immediately after the Second World War, together with progress in the South African chemical industry brought about by wartime needs, did result in a marked upgrading of the functions of chemical engineers in South Africa - and a need for improved educational facilities, but there were problems here. In the highly developed countries overseas specialisation is normal, but in South Africa, with its relatively small number of specialist engineers and scientists and its wide range of problems, versatility is essential, but not at the expense of depth of education and training. This imposes severe demands on the local chemical engineer, not only during his education, but also during his earlier professional life, as responsibilities come very early in his career. Much of the record of post-war chemical engineering education in South Africa is therefore connected with the way in which these demands can best be met.

Chemical engineering education at university level received its greatest impetus in the late nineteen-fifties and early nineteen sixties, when, as has been mentioned earlier, autonomous departments of chemical engineering were set up at some of the major universities. It is from that time that the present modern courses, facilities and active research in the field of chemical engineering began. Courses became more mathematically and theoretically based and far less descriptive. Since then there has been constant evolution to keep up with trends in chemical engineering world wide and at the same time to meet the specific needs of South Africa.

At the University of the Witwatersrand the "Industrial" and "Metallurgical" Options had been dropped in 1939 when a degree in Metallurgy was introduced. In 1976, however, two options were again introduced in view of the great and growing importance of the minerals industry. These options were a straight Chemical Engineering Option and a Minerals Process Engineering Option. The latter option is the joint responsibility of the Department of Chemical Engineering and the Department of Metallurgy.

At the University of Natal a four year course leading to the degree of Bachelor of Science in Engineering (Chemical) was introduced in 1961. Previous to this the Department of Chemistry had offered a post-graduate course open to science students who had obtained a three year Bachelor of Science degree and who wished to make a career in industry. This course entitled successful

CHEMICAL ENGINEERING IN SOUTH AFRICA

candidates to be awarded the degree of Bachelor of Science in Chemical Technology. A Master of Science course in Chemical Technology was also available.

At the University of Pretoria a four year degree course for the degree of Bachelor of Science in Chemical Engineering has been available since 1959.

The University of Stellenbosch formed a Department of Chemical Engineering in 1970 and in 1981 added to its curriculum an option in extraction metallurgy. The chemical engineering degree course is structured on similar lines to those at the other Universities.

Developments at the University of Cape Town have already been mentioned.

Of the five universities mentioned, Pretoria and Stellenbosch teach in the medium of Afrikaans, one of the official languages of South Africa.

The number of graduate chemical engineers, or the equivalent, produced annually in South Africa has always been small by the standards of large industralised countries overseas. Until 1937 less than ten graduates were produced per annum. During the war years and immediately after, numbers were naturally variable but the average was about twelve annually. With the return of ex-servicemen and the provision of special educational facilities the numbers for the years 1949 through 1952 rose to 121 or an average of just over thirty each year. Graduation lists of those years contain the names of many men who made a very significant contribution to the development of chemical engineering and the chemical industry in South Africa in the following decades. Numbers stabilized in the nineteen fifties to about twenty five graduates a year and then rose significantly for the next fifteen years, as the chemical industry developed, to about fifty annually. The world-wide swing away from science and engineering was felt in South Africa in the mid seventies. Enrollments dropped severely and annual graduation numbers fell to around twenty-five. There is now a very marked upward swing in enrollments, but even so less than half the demand for chemical engineers is likely to be met by local graduates.

In professional matters chemical engineering in South Africa has always been closely associated with Britain, particularly as in the earlier part of this century most of the educators and engineers came from either England or Scotland. In 1949 a South African Committee of the Institution of Chemical Engineers, London, was set up. In 1952 a South African Branch of the same Institution was formed, the inaugural meeting being held in Johannesburg on the

24th March of that year. This was the first full overseas Branch of the Institution to have been formed. The branch was closed down in 1964 when the South African Institution of Chemical Engineers was established as a fully autonomous South African Institution and thus eligible to take full part in the representation of chemical engineering interests to the various statutory authorities, chief of which is the South African Council for Professional Engineers which legislates for and controls professional membership and activities of engineers in South Africa.

Up to the late nineteen forties the South African chemical industry in general was not really ready to employ fully qualified chemical engineering graduates in a meaningful way. A result was that many graduates complained that industry seemed unwilling or unable to employ them to their full potential - a perhaps not unfamiliar cry. Thus a rather high proportion of the better qualified graduates emigrated, though some did return later and with the benefit of much valuable experience.

However, by the early nineteen fifties some of the more forward looking employers in the chemical industry were beginning to realise that chemical engineers possessed skills that could be of advantage to them and that they should not restrict the employment of chemical engineers to jobs such as senior plant operators, industrial chemists and the like. All the same, it was almost a decade before the chemical engineer specialising in design could find satisfying employment.

About this time, too, it was realised that in developing the country's natural resources the importance of chemical engineering should not be underestimated if a reasonable degree of self sufficiency and industrial development were to be achieved. That the mineral wealth of the country is one of its greatest assets had already been shown in the case of precious metals and uranium. In these areas chemical engineers were already employed essentially with a view to improving process efficiency and product recovery: a great deal of original work was done in the fields of flotation, ion exchange and adsorption in particular. But there were many base metals including copper, manganese, nickel, chromium and beryllium requiring further exploitation. Up-grading ores in the country of origin rather than exporting unrefined ore is sound economic policy although tariff barriers and other political restrictions do in fact sometimes make this impossible. There may also be special technical reasons which makes upgrading at source impracticable. It was recognised, however, that in many instances production of ore concentrates on an appreciable scale might be possible. This would reduce dependence upon imported materials and might result in an expanding export market. This thinking was in line with the country's desire to reduce its dependency on overseas countries which had so far largely dominated the mining as well as the chemical industries.

One result of this development has been that chemical engineers are now being integrated much more closely into research, development, commissioning and operation of processes in the minerals industry. This is particularly the case in the field of extraction metallurgy and more particularly in hydrometallurgical and allied processes. There have been many improvements in processes, particularly those dealing with ores that are difficult to treat. Highly sophisticated chemical engineering techniques have been necessary to overcome difficulties arising from the nature of the local raw materials, labour, ambient conditions such as low humidity and high temperatures conbined with high altitude, or to meet stringent quality specifications, or to elminate effluent nuisances, or reduce water usage.

The majority of chemical engineering graduates are now employed by the chemical industry either directly or indirectly. Today a considerable amount of plant design and construction is carried out in South Africa. In the case of 'Coalplex' referred to earlier 77% of the total capital cost of R229m was spent in South Africa. The corresponding figure for SASOL II and SASOL III combined is more than 60% : 'the total capital cost of SASOL II alone is estimated to be R2 500m. These figures are fairly representative of the proportion of local construction and fabrication of large projects. In cases where high pressure equipment or highly sophisticated machinery is involved the "local content" is, of necessity, lower. For example, in the case of the 650 tonnes per day nitric acid plant two thirds of the capital cost was spent abroad. Nevertheless, there are a few firms who do manufacture chemical process plant equipment in sophisticated materials to the most stringent overseas specifications and whose work is accepted by leading overseas insurers.

The challenges for the chemical engineer in South Africa are considerable. Ingenuity and initiative are pre-requisites. In such an industrially young country as South Africa use is made generally of only the latest or most efficient technologies available : it is often necessary to marry several different technologies to locally devised ones to give an integrated process that will provide an acceptable solution for local production. These constraints and challenges demand particular aptitudes of the local chemical engineer. The record of the history of chemical engineering and the training of chemical engineers in South Africa is largely a record of how these challenges have been met. The policy of self sufficiency combined with the need for liquid fuels - of which methanol is currently a favoured candidate - will ensure that further challenges are in store for South African chemical engineers.

ACKNOWLEDGEMENT

The authors wish to express their thanks to all the firms, particularly AECI Limited, organisations and individuals who so kindly supplied or confirmed the information given in this paper. Extensive reference was made to the calendars and annual reports of the universities and colleges cited.

ETHYLENE AND ITS DERIVATIVES: THEIR CHEMICAL ENGINEERING GENESIS AND EVOLUTION AT UNION CARBIDE CORPORATION

Arthur E. Marcinkowsky
George E. Keller, II
Union Carbide Corporation
P. O. Box 8361
South Charleston, WV 25303

SUMMARY

This paper deals with the commercialization of ethylene and its derivatives by Union Carbide Corporation (UCC) and the people who pioneered in the development of this petrochemical industry. In 1914 the basic chemistry of ethylene was reasonably well understood, but virtually no engineering and scale-up know-how existed. A part of this presentation will therefore deal with the engineering aspects of the early production of ethylene. Prior to the commercialization of ethylene as a chemicals feedstock most of the simple chemicals were produced from acetylene which owes its origin approximately 25 years earlier to Willson's calcium carbide discovery. Ethylene prior to ca. 1920 was regarded more as a high priced chemical whose chief source was from the dehydration of ethanol, although it was known that gases from oil cracking, etc., contained significant amounts of it. The chief problem in those days was that nobody knew how to concentrate ethylene, hence one of the first major achievements of the early pioneers was to develop an effective cryogenic separation process for its recovery. A second major development was the commercial scale cracking of gas achieved at Clendenin, WV, using primarily propane and butane recovered from the abundant natural gas of West Virginia. The production of ethylene-derived chemicals, which launched the present day petrochemicals industry, was initiated by Dr. George O. Curme, Jr., and a small group of collaborators. In order to achieve our objectives of this presentation the origin of UCC will be briefly reviewed together with what UCC is today. Finally we will deal with some current programs which may be the sources of the processes of the 21st century.

INTRODUCTION - THE ALIPHATIC PETROCHEMICAL INDUSTRY

In its broadest sense a petrochemical is any chemical produced or derived from petroleum or natural gas, other than gasoline and fuel oils, and the aliphatic petrochemical industry comprises those chemicals produced primarily from straight chain hydrocarbons. These feedstocks are mostly saturated hydrocarbons, C_2 and greater chain lengths, which must first be cracked to produce the unsaturated olefins that serve as the primary building blocks of the petrochemical industry. The petrochemical industry products, those produced in large volume, are therefore derived from the olefins -- ethylene, propylene, butenes and to a considerably lesser extent the higher molecular weight olefins and acetylene. Of these various primary building block chemicals ethylene is consumed in greatest amount and so holds the title of "King of the Petrochemicals". For example, the 1980 domestic U.S. consumption of ethylene (1), is projected to be 35 billion lbs. This represents approximately 35% of world consumption.

The petrochemical industry, differentiated from the much older coal-tar chemicals industry, started in the early part of the 20th century. It began to take shape around 1914, but the pace of activity really did not hit full stride until the early 1920s. The purpose of this presentation is to outline the role Union Carbide played in the launching and development of the olefin-based petrochemical industry. Major emphases in this paper will be placed on ethylene as the principal feedstock, its derivatives, the chemical engineering which was required to produce the derivatives and the development of the necessary markets. The other olefins, though used in large volumes today, played a secondary role to ethylene and its derivatives in the early development of the petrochemical industry. As the story unfolds it will become clear that the early technology pioneers at Union Carbide helped lead the way in developing the petrochemical industry as we know it today.

As stated in the foregoing paragraph, this paper is intentionally restricted in scope. Many of the accomplishments of Union Carbide as a corporation cannot be presented here because its involvement in high technology over many years is simply too diversified. Even as early as 1933, the wide range of the Corporation's activities was made evident at the Chicago World's Fair, where Union Carbide alone was able to supply for display purposes more than half of the chemical elements known at that time. It is interesting to note that today, with all its many interests, Union Carbide works with over 90 of the 107 isolated chemical elements.

TABLE I
UNION CARBIDE CORPORATION: - VITAL STATISTICS*

	1977	1978	1979
New York Stock Exchange Symbol - UK			
Worldwide Sales	7.036×10^9	7,869.7	9.18
Net Income	385.1×10^6	394.3	556.2
No. of Shares Outstanding	6.4533×10^7	6.51	6.62
Market range/share	$40-62 3/8	33 58-43 1/4	34-44 1/2
Equity/sh	$52.79	55.92	61.09
Numbers of Employees	113,669	113,371	115,763
Product Class Sales			
a. Chemicals and Plastics	2.8×10^9	2.9	3.35
b. Gases and Related Products	1.1	1.26	1.43
Metals and Carbons	1.2	1.4	1.77
c. Batteries: Home & Auto Products &	1.2	1.35	1.60
Specialty Products	0.7	0.93	1.03
Executive Remuneration (Direct Salary)**			
Chairman and Chief Ex. Officer	$354,960	370,000	387,500
President and Chief Operating Officer	295,195	315,000	338,333
Executive V.P.	221,089	244,705	255,294
Executive V.P.	207,796	216,667	254,046

* Annual Stockholders Reports
** Notice of annual meeting of stockholders

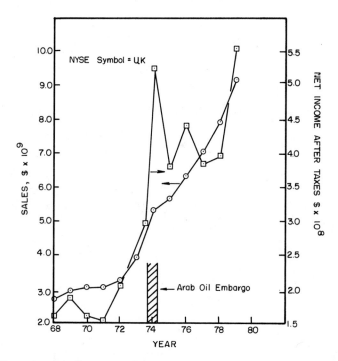

Fig. 1. Ten year sales and income summary Union Carbide Corp. and consolidated subsidaries.

UNION CARBIDE TODAY

To get a perspective on the past it is worthwhile to know where we stand today. Looking at Union Carbide as a multinational company of the 1970s will make the historical events which led to its present size and diversity understandable both in terms of a corporate evolution and the origin, growth and development of the olefin-based petrochemical industry. The fact is that Union Carbide was already a well established organization when the amazing petrochemicals industry began approximately 60 years ago.

Union Carbide Corporation, with headquarters at 270 Park Ave., New York*, is among the nation's 25 largest companies, with 1979 sales standing at over nine billion dollars. Union Carbide is engaged in three major lines of businesses: chemicals and plastics; gases, metals and carbons; and consumer and related products. Although primarily a manufacturer of basic products, Union Carbide also makes products sold directly to the consumer such as EVEREADY dry cell batteries, PRESTONE antifreeze, GLAD plastic household wrap and bags, and SIMONIZ waxes and polishes to name just a few of the more familiar items.

Today Union Carbide Corp. employs more than 100,000 people, operates about 500 plants, mines and mills in over 30 countries and has close to 115,000 stockholders (2,3). Table I presents some vital statistics of Union Carbide based on its 1977, 1978 and 1979 record as published in its annual stockholder's report. To get a somewhat more comprehensive understanding of Union Carbide, Figure 1 illustrates the 10-year sales and income performance of Union Carbide and its consolidated subsidiaries.

THE ORIGIN OF UNION CARBIDE

Union Carbide Corporation assumed its present name on May 1, 1957. Prior to this date the company was known as Union Carbide and Carbon Corp., which was formally incorporated in 1917 through the merger of four companies and their subsidiaries which had certain common interests: (1) Linde Air Products Co., incorporated in Ohio in 1907; (2) National Carbon Co., Inc., incorporated in New York in 1917 to succeed National Carbon Co., incorporated in New Jersey in 1899; (3) Prest-O-Lite Co., Inc., incorporated in New York in 1913, and (4) Union Carbide Co., incorporated in Virginia in 1898 (4). A highly simplified

---

*The headquarters of Union Carbide are scheduled to be moved in 1981 to Danbury, CT, where ground breaking for the new facility took place in 1978.

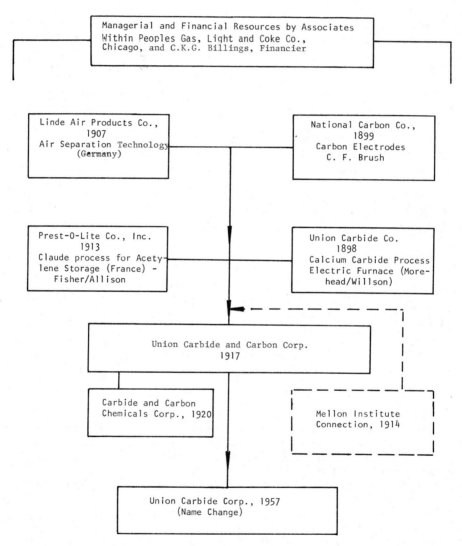

Fig. 2. Formation highlights: Union Carbide and Carbon Corp., 1917

schematic of the 1917 merger is given in Figure 2. The name, Union Carbide and Carbon Corp., and its original significance are directly traceable to two earlier events that were to have a profound effect in shaping the Corporation's course.

The first event took place in 1876, when Clevelanders crowded into the Public Square to see Charles F. Brush's brilliant carbon arc street light change night into day. This invention sparked the formation of a company in 1886 to make various carbon products - including carbons for electric arc street lights and later carbon electrodes for electric furnaces. The company, which was the forerunner of the Corporation's present Carbon Products Division, brought to the Corporation the famous EVEREADY trademark. In 1890, it produced the world's first commercial dry cell battery and, in 1894, built one of the first industrial research laboratories in America. The work of Charles Brush led eventually to the formation of the National Carbon Co. in 1899.

The second event, which was to supply the words "Union Carbide" took place in North Carolina in 1892. In the years following the Civil War, a Confederate Major, James T. Morehead, who had part of his jawbone shot away at Gettysburg (5), gained a modest fortune in various small industries in Spray, North Carolina. He became involved with an enthusiastic Canadian inventor and electrical engineer, Thomas L. Willson, in a scheme for making aluminum from its oxides in the intense heat of the newly invented electric furnaces just beginning to emerge in Europe. Their desire, no doubt, was inspired by the great discovery of Charles M. Hall's aluminum production process of 1886. They organized the Willson Aluminum Co. and built a small plant at Spray. This plant never produced any aluminum. However, in 1892 in an attempt to produce calcium metal, lime and tar were heated together and produced what we now know as calcium carbide, which eventually became the preferred commercial route to producing acetylene (5,6). Back then, however, the gray-looking material which came out of the furnace was unknown to the furnace man, who water quenched it according to standard procedure. It must have been one of the biggest surprises of his life, seeing voluminous quantities of gas evolving and perhaps even bursting into a flaming torch. What a dramatic way to make a great discovery!

Unfortunately, the Major went broke soon after this discovery, largely because his plant burned down and his creditors sold him out, so he packed up some of the manufactured calcium carbide and went north to sell the discovery. Through his persuasiveness he succeeded in interesting some New York capital in forming the Electro Gas Co. to promote acetylene for city and home lighting. He sold them everything except the chemical rights to calcium carbide for cash and stock. From here, the

Major went on to develop power dams on the James River at Holcomb Rock, VA, and on the Kanawha River in West Virginia to supply his ferrometals alloy business, which became very successful. To conclude the colorful story on Major Morehead we find him selling all the plants and patents on his electric-furnace developments in 1906 to Union Carbide (1898), the successor of Electro-Gas. He died at 67, leaving a substantial estate, including $200,000 to his widow.

Electro Gas Co. was never really successful and basically failed to achieve its intended objectives. However within Electro Gas was a group of men known as the Chicago Associates whose common ties were in the Peoples Gas, Light and Coke Co. of Chicago. These people were interested in acetylene as a possible substitute for oil in enriching their company's city gas. This idea never worked but the search was so successful in other directions that by 1898 the Peoples Gas group took over Electro Gas and formed Union Carbide Co. Union Carbide became quite successful selling acetylene lamps, and sales peaked in 1909 when there were 235 U.S. towns glowing nightly from these lamps. This business faded when the electric lighting era entered the scene. Next, Carbide developed the miner's acetylene safety lamp and took the lead in importing from France a new oxygen-acetylene technique for welding and cutting metals, which resulted in setting up the Oxyweld Co.

In making calcium carbide the furnaces could not be kept busy continuously, and because power was contracted for on a continuous basis, slack time represented a sizeable loss. Major Morehead's electroalloys business was bought and the Electro Metallurgical Co. (Electromet), a subsidiary of Union Carbide, was set up to utilize the power and turn the loss into a significant profit. With oxyacetylene welding growing in volume, Carbide needed large quantities of oxygen to complement its product and began to acquire stock in the Linde Air Products Co. which had been organized in 1907 by a group of Clevelanders for U.S. exploitation of the German Linde Process for the fractional distillation of liquid air.

Union Carbide soon discovered that its biggest customer for calcium carbide was the Prest-O-Lite Co., an Indianapolis business built up by a bicycle repair shop owner, Carl G. Fisher, and a businessman, James A. Allison. The business was fashioned around the Claude process, imported from France, for storage of acetylene gas in cylinders which contained a porous filler impregnated with acetone - then and still the safest packaging method for touchy acetylene. Through good business acumen they acquired the sole American rights to this process. Prest-O-Lite at this point started using acetylene in small cylinders for lights on their bicycles, and later in automobile lights.

Carbide bought into Prest-O-Lite to keep its finger on both streams of the oxyacetylene business. At about the same time, National Carbon Co., which had been making carbons for street lamps for years, had swung into making the big carbon electrodes in quantity for Carbide's growing furnace market.

It will be well to pause here and note the interplay of the various business ventures mentioned earlier. The oxyacetylene process was the big opening wedge for the Carbide business. The reason why Union Carbide was able to take advantage of it was because of a strategic position obtained by buying into the related companies. The big problems in oxyacetylene welding and cutting were the actions of heat on various metals and the need to develop special welding rods to effect welds on various metals. To address these problems Carbide had a ready-made research organization in the Electro Metallurgical Co.

Union Carbide and Carbon Corporation came into being in 1917 through the merger of the four companies cited above. These four companies had been interdependent for many years and a delicate balance existed between them which seemed worthwhile to solidify. National Carbon was losing electrode business and needed both financial and managerial help while Linde needed more capital. Prest-O-Lite's Carl Fisher wanted to proceed with his Miami Beach development project, so he was a willing listener. Moreover Union Carbide already owned 40% of Prest-O-Lite and 30% of Linde. The financial man behind the scene was Cornelius Kingsley Garrison Billings, a wealthy Chicago business tycoon with a reputation of sound judgement. The story goes that everybody talked and had their say, then Billings would say a few pertinent words to the point and the decision was accepted.

In summary, Union Carbide and Carbon Corporation came into being because of mutual interests among the four companies, the efforts of a powerful group of associates with common ties through the Peoples Gas Light and Coke Co., and the financial support of C. K. G. Billings. Naturally acquisitions, mergers and spin-offs, in addition to internally generated new business ventures, are an on-going activity for a large corporation. Within Union Carbide such activity was, and is, no exception. A complete listing of all the acquisitions, mergers and spin-offs are listed by such publications as the Moody Industrial Manual (4).

THE MELLON INSTITUTE CONNECTION

Research in the early 20th century was not conducted in the deliberate manner it is today. Most companies did not possess the means to set up viable research departments as they exist today, which is the reason the Mellon Institute was founded. We introduce here the special relationship which the predecessors of Union Carbide Corp. started with the Mellon Institute in 1914, and how this relationship evolved eventually to give birth to the Carbide and Carbon Corp. in 1920, and as a consequence thereof, the ethylene-based petrochemical industry.

The Mellon Institute, originally known as the Mellon Institute of Industrial Research (7), was established in 1913 by Andrew W. Mellon, and his brother, Richard B. Mellon, to implement Dr. Robert Kennedy Duncan's (8) ideas on cooperative research. The institute was affiliated with the University of Pittsburgh and was initially housed in a temporary wooden laboratory. Its first permanent building was formally opened in February, 1915. Figure 3 is a picture of a laboratory at the Mellon Institute known as "The Shack". This is where some of the original work by Dr. George Curme and his associates, to be discussed below, was carried out. Apparently, several of these shacks existed at the time. The Mellon Institute has been very successful over the years, and in 1967 the Mellon Institute and Carnegie Institute of Technology merged to form the Carnegie-Mellon University. The Institute is now known as the Carnegie-Mellon Institute of Research.

The Prest-O-Lite Company established its "Illumination Fellowship" at the Mellon Institute of Industrial Research on November 15, 1914. The purpose of this fellowship was to find sources of acetylene other than calcium carbide, which was controlled solely by Union Carbide. Dr. George Oliver Curme, Jr. was hired for this assignment. He held fellowship No. 37, for which the Prest-O-Lite Co. contributed the sum of $3000 per year, including an apparatus fund (9). While the Prest-O-Lite Fellowship work was progressing satisfactorily and Curme was demonstrating that acetylene could be obtained from electric arc cracking of petroleum, the Union Carbide Co. also became interested in research. On July 1, 1915, Union Carbide Co. established its Organic Synthesis Fellowship. The objective of this fellowship was the utilization of acetylene or calcium carbide in fields other than lighting and oxyacetylene welding and cutting. A fund of $6,000 for one year was set up to finance the study. The Linde Fellowship at Mellon resulted because of a need to find uses for pure nitrogen. The increasing commercial importance of oxyacetylene cutting and welding had increased the demand for oxygen, causing enormous volumes of pure nitrogen, produced along with the oxygen by cryogenic distillation, to be wasted for lack of a commercial outlet. The Linde Fellowship was to explore the new field of nitrogen fixation and possibly establish a market for its surplus nitrogen. Thus the Prest-O-Lite, Linde, and Union Carbide fellowships became a part of the Institute early in its history.

ETHYLENE AND UNION CARBIDE 303

Fig. 3. Laboratory (known as "The Shack") at Mellon Institute in Pittsburgh, PA, where some of the original chemical processes of the huge synthetic organic chemical industry were developed. Left to right -- Glenn D. Bagley; George O. Curme, Jr.; Henry R. Crume (brother and co-worker); J. Compton; others unidentified.

Following the 1917 merger which produced Union Carbide and Carbon Corp., the three independent fellowships originally at Mellon were somewhat at cross purposes. With the merger the need to find an alternate source of acetylene disappeared. Furthermore Linde had not found the much-sought-after market for nitrogen. As a result, the three fellowships were combined as one unit under Dr. Curme, who became the senior fellowship holder. The purpose of the combined fellowship was redirected toward finding uses for ethylene, which was created in abundant yield during the production of acetylene by hydrocarbon cracking.

The Union Carbide-Mellon Institute connection has been maintained continuously up to the present and now serves as the Union Carbide toxicological study center. The progess of this fellowship from 1914 through the formative years of the petrochemicals industry development will be seen in subsequent paragraphs.

PROFILE OF GEORGE CURME (10, 11)

The senior Union Carbide fellowship holder at Mellon was Dr. George Oliver Curme, Jr. At the time of his promotion in 1917 he was 29 years old, a three-year man in research at Mellon and exceptionally well educated. One year after the fellowships were combined he wrote, "Starting with a plentiful supply of ethylene and acetylene, and the necessary by-products obtained in the manufacture of these substances, a huge chemical industry can be built up capable of absorbing thousands of tons of the products annually. The Union Carbide Corporation, with control of the Linde process, the Carbide process, the electrothermic acetylene process, and with its knowledge and control of electric power projects, is in an exceptional position to exploit this field." How true this view was will become clearer as we observe the growth of Union Carbide's chemical activities.

Dr. Curme was born in Mount Vernon, IA, in 1888, son of a renowned grammarian, in deference to whom Dr. Curme throughout his life maintained the "junior" following his name. He was an alumnus of Northwestern University, where he was graduated with the degree of B.S. in 1909. After that Dr. Curme did graduate work in organic chemistry at Harvard. In 1913 he received the Ph.D. degree from the University of Chicago, after which he did further graduate work at Kaiser Wilhelm Institute and the University of Berlin in Germany, working with Fritz Haber and Emil Fischer. The story has it that he switched from Haber to Fischer for his second semester because Haber was always absent from the laboratory. It was not until after the war that Curme found out that Haber was directing the buildup of Germany's ammonia industry in preparation for World War I. Upon his return from Germany, Dr. Curme took employment with the an established company but soon left to join the Mellon Institute, having had made contact with them earlier, thereby starting his association with Union Carbide Corporation.

The honorary degree of D.Sc. was conferred upon him by Northwestern University in 1933 and by the University of Chicago in 1954, whose alumni association also presented him with its highest medal for his contributions to the nation. Internationally known in his field, Dr. Curme had in his name numerous patents on the production of synthetic organic chemicals. From his original research have come hundreds of familiar products based on petrochemicals such as synthetic rubber, plastics, polyesters and other synthetic fibers, agricultural chemicals, and car care products such as antifreeze. He was a vice-president of Union Carbide from 1948 until he retired in 1955, and a Corporate director from 1952 to 1961.

His accomplishments earned him many awards, including the Chandler Medal from Columbia University in 1933, the Perkin Medal from the Society of the Chemical Industry in 1935, the Elliott Cresson Medal from the Franklin Institute in 1936, the National Modern Pioneer Award from the the National Association of Manufacturers in 1940, and the J. Willard Gibbs Medal of the American Chemical Society in 1944. He was elected a member of the National Academy of Sciences in 1944, and a Fellow of the New York Academy of Science in 1950. In 1952, Dr. Curme received an honorary membership in Phi Lambda Upsilon, the national honorary chemical fraternity.

During World War II, he was a member of the five-man planning board of the office of Scientific Research and Development that was responsible for the technical and engineering aspects of the atomic bomb project, and also served as a consultant to the National Defense Research Committee of this group. In addition, he was a consultant to the War Production Board.

Dr. Curme was a member of numerous technical and honorary societies, including the American Chemical Society, American Institute of Chemical Engineers, Society of Chemical Industry, American Association for the Advancement of Science, and the National Academy of Sciences. He was vice-chairman of the Engineering Foundation in 1960 and was a member of The Chemists' Club, the University Club, Phi Beta Kappa, and Sigma Xi.

Dr. Curme died on July 29, 1976, at 87 years of age. His contributions to the development of the petrochemicals industry were profound.

ETHYLENE AND ETHYLENE DERIVATIVES AND THEIR EARLY HISTORY

It is well to remember that while ethylene and its primary derivatives were well known chemicals prior to 1914 no commercial or large scale production for any of them existed (12,13). The chief reasons for this situation were that much of the associated chemical engineering was lacking, proven commercially available feedstock had not yet been identified and no product markets existed.

A history of the discovery and early studies of ethylene along with some of the first ethylene derivatives is given in considerable detail by Miller (14), and Curme et al (16-19). Becker, around 1650, was apparently the first to observe that on heating a mixture of alcohol and sulfuric acid, an inflammable gas is formed which, it turns out, is impure ethylene. This work was rather crude and credit for producing pure ethylene, studying and reporting its properties and converting it into its first derivative generally goes to the four famous Dutch chemists, Deimann, Van Troostwyk, Bondt and Louwrenburgh - men whose names hardly anyone can recite. The derivative, originally known as "oil of the Dutch Chemists", was produced by reacting chlorine with ethylene and is now known as ethylene dichloride. In 1795 this work was communicated to the Paris Institute by de Fourcroy, who at that time introduced the word olefin, which eventually emerged as the generic term for unsaturated hydrocarbons. The existence of a double bond in ethylene and a triple bond in acetylene was first clearly elucidated by Erlenmeyer (14).

An overview of ethylene and its major derivatives is presented in Figure 4 and Table II. The percentage contributions of the chief sources of ethylene to the total U.S. production in 1976 are shown at the top of Figure 4. At that time approximately 80% of the ethylene production was still derived from ethane and propane, the preferred sources because of their high cracking efficiency. The remainder of the production was petroleum derived. In the future the feedstock distribution will likely shift more toward petroleum fractions because of a projected natural gas shortage. Typical weight cracking yields to ethylene from several precursors are: ethane 76%, propane 40%, naphtha 33% and deasphalted heavy crude oil 30%.

In Table II the six principal first line derivatives (those obtained in one reaction) of ethylene are listed showing the earliest recorded date of synthesis and identification, the commercialization date and the percentage of ethylene consumed by each product for 1960, 1970 and 1980 (estimated). For ethylene oxide production, both the chlorohydrin and silver catalyzed processes are shown even though all ethylene oxide is currently produced by silver catalysis. The chlorohydrin process is shown

TABLE II
SIX PRINCIPAL PRODUCTS OF ETHYLENE

Product	Synth. & Ident. Date	Commercialization Date	Ethylene Consumed(e) % (1960)	1970	1980 (est.)
Polyethylene	1933 (ICI)	1939(a), 1943	26	36	45
Ethylene Oxide	1859 (Wurtz) 1931 (Lefort) Ag^+ Catalyst	1925(b) (UCC), 1937 (UCC) Ag^+	29	21	17
Ethyl Alcohol	1825 (1855)(c)	1930 (UCC)	21	8	4
Ethylbenzene (Styrene)	1850 (1831)	1943(d) (1935)	10	9	8
Ethyl Chloride	1856 (Reynoso)	1922	12	14	13
Ethylene Dichloride (Bromide)	1795	1925 (CI)			
Others			2	12	13

(a) Commercial production in England, and 1943 in US by UCC and DuPont.
(b) Chlorohydrin process, now obsolete.
(c) First preparation of ethyl alcohol from ethylene and H_2SO_4 (1825). First published paper on ethyl alcohol from ethylene synthesis and H_2SO_4 (1855).
(d) Prior to this data all ethylbenzene was obtained from coal-derived processes.
(e) A. L. Waddams, "Chemicals From Petroleum," 4th Ed., John Murray, London (1978).

NOTE: Synthesis and commercializtion data taken from (1) Kirk-Othmer, "Encyclopedia of Chemical Technology", and (2) S. A. Miller (Ed) "Ethylene and its Derivatives", Ernest Been LTD, London (1969).

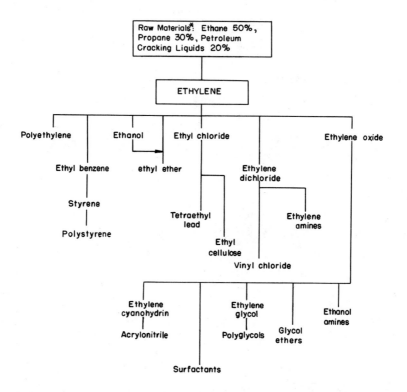

* J.R. Dosher, Chem. Eng. Prog. Sept. (1976)

Fig. 4. Ethylene source and major derivatives.

for its historical interest in regards to ethylene oxide. Although it is still practiced for the production of propylene oxide.

EARLY ATTEMPTS AT DEVELOPING AN ETHYLENE DERIVATIVES INDUSTRY

The phenomenal growth of ethylene-based industrial organic chemistry, which today is world-wide in extent, was pioneered by Union Carbide at the Mellon Institute after 1917. In retrospect, this success stemmed primarily from the capability to produce pure ethylene in large quantities by cryogenic distillation which, as mentioned earlier, was practiced by the Linde Air Products Co. for air separation. In addition to a proven separation technology, four requirements had to be satisfied before large scale production became feasible. The requirements were: 1) a knowledge of the underlying chemistry, 2) the engineering skills to exploit the chemistry on a large scale, 3) the availability of an adequate supply of a chemically pure raw material, and 4) a market demand.

The chemistry of ethylene and its derivatives, which began in 1795, was developed during the 19th century and so was in hand. The second and third requirements were not met until the early part of the 20th century. The lack of sufficient engineering skills and an adequate supply of raw material doomed to failure several early attempts to develop an ethylene-based derivatives industry. The fourth requirement was satisfied as products became available and chemical salesmen got into the act. A brief review of several earlier attempts at commercialization follows.

Ethanol manufacture via ethylene hydration was effected in France around 1860 based on work by Berthelot involving the absorption of ethylene in concentrated sulfuric acid. Allegedly, one liter of synthetic ethanol from ethylene was shown at the London Industrial Exhibition at that time. A company was formed in 1862 to promote the idea, but nothing came of it because of adverse economics. The ethylene level in the feedstock, which was a crude coal carbonization gas, was too low for profitable ethanol production.

At the turn of the century P. Fritzsche came to the United States from Germany and established a plant for making ether from a mixed gas containing approximately 20% ethylene. By utilizing ethylene as a starting material, Fritzsche avoided the excise tax levied on ethyl alcohol which served as the ether source at that time - a decided economic advantage. The process was crude and the scale small, but the heavy ethanol tax made the venture profitable until 1907 when Congress enacted a law permitting the sale of denatured ethanol without tax. Fritzsche's company then folded.

In the 1920s other attempts at hydrating ethylene by absorption in concentrated sulfuric acid were made in England by the Skinningrove Iron Co., using coal gas containing about two percent ethylene, and in France by the Compagnie des Mines de Bethune using a coal gas fraction containing about 20-30 percent ethylene. Neither of these processes could compete with the production of ethanol by fermentation.

The first commercial plant for the manufacture of synthetic ethanol was started up in 1930 by Union Carbide at South Charleston, WV. A great deal of national publicity attended this startup in view of the 100 years of history behind it and the unsuccessful efforts in Europe during the preceding ten years. At the time, the Commissioner of Prohibition was Dr. Doran, a competent chemist with considerable interest in this project from the standpoint of a scientific man. He personally took part in the government's inspection, which was required before production could be authorized.

While this presentation deals primarily with the initiation and development of a commercial industry based on ethylene and its derivatives, it is of interest to review the commercialization of isopropanol manufacture via propylene hydration. The availability of a cheap and abundant supply of propylene from petroleum cracking and the pressing need for an alternative method for producing acetone from isopropanol for use in nonflammable, cellulose acetate airplane dopes provided the economic incentive for process development. This process was based on absorption into sulfuric acid followed by hydrolysis and distillation. It was first worked out by Ellis and Cohen in the Ellis laboratory at Montclair, NJ, and pilot tested in 1917. The process was subsequently acquired by Standard Oil Company of New Jersey and put into production in 1920. Union Carbide entered the isopropanol market in 1929.

UCC COMMITMENT TO BUILD

So far we have focused on the origin and evolution of Union Carbide, the path that led to the Mellon Institute, the research effort under Dr. Curme, the knowledge available on ethylene and its derivatives before industrialization, and the early attempts at developing an ethylene derivatives industry. We now examine the developments within Union Carbide which eventually led to the commitment to build plants to support a petrochemicals business. As the events unfold it will be seen that the commitment went through a certain evolutionary process as the various pieces of the jigsaw puzzle fell into place. The salient features are the scale-up effort for pure ethylene production which took place at

Buffalo, NY; the search for an adequate raw material source, which led to West Virginia, and the purchase of the Clendenin Gasoline Co. in 1920; and creation of a marketing and sales effort to identify customers and product outlets. Then along technical lines large scale processes had to be engineered and proved. Expansions had to be planned which in turn led to the move to South Charleston, WV. Now from hindsight it is quite clear once expansion went beyond South Charleston, the ethylene derivatives petrochemicals industry was firmly established. Adequate markets were in place and growth patterns were quite well established. In short a gigantic new industry was ready to prove itself.

THE BUFFALO PLANT AND RELATED EVENTS

The first ethylene production expansion beyond the Mellon Institute laboratories was undertaken at Buffalo, NY, at the Linde Air Products Plant. The plan was to build a small commercial plant for the manufacture of pure ethylene which would be used in both process development and new product research at Pittsburgh. Prior to this time pure ethylene was produced only on a lab scale at Mellon. Plant construction started in the summer of 1918 and was completed in 1919. Many unforeseen difficulties were encountered during the plant start-up, and the unit, at first, did not produce ethylene. After long months of hard work the first ethylene was shipped to Pittsburgh in January, 1920, with additional shipments following in February and March.

At the time the plant was started, knowledge of the details of handling hydrocarbon feedstocks for the processes was startingly primitive. For instance, there was no record that pure ethylene had ever been compressed, and there was a fear that dangerous spontaneous decomposition similar to that experienced with acetylene would occur. Prior to this Union Carbide effort, the largest scale ethylene manufacture had been its preparation from ethanol for the manufacture of mustard gas, and this did not involve any compression. To check the effect of compression, a small compressor was set up behind a barricade in an open field, and the possibly hazardous operation was tried for the first time and demonstrated to be safe.

The method of operation at the Buffalo plant was to run for several weeks cryogenically purifying ethane, compressing it and storing it in cylinders. After several hundred cylinders were charged, the ethane cracking operation was started. The furnace product containing ethylene was also compressed and charged into cylinders. The ethylene was then separated and recovered by cryogenic distillation. The method was crude and inefficient. Usually, before the operation had lined out, all of the starting material was exhausted. Much time was also wasted in the production of ethane because the original thinking was that pure ethane was needed for the cracking to be successful.

The signing of the World War I armistice put the entire organic chemicals program into a tenuous position, and the possibility of its cancellation seemed high. The times did not bode well for the ethylene project. Aside from the technical uncertainties involved in the project, the general post-war economic and political climate did not look encouraging. The Corporation was being reorganized following the war and no one in the organization knew anything about the business of organic chemicals manufacturing. It was only the farsighted vision and enthusiasm of Mr. E. F. Price, a vice president, which kept the organic chemicals program in a favored position in the high echelons of the corporation and insured attention and support. The major decision to keep the ethylene project alive was reached in December, 1919, following a conference at which Dr. Curme and J. N. Compton presented their economic assessment of ethylene derivatives manufacture. One of the pivotal points in their presentation was the great market expectations for diethyl sulfate, according to Compton (8). Later he said, "The subsequent course of development of diethyl sulphate was disappointing, but some of us will always do it honor and reverence because of memories of the days when it was pointed out with confidence as the cornerstone of the magnificent edifice which arose in our imaginations at that time, and which otherwise might not have had even an imaginary cornerstone to start from."

Pending the construction of a new chemical plant, which was finally located in Clendenin, WV, the Buffalo plant was to be kept in operation, not only to gain additional technical information, but also to continue to furnish cylinder ethylene for the operation of the chemical units at the Mellon Institute.

It was particularly fortunate that the decision to proceed was made no later than it was and that some momentum had been generated in the early months of 1920, because disaster struck the project in May when a fire destroyed the ethylene plant at Buffalo. The damage made restoration or further operation impossible, thus giving additional incentive for early action elsewhere. Eventually all ethylene production and development work were moved to Clendenin. It may be concluded that the major developments at Buffalo were the furnace operation for ethane cracking, the proving of the cryogenic fractionation method for ethylene purification and the handling and compression of ethylene on the large scale essential to ethylene derivatives manufacture.

ETHYLENE AND UNION CARBIDE

THE MOVE TO CLENDENIN, WV - 1920

After the decision to expand production facilities was reached it was necessary to address immediately the question of finding a large and abundant supply of feedstock material from which the ethylene would be produced. A study of this problem led to the idea of using natural gas, or more precisely the ethane, propane, and butane in natural gas.

A survey was therefore conducted of the various gas fields in the United States, and samples were taken and analyzed at the Mellon Institute to assemble information as to the relative merits of gas from these fields. The study indicated that gas from West Virginia was as good as or better than any other gas in the country, and estimates indicated ample reserves for a permanent chemical plant there. West Virginia also had the advantage of convenient location, being one of the most favorable shipping points of all the territories considered. It was not far from the center of manufacturing activities in the United States and only about 300 miles from the center of population at that time.

In studying the West Virginia situation, contact had been made with the Clendenin Gasoline Co., a small independent organization operating a plant at Clendenin on the Elk River about 20 miles above Charleston. The plant had contracts for purchasing natural gas from the surrounding wells, processing it for the removal of gasoline by an absorption process, and selling the stripped gas to the United Fuel Gas Company. Such a plant seemed to be an ideal source for the desired hydrocarbon raw materials. The property adjoining the original plant was available for lease for the erection of the necessary new facilities and, in general, the location and the town seemed suitable for the desired purpose. This company was purchased in 1920.

As the technical personnel of the new corporation became more familiar with the techniques of the natural gasoline industry, they were immediately struck by the crudity and inefficiency of certain phases of the industry and of the entire allied petroleum industry. Most striking was the apparent ignorance in the gasoline industry of the principles and practice of rectification, which were not only well known to the Linde Company through its work on liquid air, but also to the chemical engineering profession in general, especially in its application to the separation of coal tar derivatives and the manufacture and purification of ethyl alcohol.

The constituents of natural gas most desirable as raw materials for the new industry were ethane, propane and butane.

Curiously enough these specific constituents were absolutely wasted and allowed to escape to the air in some early natural gasoline plants. In these plants, natural gas was passed through the heavy absorbing hydrocarbon oil, which tended to pick up selectively the heavier constituents contained in natural gas together with decreasing proportions of propane, ethane, and even a little methane. (This technology will be discussed in more detail in a later section.) However, these lighter constituents could not be incorporated in the natural gasoline finally offered for sale. Therefore, in order to reduce the vapor pressure of the "wild" gasoline and make it shippable, the mixture containing the undesirable lighter constituents was subjected to a process called "weathering", which in many places even as late as 1920 consisted simply in exposing the gasoline to the atmosphere in an open tank. By allowing the undesirable gases to pass off, which in themselves were of value, important quantities of valuable gasoline were also vaporized. In the better plants, the weathering process was modified to permit the collecting and recompression of the weathered gases for their use as boiler fuel. In absolutely no case was there any thought of applying the obvious expedient of rectification, which the Carbide and Carbon Chemicals Corporation immediately proceeded to use to good advantage.

One of the earliest technical developments of the Corporation was a special rectifying device, later known as the "stabilizing column," which was designed to carry out the weathering process with efficient separation of natural gasoline from lighter constituents and the delivery of the latter for use in the manufacture of olefins. It was also recognized that this procedure might be used in the plants of other companies for the improvement and standardization of their natural gasoline. The process and equipment were covered by patents, and it was found that several companies, especially some of the more progressive ones in the mid-continent field, were interested in use of the stabilizing column under royalty agreements. Its advantages were so obvious that its application spread rapidly, but attack was made on the validity of the patents and, after several years of litigation, they were found not valid. The term "stabilizing column", which was originally coined at Clendenin for this device, scattered throughout the literature of petroleum technology and in almost every article on natural gasoline. The device has been of great importance to the Corporation in its applications.

Clendenin in 1920 was pretty much isolated. In certain seasons it was practically impossible to drive a car from Clendenin to Charleston. Even in the most favorable seasons this was possible only by a road which forded streams and wound circuitously in and out of hollows to pick the most favorable natural path. The best connection between Clendenin and Charleston was by railway. There were two passenger trains per day in each direction, one leaving Charleston at about 7 am and the other at about 4 pm.

Most of the men connected with the program elected to live in Clendenin in various boarding houses or in the one hotel. Dr. Curme chose to live in Charleston which involved the most severe routine sustained by anyone concerned. He faithfully travelled to Clendenin every day, including Sunday, catching a train before daybreak in winter and usually returning after dark. What made it even worse was that the trains usually ran one to four hours late.

Construction work at Clendenin was completed and operations started in the summer of 1921. The completion was unusually timely since the post-war business boom lapsed at just about this time and future business prospects were extraordinarily uncertain. Fortunately the Corporation was committed by having gone ahead at Clendenin and did not hesitate to spend the additional money required to operate the plant even in a period when routine expenditures were curtailed to the limit and less farsighted executives would have been tempted to suspend an operation which gave no promise of immediate profit.

As an addendum to the manufacture of chemicals, the idea of selling bottled gas for household use developed. Originally a mixed hydrocarbon gas under the trade name PYROGEN was sold. It was later dropped and PYROFAX, containing primarily propane, was put on the market. Gas selling provided good income for the young company before the chemicals business really got started. Eventually the gas business was sold because of other priorities. Today this business is known as the LPG industry.

Union Carbide was noted from the very beginning for the complicated refining systems that the engineers devised. It is a possibly apocryphal story that Dr. Nelson Wickert wryly observed, "Curme invents messes and we fractionate them into good solvents" [20].

GROWTH AT CLENDENIN

The company stayed in Clendenin for almost five years. During those years the pioneers built up and tore down production units at an amazing rate while the natives of Clendenin looked on with a growing air of mystification wondering how a company could succeed with more than 20 men when the former gas company could hardly make it with only six men. The men were dealing with what gas and oil men called "wild gases". Explosions could almost be predicted from the time a reaction was started. In later years when Curme was asked what was produced at Clendenin he would reply, "Mostly mistakes," which summed up the trials of this early period.

Following the move to Clendenin economic and engineering evaluations were carried out on a fairly continuous basis. In September, 1922, an exhaustive report was released which dealt with the prospects of manufacturing chemicals from ethylene and propylene. In January, 1923, a second series of economic studies was completed in which attention was concentrated on alternative methods of obtaining hydrocarbon raw materials for producing ethylene and propylene. These studies indicated clearly that the most promising method of obtaining raw materials would be the installation of stabilizer columns in existing natural gasoline plants to produce ethane-propane-butane mixtures. The studies also proposed the construction of cross-country pipelines to transport natural gas from distant sources to the manufacturing sites and the installation of equipment to separate and recover ethane and propane to provide a raw material source independent of the natural gasoline supply. Such arrangements required entering into long-term contractual arrangements with the gas companies, under agreements sufficiently profitable to them to be attractive. This report also addressed, for the first time, the possibility of locating expansion efforts in the Charleston, WV, district.

One of the major achievements at Clendenin was the development of the cracking furnace. Another was the demonstration that chemical processes functioned in a sheet-iron shack quite as well as under the best laboratory conditions with the processes running around the clock in a steady state manner. In retrospect, however, it seems the most important achievement of the Clendenin era was in the development of new chemical processes on a pilot plant scale and the production of new chemicals at sufficient volume to permit evaluation for commercialization purposes.

MARKETS FOR NEW CHEMICALS

The work at Clendenin was highly successful in turning out new chemicals. Processes were shown to be successful and the necessary chemical engineering innovations for scale-up purposes kept pace with the process advances. New chemicals began to build up at a remarkably fast rate. However, in the 1920s and earlier when the Carbide and Carbon Chemicals Corp. launched its efforts to build a petrochemicals industry with olefins, principally ethylene, as the raw material, no ready markets existed. Up to this time the chemical industry was essentially a coal-tar-derivatives business with a few chemicals produced from acetylene, e.g., by the Niacet Chemical Corporation. The reason was perfectly logical: Before this time none of these emerging chemicals were available so no markets or uses for these materials existed. In those days for the Carbide and Carbon Chemicals Corp., the only outlets were for acetylene used in the oxyacetylene welding business, propane sold as PYROFAX for heating and perhaps a few other chemicals.

Although PYROFAX sales were of real importance to the struggling young chemical company, it was still not a real synthetic chemical. It must be noted however that the first cash income for the chemicals co. proper came from the sales of propane and natural gas stabilizing equipment.

The first synthetic chemical product sold on a really large scale was CELLOSOLVE solvent, which found a good market in the lacquers industry. Another major product turned out to be ethylene glycol which originally found its major use in antifreeze and as a low temperature solvent for dynamite. To improve the quality of ethylene glycol as an antifreeze solvent, anti-corrosion and antifoaming additives had to be developed. Ethylene was also sold in 1925 as a ripening agent for fresh fruit and provided a fairly important source of income. A third product first sold in 1925 was ethylene dichloride produced as a by-product of the ethylene glycol operation. It found its biggest outlet as a major dry-cleaning fluid and then later as the precursor for vinyl chloride.

In 1929 dichloroethyl ether became a tonnage quantity product as CHLOREX solvent for the refining of lubricating oils. This CHLOREX process led the way to today's high-quality lubricants.

An interesting story concerning ethylene oxide is worth relating. In 1929 its polymerization and toxicological properties were not too well known, and at first the material was sold as a fumigant to sterilize surgical bandages and supplies. One day a professor at Vassar called about an invasion of silverfish into her library. Could Carbide help? Two chaps named Doley and Leonard rushed to the rescue (some said they were more interested in the undergraduates than the silverfish) and carted in buckets full of ethylene oxide which they proceeded to splash around the library. This unintentional toxicological study killed all the silverfish and made both boys pretty sick; it was fortunate there was no fire and no explosion. In 1930 the U.S. Department of Agriculture showed that ethylene oxide was a promising fumigant for foodstuffs.

In attempts to find markets for their products Messrs. J. A. Rafferty and Curme would fill their satchels with four-ounce bottles of chemicals eked out at Clendenin and peddle them around the country. Months later they would return and find that "the sons of guns" hadn't taken them off the shelf. A flourishing but short-lived business turned up in isopropyl alcohol which was in sharp demand as a special denaturant in the early prohibition era with prices hitting $5.50/gal.

In January 1924 the first sales department for the Chemicals Co. was created with R. W. White from Linde Co. becoming its first General Sales Manager. Dr. J. G. Davidson was borrowed from Curme as a technical advisor to White. Davidson soon became Manager of Chemical Sales, the first chemist-salesman in the industry and the prototype of our technical representatives today. Davidson recognized the two major problems to be solved. He had to continue creating markets as research turned out new products and he had to help the customers with their technical problems. He solved both problems by training chemists and chemical engineers to be salesmen - men who could recognize potential markets and teach the customers to use the products correctly. This chemical knowledge of our salesmen backed by technical service from Mellon revolutionized the selling methods of the chemical industry.

Marketing chemicals in the twenties was certainly different from today's situation. Prices were high and known applications were limited. Davidson commented more than once that, "If these men had to work on commission they would starve to death". Sometimes sales, not commission, averaged as little as $300 a month per territory during the beginning of the chemicals business. Another important device used by Carbide in those days was to provide credit to customers.

The emerging auto industry proved to be an excellent market for chemicals and this special relationship has continued to the present. Union Carbide and the auto industry have had special ties dating as far back as 1904 when the Prest-O-Lite Co. first started selling cylinder acetylene for automobile lights. The Prest-O-Lite Co. started on a very modest scale in Indianapolis, but was swept along with the rapid upswing of the automobile industry, with which they kept pace by developing the first important nation-wide automobile service system. Within a few years, the Prest-O-Lite Co. had become the largest single consumer of calcium carbide and probably also of acetone in the United States.

CARBIDE AND CARBON CHEMICALS CORPORATION

The Carbide and Carbon Chemicals Corporation which presently is part of the Chemicals and Plastics Division (reorganized 1979) was formed and incorporated under the laws of the State of New York in 1920. This company started as a division of the Linde Air Products Company. Mr. M. F. Barrett, then vice president of the Linde Co., was put in charge. He appointed Mr. J. A. Rafferty as its full time manager. Mr. H. E. Thompson was appointed principal technical advisor to Mr. Rafferty and given responsibility for all of the gas separation development work and for general engineering. Arrangements were also made to use the engineering, drafting and construction division of the Linde Co.

The technological developments responsible for its formation actually started in 1914 as stated earlier, gaining what might be called a single focus when the four separate companies listed earlier united to form the Union Carbide and Carbon Corporation in 1917. This move consolidated the three fellowships into the single pursuit of developing chemicals based on ethylene. After about three years of work, new compounds of commercial potential had piled up at such a rate that the Carbide and Carbon Chemicals Corp. was created to handle them.

MOVE TO SOUTH CHARLESTON - 1925

By 1925 the manufacture of chemicals at Clendenin had grown to the point that new and expanded facilities were needed. A contract with dynamite makers for three million pounds of ethylene glycol a year was encouraging and so too were the CELLOSOLVE sales, as they caught on as solvents for nitrocellulose.

In 1923 a business study report, mentioned earlier, gave attention for the first time to possible locations in the Charleston district. Eventually the idle property of the Rollin Chemical Co. at South Charleston, which had fallen on hard times, was leased at exceptionally favorable terms. The lease included buildings and a modern steam power plant, all of which promised a saving of considerable magnitude as compared with the cost of a completely new plant.

The initial contract at South Charleston called for taking possession as of November 30, 1923, and covered a five year period with option to renew and option to purchase. Early in 1924, a construction crew was at work cleaning out the old buildings and making the site ready for occupation.

At this point it may be of interest to provide a picture of the conditions of South Charleston and of the major industries there in 1925. The town was essentially a wartime boom town because of the Naval Ordnance Plant and of the prosperous condition of the chemical industry during and immediately following the First World War. After the closing of the government plants and the severe post-war depression of the early 1920s, South Charleston had become a forlorn and desolate community. The Rollin Chemical Company had been managed by the Rollin brothers who dissipated their reserves and mortgaged their future in the later years of the war to go into a new and foreign field, namely the manufacture of chlorine, caustic soda, carbon bisulphide, carbon tetrachloride and monochlorobenzene. Their incentive for doing this was the success attained during the war years in the identical project by their neighbor, the Warner-Klipstein organization. The Rollin Company succeeded in getting its plant into operation just in time to miss the wartime profits and was left with a useless chlorine-derivatives plant and a barium products plant badly crippled by the business depression. Colonel J. J. Riley had become interested in the plant and was persuaded to undertake its rehabilitation. While the barium business was a sound one, the idle property originally used for chlorine products was a considerable drag on the company and the proposal to lease it to the Chemicals Corporation, even at a low rental, was an important factor in enabling Colonel Riley to put his business on the road to solvency under the new name of Barium Reduction Company. In 1925, the Rollin property was a sadly rundown plant but still the most important in South Charleston.

Next door to this property was the small plant of E. C. Klipstein and Sons Company, occupying a frontage of only about 150 feet and dependent upon the adjoining Warner Klipstein establishment for its steam, water and chlorine. This also was a war plant, extravagantly but badly housed, and at one time extremely profitable as a producer of sulfur black, anthraquinone and allied products. This company also had fallen on bad times and was struggling to exist. It continued to struggle for five or six years, with one drastic change of management, steadily deteriorating physically, and finally succumbing to an offer of the Calco Chemical Company for all of its assets.

Perhaps the saddest business failure of the whole group was the Warner Klipstein Plant, originally owned jointly by the Warner Chemical Company and E. C. Klipstein and Sons Company but acquired at some period by the Warner Company when the obligations of the Klipstein Company to the partnership became too great. This plant had been an important producer of chlorine and chlorine products during the war period, and the company had control of the Nelson cell, which had been the most widely used cell in wartime expansion such as the Edgewood Arsenal Plant. It had, however, fallen into misfortune in the early 1920s, and in 1924 its future looked very dark. It had a production capacity of about 20 tons of chlorine per day, but was able to sell only about half this quantity.

To this organization, the proposal by the Chemicals

Corporation to sign a contract for their full requirements up to ten tons of liquid chlorine per day was looked upon as a life saver which would mean the difference between continued operation and possible total extinction. As in the case of the Rollin Chemical Company, Warner Klipstein changed its name and became Westvaco Chlorine Products Company, with prospects of a new lease on life.

It was now proposed to add to this aggregation an infant company, smaller and weaker than any of them and dependent upon the strongest of the group, Barium Reduction Company, for some of the essentials of its existence.

In addition to the chemicals manufacturing organizations, a small glass plant adjoined the east side of the Barium Reduction Plant. This, the Dunkirk Glass Company, later sold out to Libby Owens. Beyond their property was a yard of the Hamilton Lumber Company. In the Kanawha River, extending from the far end of the plant of Westvaco and far past the limits of the developed South Charleston industrial area to the east, were the 80 fertile acres of Blaine Island, which was then devoted to farming with a reputation for producing the best and most easily stolen melons in Kanawha County. At its upper end there was a small amusement park and bathing beach connected with the mainland by a light pontoon bridge. At its lower end was a shed used by the farmer in the summer for sheltering his team.

This picture of South Charleston in 1925 may be contrasted with the picture in 1939. Barium Reduction Corporation remained a self-supporting and reasonably prosperous organization, having sold additional acreage to the Westvaco Chlorine products which expanded until it was one of the largest chlorine plants in the world. The Klipstein organization was entirely out of the picture after its sale to Calco and the subsequent acquiring of the property by Union Carbide. The Dunkirk Glass Plant was demolished and removed as junk, and the property was acquired by Union Carbide and was used at the time as a railroad siding and coal-storage yard. The Hamilton Lumber Company likewise was completely liquidated and its property occupied by UCC's shipping department, with a small corner allotted to the Prest-O-Lite Company. Finally, the mile-long industrial development of Blaine Island provided an extension of the operations of the Chemicals Corporation. This island is larger and more important in area and in production than was the entire aggregation of plants at South Charleston in 1925, excluding the government Naval Ordnance and Armor Plants.

Today the UCC South Charleston Plant produces more than 500 different chemicals and plastics. Most of the materials produced are intermediates and are either used in other processes or sold to customers who in turn use them to make finished products. Total land area is 230 acres, all of it developed, and it employs 1,763 people (1976 figures).

EXPANSION BEYOND SOUTH CHARLESTON

When political considerations do not intrude, the locations of ethylene and derivatives plants are determined primarily by two factors: proximity to abundant and cheap feedstocks, and proximity to markets. The south-central West Virginia area was chosen for initial commercialization because of the first factor, although the location was also within 500 miles of over half of the country's population. All ethylene production was carried on at South Charleston for about ten years until, in 1935, a plant was started up at Whiting, IN, to fill the burgeoning demands of the mid-west. The feedstock for this plant was gas from a nearby refinery. Major gas discoveries in Texas provided the impetus to locate a plant at Texas City in the late '30s. In this case the low price of feedstock overcame the problem of relatively long distances for delivery of products to customers.

During World War II Union Carbide and Goodrich operated a plant for the government at Institute, WV, just a few miles from South Charleston, to produce synthetic rubber from butadiene and styrene. The butadiene was originally made from grain-derived ethanol, but after the war, when the two companies bought the plant, Union Carbide installed a cracking plant which then supplied butadiene to Goodrich. Styrene was also supplied by a chemically circuitous route: Benzene from the local cracking plant was alkylated with ethylene to form ethylbenzene, which in turn was oxidized to acetophenone, hydrogenated to the corresponding alcohol, and dehydrated to styrene.

By the late 1950s a substantial price differential had developed for concentrates in favor of the Gulf Coast over West Virginia, and an inexorable shift southwestward in the center of gravity for Union Carbide's olefin industry began. A second cracking plant was built at Texas City, and another was commissioned at a new site at Seadrift, TX, in 1954. The only non-Texas plant built during this decade was at Torrance, CA, to service West Coast needs.

In the decade of the '60s a second cracking plant was built at Seadrift, a third at Texas City, and yet another at Taft, LA. Concurrent with the commissioning of these units, all olefin production was terminated in West Virginia and Indiana. Finally, in 1972 a large cracking plant was built at Ponce, PR, to take advantage of the then-low price of foreign naphtha. Later, the Taft unit was doubled in size. Thus, in a period of 40 to 50 years the area which had spawned the ethylene-based chemical industry saw its child grow to adolescence, leave home for the sunny South, and then continue to grow there to the robust maturity it exhibits today.

PRODUCTION AND GROWTH OF ETHYLENE

In a previous discussion the history of ethylene and six of its major derivatives was briefly reviewed. We also saw how ethylene came to the attention of the early Union Carbide pioneers and noted the evolution of certain ethylene-based processes. We noted how the search for an adequate source of feedstock at commercial scale levels that led to Clendenin, West Virginia, and reviewed the early growth of the UCC Chemicals venture. In this section we wish to briefly consider the growth of ethylene production over the years not only by Union Carbide, but in the United States and worldwide. By considering the growth in ethylene consumption one can, to some extent, arrive at an appreciation of the evolution of the entire aliphatics chemical industry.

It was pointed out that at the beginning of the commercialization efforts most of the chemicals intended for large-scale production were well-known laboratory chemicals and in some instances, some synthesis work to make a few select derivatives had already been attempted. However, no gigantic industry was perceived by anyone prior to the Union Carbide-Mellon connection. What the early Carbide pioneers undertook was to develop the engineering and chemistry needed for tonnage-scale production and in support of this production establish the necessary sales and markets.

Ethylene first became available as a commercial raw material in 1922. Prior to this time all ethylene production was used in-house. By 1930 US production stood at approximately 35 million lbs per year. In comparison the extrapolated 1980 production figure is approximately 35 billion lbs. In the 50 year interval from 1930-80 ethylene growth averaged out to 14.8 percent per year, which is shown in Figure 5. World-wide ethylene production in 1978 stood at approximately 85 billion lbs. These figures clearly show the present gigantic size of the ethylene-based chemicals industry, which establishes ethylene as the undisputed king of the petrochemicals business. In terms of other chemicals produced in the US, ethylene ranked number 5 in 1977 with a production of 25.43 billion lbs following in order of size, sulfuric acid, lime, ammonia, and oxygen (21), whose values in billion lbs/yr were: 71.64, 37.85, 35.15, and 32.53 respectively. From the same source the estimated 1978 ethylene production is 28.13 billion lbs, which represents rank number 6,

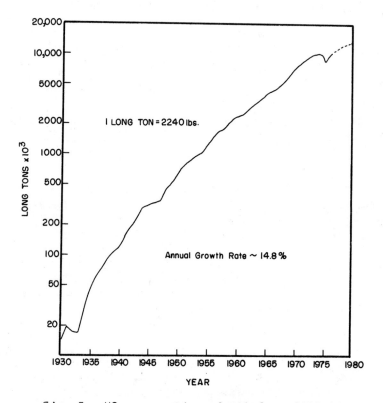

Fig. 5. US consumption of ethylene 1930-80.

ETHYLENE AND UNION CARBIDE

nitrogen moving up to 5th place at 28.41 billion lbs.

Purchase price of ethylene as a function of year from 1950-78, supplied by the Union Carbide Purchasing Department, is given in Figure 6. The cost is represented in both actual price and price based on 1975 dollars. It is interesting to note the significant price changes that have taken place since the Arab oil embargo. The price of ethylene from 1950 to the mid '70s stayed relatively constant primarily because of the increasing scale of new plants as they came onstream and the retiring of the older, less efficient production units and the stable price of raw materials during this period. The only sure thing which can be deduced from these figures is that ethylene will continue to rise more rapidly in cost in the next 10 years than in the previous decade. As more of the produced ethylene is derived from petroleum, the price of ethylene will eventually be keyed to the price of petroleum.

OVERVIEW OF CHEMICALS PRODUCED DURING THE PERIOD 1922-1960

Earlier, for illustrative purposes we discussed the development of markets for a few of the first chemicals produced by the Carbide and Carbon Chemicals Co. From the very beginning each succeeding year saw new chemicals added to the market list. In Table III a fairly complete tabulation is given of those chemicals sold in quantities of at least 55 gallon drums. Formulation products have been omitted. Other chemicals sold at one time or another in small quantities cannot be identified. One interesting observation of Table III is that a very large fraction of the listed products were firsts in this country and in many instances also firsts on a world basis. Figure 7 illustrates a chemical synthesis scheme which was included in a report issued in 1928. Presumably the proposed spectrum of chemicals by the specified processes were not all translated into commercial production so that this schematic is included for its historical value to show the thinking at that time.

Table II lists the six largest volume derivatives of ethylene. The table is set up to show the first synthesis and identification date which will put each chemical into its proper historical perspective. The commercialization date lets us see how long it took to establish a business with these chemicals. Ethylene oxide, identified in 1859 by Wurtz, became an item of commerce in 1925 via the chlorohydrin process. This process is now obsolete having been replaced by the unique silver-based heterogeneous catalyst process first discovered by Lefort and commercialized by Union Carbide. The information on ethyl alcohol refers to synthetic alcohol obtained via ethylene hydration. The percent of ethylene consumed for each product is

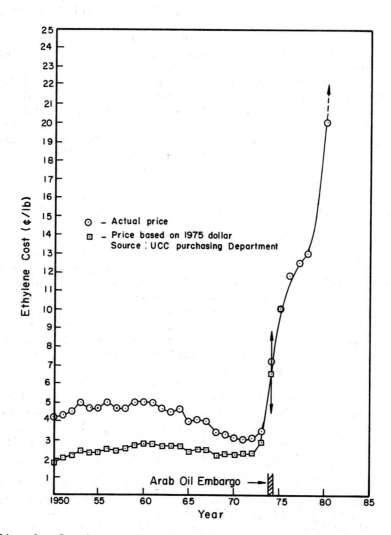

Fig. 6. Purchase price of ethylene as a function of year.

TABLE III

DATES OF FIRST SALES OF CHEMICALS IN LOTS OF AT LEAST 55 GALLONS
IN THE USA BY UNION CARBIDE CORPORATION BETWEEN 1922-1960*

1922- Propane (PYROFAX)

1924- Acetaldehyde
Paraldehyde
Aldol
Crotonaldehyde   Niacet
Acetic Acid   Co.
Glycol Diacetate

1925- CELLOSOLVE Solvent (a)
Ethylene Glycol
Ethylene Chlorhydrin (a)
Ethylene Dichloride (a)
Ethylene (a)
Methane
Ethane
Isobutane

1927- CELLOSOLVE Acetate (a)

1928- CARBITOL Solvent (a)
Butyl CARBITOL (a)
Butyl CELLOSOLVE (a)
Diethylene Glycol (a)
Triethylene Glycol (a)
Triethanolamine (a)
n-Butene
Acetic Acid

1929- Acetone
Isopropanol
Dichlorethyl Ether (a)
Dioxane (a)
Ethylene Oxide (a)
Methyl CELLOSOLVE (a)
Vinyl Chloride (a)
Propylene
Butylene

1930- Vinyl Resins (a)
Ethanol (b)

Methanol
Isopropyl Ether
Ethyl Ether
CARBOXIDE Fumigant (a)

1931- n-Butyl Acetate
n-Butanol (b)
Butyraldehyde (b)
Propylene Oxide (a)
Propylene Dichloride (a)
Propylene Chlorhydrin (a)
Propylene Glycol (a)

1932- Ethyl Acetate
Ethyl Acetoacetate (b)
Acetoacetanilide
Methyl Acetate
Methyl Acetone
Monoethanolamine (a)
Diethanolamine (a)
Diethyl Sulfate (a)

1933- Acetic Anydride (b)
Isopropylacetate
Butryic Acid
Methyl Amyl Acetate (a)
Methyl Amyl Alcohol (a)
Methyl Isobutyl Ketone
  (a)

1934- n-Hexanol (a)
2-Ethyl Hexanol (a)
2-Ethyl Butanol (a)
2-Ethyl Hexyl Acetate (a)
Butyl Ether

1935- Ethylene Diamine (a)
Diethylene Triamine (a)
Triethylene Tetramine (a)
Ethyl Silicate (a)
Morpholine (a)

TABLE III
(CONTINUED)

1936- Methyl Acetoacetate
o-Chloroacetoacetanilide
Dichloroacetoacetanilide

1937- Butyl CARBITOL Acetate
(a)
CARBITOL Acetate (a)
Diisobutyl Ketone (a)
Isopropanolamine (a)
Diisopropanolamine (a)
Triisopropanolamine (a)

1937- Dipropylene glycol (a)
Ethyl Butryic Acid (a)
Dioctyl Amine (a)
Ethyl Silicate (a)
Methyl n-Amyl Ketone (a)
Thiodiglycol (a)
Methyl CARBITOL (a)
Methyl CELLOSOLVE
Acetate (a)
TERGITOL Anionic
Surfactants
Triethylene Glycol
Dichloride

1938- CELLOSIZE WS (a)
Diacetone Alcohol (a)
Diethylaminoethanol (a)
Diethyl CARBITOL (a)
Ethyl Butryaldehyde (a)
Ethyl Hexaldehyde (a)
Ethylhexoic Acid (a)
n-Hexaldehyde (a)
Hydroxylethyl Ethylene
Diamine (a)
Isophorone (a)
Mesityl Oxide (a)
Methyl Amyl Carbinol (a)
Propylene Diamine (a)
Tetramine Pentamine (a)
VINYON Fiber (copolymer
of vinyl chloride and
vinyl acrylate)
VINYON N (copolymer of
vinyl chloride and
acrylonitrile)

Ethyl Propylacrolein
Propionic Anhydride

1939- Butyl Amine
Dibutyl Amine
Dichlorisopropyl Ether
(a)
Polypropylene Glycols
(a)
Tetraethylene Glycol
(a)
Butyric Anhydride
Dicyclopentadiene
Methyl Propyl Carbinol
Piperazine
Polyethylene glycol
300 (a)

1940- Tetraethanol
Tetraethylene
Pentamine (a)
Hexyl Ether (a)
Pentaethylene Hexamine
(a)
Amine 220 (a)
Butadiene

1941- Dimethyl Ethanolamine (a)
Glyoxal (a)
Acetoacet-o-Toluidide
Butyl Chloride

1942- 2-Ethylhexanediol (a,c)
Methyldiethanol Amine
(a)
2,4-Pentanedione (a)
Polyethylene Glycol
600 (a)
UCON Lubricants
Trichloroethane

1943- CARBOWAX Methoxy
Polyethylene
Glycols 350 and 550
Styrene

1944- Allyl Alcohol
Allyl Chloride
Epichlorhydrin

TABLE III
(CONTINUED)

1946- Dipropylene Triamine (a)
Methyl Morpholine (a)
Monoethylhexylamine (a)
Hexylene Glycol

1947- Ethyl Butyl Ketone (a)
Ethoxy Triethylene
 Glycol (a)
2-Methyl Pentanol-1 (a)
Pyruvic Aldehyde (a)
Sorbic Acid (a)
TERGITOL Nonionic
 surfactants

1948- Diisobutyl Carbinol
Isopropyl Amine (a)
Acetoacet-o-Anisidide
Acetoacet-p-Phenetidide
Acetophenone

1949- Polypropylene Glycols
 425 and 2025 (a)
Phenyl Methyl Carbinol
 (a)
Hexyl CELLOSOLVE (a)
Butanediol
Diethyl Amine
Ethyl Acrylate
Tetrahydrophthalic
 Anhydride
Triethyl Amine

1950- Methyl Ethyl Pyridine (a)
Polyethylene Glycol
 Amine 175 (a)
Allyl Cinnerin (a)
Acetoacet-p-chloranilide
Acrolein
Butyl Acrylate
Diethylaminoethyl Amine
2-Ethylhexyl Acrylate
Isodecanol
Nonyl Acetate

1951- Ethylhexyl Chloride (a)
Methyl Monoethanolamine
 (a)

Isopropenyl Acetate (a)
Vinyl Butyl Ether (a)
Amyl Alcohols
Butyl Octanol
Calcium Acrylate
Di(Methyl CELLOSOLVE)
 Maleate
Ethoxypropionaldehyde
Ethyl Formate
Hydroxyethyl Diethylene
 Triamine
Methyl Acrylate
Propionaldehyde

1952- Butoxy Ethoxy Propanol
 (a)
Diisopropyl Ethanolamine
 (a)
Ethoxy Ethoxy Propanol
 (a)
n-Ethyl Ethanol Amine (a)
Alpha-Picoline (a)
Gamma-Picoline

1953- Vinyl Butyrate
Acrolein Dimer
m-Amino Phenylmethyl
 Carbinol
Butyl Acetanilide

1954- Diethylamine Propylamine
 (a)
1,2,6-Hexanetriol (a)
Dimethylamino
 Propylamine (a)
Vinyl Ethylhexoate (a)
Methyl Isoamyl Ketone (e)
Tetra (Ethylhexyl)
 Silicate (a)
Decyl Acrylate
Ethylene Cyanohydrin
1,5-Pentanediol

1955- Di(2-Ethylhexyl)
 Phosphoric Acid (a)
Dioxolane (a)

TABLE III
(CONTINUED)

Ethylbutyl Ethylhexyl
  Silicate (a)
Ethoxydihydropyran (a)
N-Ethyl Morpholine (a)
Dihydroxyethyl
  Diethylene Triamine (a)
Acrylic Acid, Glacial
Glutaraldehyde
n-Hexyl Amine
Isooctoic Acid
Methacrolein
p-tert-Butyl Phenol
Tridecyl Alcohol
N-Methyl Butyl Amine (a)

1956- Aminoethyl Hydrogen
  Sulfate (a)
Butyl Glycidyl Ether
Calcium Acetate
  Propionate
CELLOSIZE WP-4400
Decaldehdyes, Mixed
Diallyl Amine
Hydroxylethyl Piperazine
  (a)
Methyl Pentanoic Acid (a)
n-Pentaldehyde (a)
N-methyl Piperazine (a)
Triisooctyl Amine (a)

1957- Imino-bis-Propyl Amine
  (a)
Aminoethyl Piperazine (a)
Triethoxy Butane (a)
Isodecanoic acid (a)
Methyl Butanol (a)
2-Methyl-2-Ethyl
  Dioxolane (a)
Acrylic Acid, 30% Aqueous
Calcium Propionate
Dibutyl Tin Dichloride
Dimethyl Morpholine
Di-n-Propyl Amine
Methyl bis Aminopropyl
  Amine

n-Propyl Amine
Sodium Propionate
Tetrahydrobenzaldehyde
Valeric Acid

1958- Methyl Propyl
  Propanediol (a)
Cyanoethoxyethyl
  Acrylate (a)
Dipropylene Triamine (a)
POLYOX Water Soluble
  Resin 35 (d)
Potassium Sorbate (a)
N,N,N',N'-Tetramethyl-
  1,3-Butane diamine
  (a)
1,1,3-Trimethoxybutane
  (a)
Butylene Oxides
1,3-Diaminopropane
2,3-Dichloropropene-1
Diethylcyclohexane
Methacrylonitrile
Methyl Aminopropyl
  Amine
Nonyl Phenol
Thioxane
1,2,3-Trichloropropane

1959- Epsilon-Caprolactone (a)
2-Methyl-2-Ethyl-
  1,3-Propanediol (a)
Styrene Oxide
1,2,3-Trichloropropene-1
UCAR Butylene Oxide 12
UCON Diesters JL-1 and
  JL-3
3-Amino propanol
Butylene Glycol
2-Cyanoethyl Acrylate
Ethyl Butyl Amine
N-Glycidyl Diethylamine
Hexamethylenetetramine
Isobutyronitrile
Isohexanol
Isopentanoic Acid

TABLE III
(CONTINUED)

1960- NIAX Hexol LS-490 (d)
      Hexyl Ethyl CARBITOL (a)
      2-Mercaptopropanol (a)
      n-Butyronitrile
      CELLOSOLVE Acrylate
      Epoxidized Oleate
      Heptanol-2
      UCAR Bisphenol-A
      Isobutyl CELLOSOLVE

(a) First commercial production in US.
(b) New process.
(c) New molecule.
(d) New composition of matter.
(e) Second supplier.

* Chemical formulations not included.

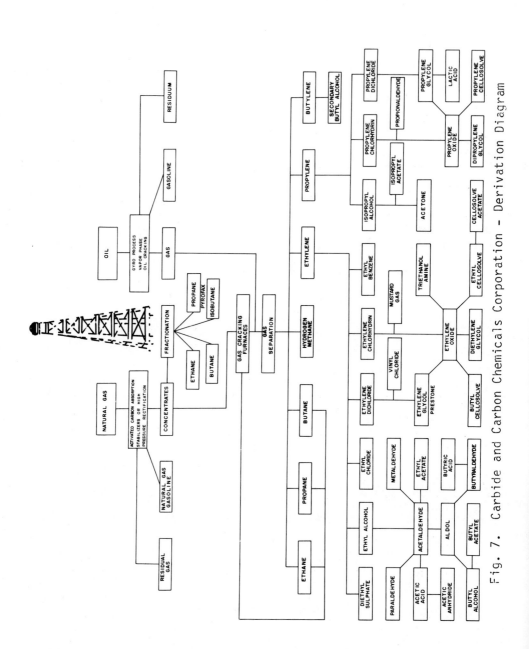

Fig. 7. Carbide and Carbon Chemicals Corporation – Derivation Diagram

ETHYLENE AND UNION CARBIDE

given in the three columns on the right. Polyethylene is gradually capturing a greater fraction of the ethylene consumed as time advances and will probably hit the 50% level in the near future.

A word about the Niacet Chemicals Corporation is in order because this organization produced several acetylene-based chemicals around the time ethylene based products were developed. The Niacet Chemicals Corporation was created in 1925 and derives its name from Niagara acetylene. It was originally owned by the then Carbide and Carbon Chemicals Corporation, DuPont and Shawinigan Power. Soon thereafter Union Carbide acquired the entire company. It was essentially a specialty chemicals organization during its tenure with Union Carbide. In 1979 the Niacet Chemicals Corporation was sold to M. R. Brannen who at the time was its operations manager.

CHEMICAL ENGINEERING AND THE INFANT CHEMICAL INDUSTRY

Chemical engineering was the primary driving force in launching the petrochemical industry. Once a chemical synthesis had been identified in the literature and tested in the laboratory, pilot scale testing had to be carried out. Containment and corrosion problems had to be dealt with very early in the program. Optimizing reaction conditions and working out a separation and purification scheme were also problems that required solutions early in the development program. In many cases several engineering problems were tackled at one time as a process was being developed. The application of engineering principles took many forms. Some of the more noteworthy accomplishments in the production of ethylene are summarized briefly below.

Feedstock Preparation

The reason for locating the first cracking plant at Clendenin, as discussed earlier, was the presence of copious quanties of feedstock - ethane, propane and butane - in the natural gas of that area. The key to economical recovery of the "concentrates", as these light paraffins were called, was the stabilizer column. The original recovery process consisted of the following:

Compressed natural gas, valued at three to six cents per 1000 cubic feet and containing a large component of concentrates and natural gasoline - C_5 and higher paraffins - was fed to an absorber. The solvent was a heavy oil. Most of the propane and even larger proportions of the heavier components were absorbed and then stripped from the solvent. Little ethane was recovered in this process.

The concentrates and natural gasoline mixture was then fed to a stabilizer column in which the two fractions were separated. The stabilizer was a low-temperature column which operated at near-atmospheric pressure. Refrigeration techniques, including Joule-Thompson expansions, were used extensively. Then-extant air-separation technology served as the technological basis for the column.

Traces of water in the feed were a nagging problem for the stabilizer. Over a period of several weeks the heat-transfer surfaces would become fouled with ice to the extent that operation was no longer possible. The column was then shut down and warmed up to clear these surfaces. To retard the rate of ice buildup, "calcium traps" were placed in the natural gas feed line to the absorber. These traps contained calcium chloride, which removed part of the water by formation of the hydrated salt. The drying capability of silica gel was known at that time but was not used.

During the period at Clendenin, other columns were developed to separate concentrates into propane (PYROFAX), propane/butane (RURALITE) and isobutane, which was sold as a refrigerant. Superatmospheric-pressure technology was developed, and some of these columns were operated without the need for refrigeration. This eliminated the water-freezing problem (but not necessarily the gas-hydrate problem) in these columns.

The high cost of separation in the absorber and the original low-pressure stabilizer provided an incentive to investigate alternative methods for recovery of concentrates from natural gas. The method siezed upon was adsorption on activated carbon. Natural-gas feed was passed through a bed of granular carbon at essentially ambient temperature. Once loaded, the bed was taken off line, and the concentrates were desorbed by treating the bed with steam or hot water. The evolved gas was then compressed and liquefied.

Based on a successful development program at Clendenin, the decision was made to recover concentrates feedstock for the new South Charleston cracking plant with the activated-carbon process. A unit consisting of three 10,000-gallon tanks of carbon was built at the South Penn Oil Co plant at Diamond, WV, about 20 miles fom South Charleston. On October 3, 1925, a specially-designed tankcar made the first delivery - 298,000 cubic feet (gas) - of mostly propane to the newly completed South Charleston No. 1 cracking plant.

The need to transport concentrates to South Charleston provided the impetus to develop the special tankcar. The car contained an inner chamber of 1 1/2-inch welded steel plate. Surrounding this was a layer of compressed cork for insulation, which was held in place by a 1/8-inch steel corset. This car was so well built that it served the South Charleston plant for over 25 years, and cars of quite similar design are in common use today transporting a wide variety of refrigerated liquids.

The activated-carbon process at Diamond was found to have an Achilles' heel: In the summer the bed temperature during adsorption became so warm that the efficiency of recovery dropped substantially. The result was that the unit was bottlenecked. Consequently, the process was dropped in favor of an improved, pressurized stabilizer which processed the raw natural gas. By the late 1920s stabilizers were located at four gas-processing stations in West Virginia.

The concentrates-carrying tankcars also became an endangered species during this period. As the feedstock appetite of the South Charleston cracking plant grew and the stabilizers proliferated, the preferred method of concentrates transport ultimately became the pipeline. Thus by the end of the 1920s the concentrates-recovery and delivery systems in use resembled closely, except in scale, those used today.

Development of the Cracking Furnace for Ethylene and Propylene

Ethylene generation via thermal cracking of ethane was originally done at Pittsburgh and then at Buffalo, as mentioned earlier. The objective of these operations was to supply feedstock to Dr. Curme at Pittsburgh for his experiments on the production of ethylene derivatives. The first reactors were fused silica tubes mounted in small furnaces.

The first cracking furnace started up at Clendenin in February, 1921. A sketch of the furnace is shown in Figure 8. The reactor consisted of 12 single-pass, copper-lined, three-inch-diameter, iron pipes mounted horizontally and heated for a length of about 15 feet. Heat was supplied by burning natural gas in the cylindrical furnace cavity. The feed consisted primarily of propane and butane since these were more easily removed from natural gas than ethane. Somewhat earlier, ethane had been thought to be the only satisfactory feed for producing ethylene, but subsequent experiments showed the feasibility of using higher paraffins too. The cracked-gas flow from this furnace was originally about 1200 cubic feet per hour, from which about 100 to 125 cubic feet per hour of ethylene were recovered. Thus the ethylene production rate from this first commercial unit was

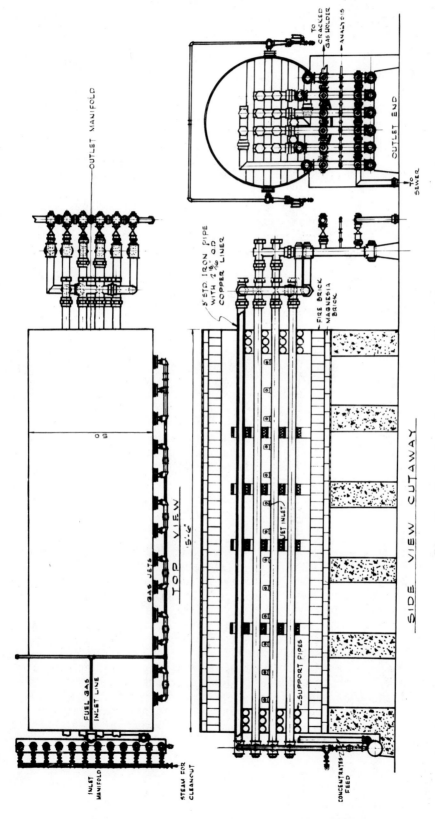

Fig. 8. Scale Drawing of the First Cracking Furnace

about seven to nine pounds per hour!

Metallurgical problems plagued the early operation. Unlined iron pipes produced too much carbon, prompting the use of copper liners. The copper liners apparently did not fit tightly, and the air gaps between the liners and the pipes caused uneven heating of the gas. Pipe life was also a problem. The average life was about one month. The first breakthrough in furnace design improvement came with the substitution of chromium-iron pipes for the copper-lined iron pipes. Originally it was felt that the new pipes would also have to be protected on the inside by copper, and so a copper-plated, one-inch-diameter pipe was placed in a small furnace and tested. At the end of the test, the pipe was broken open to determine the fate of the copper. It had apparently scaled off during the test, but no deterioration in cracking performance was detected. As a result, the presumed need for copper - like the copper plating itself - disappeared.

About the same time - 1924 - that the chromium-iron pipes were being proven out, the desirability of longer pipes became apparent. Heat-transfer calculations, made with the aid of Walker, Lewis and McAdams' "Principles of Chemical Engineering (which issued in 1923), showed that the controlling resistance was the inside gas film. Some experimentation was done using twisted ribbons of copper inside the pipes to increase the gas-film coefficient, but this approach was abandoned in favor of longer pipes, which, coupled with higher velocities, could increase the coefficient. So in July, 1924, the furnace internals were rearranged to accept a six-pass, chromium-iron reactor. Four of the old copper-lined iron pipes in parallel served as a preheater. The test was a success, and the multi-pass reactor became the basis for all subsequent designs.

Improvements in metallurgy and reactor geometry allowed the cracked-gas rate to be raised substantially - first to 7,500 cubic feet per hour and then, before shutdown of the Clendenin operation in 1925, to 15,000 cubic feet per hour. At that rate, furnace productivity had been raised by over one order of magnitude. The Clendenin plant never had more than one furnace in operation.

Concern over metallurgical problems also prompted searches for alternative methods for cracking. Early in 1924 a small, cyclic regenerative cracker, similar to the then-common water-gas reactors, was constructed. During the "blast" part of the cycle, combustion of natural gas heated the silica-brick checkerwork. Following a short steam purge, gas feed was injected and cracked. Olefin yields were not as high as could be obtained in the tubular reactor, and development work was stopped.

In 1925 cracking by blowing feed into a molten bath of lead was demonstrated. Olefin yields were comparable to those in tube cracking, but the practicality of the tubular reactor seemed greater, and lead-bath cracking development was also discontinued.

In 1924 work was begun on the design of a new cracking plant in South Charleston. Three furnaces were specified, each containing two reactors. Each reactor consisted of seven passes of three-inch-diameter chromium-iron pipes in the radiant section, with an eighth pass serving as a preheater in a convection section. The basic furnace design is shown in Figures 9A and 9B. With the addition of the convection section to the furnace, essentially all of the features of modern-day furnaces were included in a rudimentary way.

This new furnace, although designed to effect more severe cracking than the Clendenin furnace, still could not approach modern-day furnaces in severity. A comparison of cracking yields is shown in Table IV. Design conditions in 1924 produced a 40 percent per-pass conversion, compared to a present-day value of 92 percent; the per-pass yield to ethylene in present-day furnaces is about triple that of the 1924 design.

South Charleston No. 1 olefins plant successfully started up late in 1925, producing on the order of 10 million pounds of ethylene per year. Over the next few years new furnaces, stretched out to 25 feet long, were added, and the unit and its successor, South Charleston No. 2, remained a beehive of experimental activity for many years.

One development which came relatively late - 1940 - was the use of steam as a diluent in the feed to the furnace. Since the early days of Clendenin, steam was used when a furnace was shut down to remove carbon deposited during cracking. The impetus for adding steam to the feed came in an unusual way. As furnaces were lengthened and the number of passes increased, gas velocities naturally increased. As a result, erosion of the bends connecting the passes by carbon formed in the cracking reaction became a serious problem. Steam was added to the feed to suppress carbon formation. The beneficial effect of steam addition on olefin yields was an unexpected benefit.

A second and quite different type of furnace was also used at South Charleston. This furnace was called the Gyro unit and was widely used in this country prior to the development of catalytic cracking in refineries. Typically, Gyro units were fed vaporized heavier-than-gasoline fractions which were lightly thermally cracked to form some gasoline-range material. The reactor consisted of six-inch-diameter Duralloy pipes in a radiant section. The cracking temperature was much lower (670 vs. over

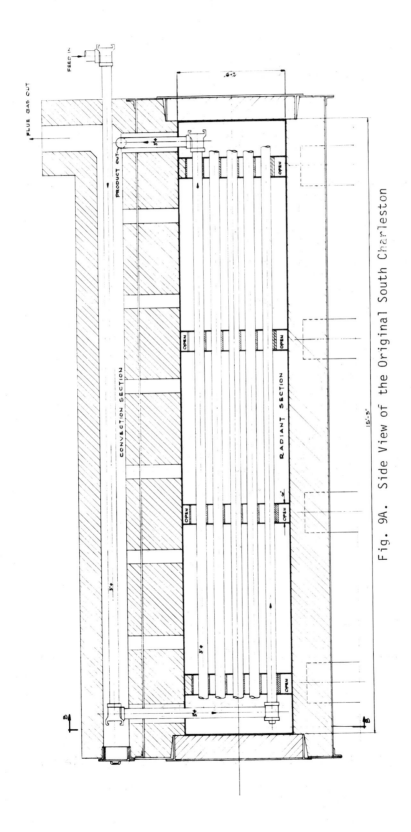

Fig. 9A. Side View of the Original South Charleston

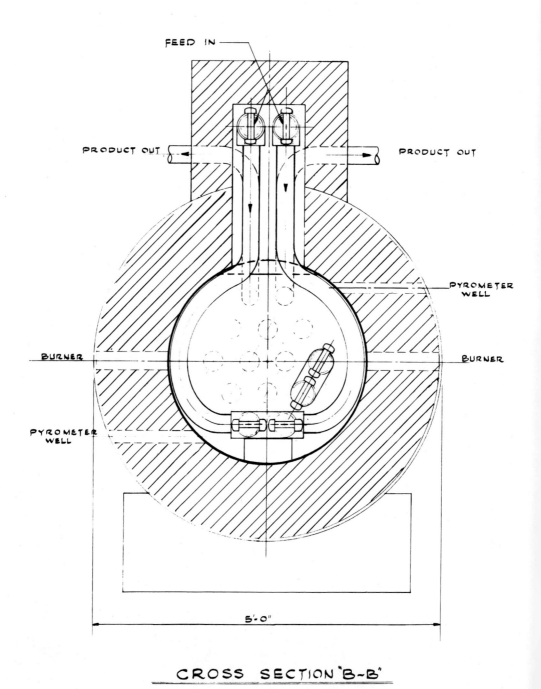

Fig. 9B. End View of the Original South Charleston

TABLE IV

COMPARISON OF SINGLE-PASS CRACKING YIELDS
IN EARLY AND MODERN FURNACES

Feed: 83 mole % propane, 17 mole % n-butane

Component	Single-Pass Yield, lb/100 lb feed	
	1924	1980
H_2	0.7	1.5
CH_4	8.4	22.6
C_2H_2	*	0.7
C_2H_4	13.1	38.1
C_2H_6	1.9	3.8
C_3H_4	*	0.7
C_3H_6	14.1	14.2
C_3H_8	56.1	7.5
C_4H_6	*	2.8
C_4H_8	*	1.3
C_4H_{10}	5.4	0.5
Heavier + CO	*	6.4
	100.0	100.0
%Conversion of Feed	40.4	92.0

*Not measured in early analyses, but the sum is probably less than about two pounds per 100 pounds of feed.

800°C) than that of the Union Carbide furnaces, and the residence times were much longer (several seconds vs. about one second).

The Gyro unit in South Charleston was originally used to produce some gasoline which was sold to local refineries. Experimentation was also done to determine the unit's flexibility. For example, higher cracking temperatures were explored to see if significant quantities of propylene could be produced, and by the early 1930s some Gyro gas was being processed at South Charleston. And in 1932 crude oil, at least that part which could be vaporized, was cracked in the Gyro unit. This event marked the first time, to our knowledge, that crude oil was cracked in such a fashion.

Later in the 1930s gasoline production at South Charleston ceased, and the Gyro unit was converted to a new use: purification of benzene and toluene. The benzene/toluene-containing fraction from the cracked gas stream was subjected to a long (several seconds) crack which reacted out almost all of the non-aromatic components. Subsequent distillation of the product produced a very-high-purity benzene, but the toluene had to be further treated with sulfuric acid to produce a nitration-grade product. The advent of liquid-liquid extraction technology rendered this method of aromatic upgrading obsolete, and the South Charleston Gyro unit was finally shut down in the early 60s.

Product Purification

Most of the separation equipment used at Clendenin came from Buffalo. The basic flowsheet for processing cracked gas was quite simple. Gas from the cracking furnace was cooled to near ambient by passing through an uninsulated pipe. Four distillation columns, each operating cryogenically and at essentially one atmosphere, then separated the gas. The first column was a demethanizer, which separated the hydrogen and methane from the remaining gas. The next column separated the C_2 fraction from C_3 and heavier materials. The C_2 stream was fractionated to produce a relatively pure ethylene stream, and the C_3 and heavier stream was also fractionated to produce an upgraded propylene stream.

Operation of the stills at about one atmosphere presented two problems: excessive energy costs and plugging by ice and hydrates. As a result, experimentation was begun at Clendenin to develop a high-pressure recovery train. This experimental work was accompanied by extensive thermodynamic and plate-to-plate calculations to provide a design package for the new South Charleston plant. The conditions chosen for the columns were quite similar to those used in modern-day olefin plants.

The icing problem was ameliorated, but not completely solved, by high-pressure operations, since the columns still operated at sub-freezing temperatures. The original method of coping with the problem was the same as with the stabilizer: The stills were run until they became fouled, and then they were warmed to remove the water. The first improvement involved the use of a glass-wool plug upstream of the demethanizer to remove "snow" which formed as the gas cooled. The glass wool reduced, but did not eliminate, the icing problem, and it was not until later at South Charleston that the addition of silica-gel driers provided the ultimate solution.

The columns at the South Charleston No. 1 unit were truly miniscule: two feet or less in diameter and about 30 feet high. In other words, both dimensions were nearly an order of magnitude less than the behemoths in a typical billion-plus pounds per year, modern-day ethylene plant. These early columns were outfitted with bubble-cap trays, reboilers mounted beneath the columns, and in some cases condensers mounted inside the columns. An additional feature of these columns was that they were housed inside a building - undoubtedly a safety hazard because of the possibility of flammable gas release in a confined space. And in fact South Charleston No. 1 met its demise on New Year's Eve, 1928, when an explosion destroyed much of the refining train. The cracking furnaces were unhurt in the explosion and continued to operate until new stills were built, and the impure olefins were consumed as best they could. When the new stills were completed, and in the new flowsheet they were double-trained, the plant became known as South Charleston No. 2.

As could be guessed from the heights of these early columns, olefin-product purities were substantially lower than typical values today. A typical 1920s value was less than 95 percent ethylene, compared to 99.9 plus percent today. The impurities were primarily methane, ethane and acetylene. The first two did not cause any particular problems in the derivatives processes, except for promoting extra losses of ethylene in purge streams. Acetylene, because of the relatively mild cracking conditions used (see Table IV), was not present in large enough concentrations to be a problem either. (It should be remembered that polyethylene and ethylene oxide via silver catalysis, two products requiring very low levels of acetylene in the ethylene feed, were not produced until the late '30s and early '40s.)

The first impetus to remove acetylene from ethylene came from a very unusual source: the desire to produce anesthesia-grade ethylene. Prior to 1931, the ethylene which was used as an anesthetic was produced from dehydration of fermentation ethanol, and the acetylene level in this product was vanishingly small. To qualify ethylene from cracking for anesthetic use, three methods of acetylene removal were studied. The first was absorption in acetone. Acetone was known for many years to be a good absorbent for acetylene and was used as part of the "filler" in acetylene cylinders, the Claude process discussed earlier. An absorber-stripper process was developed and installed in 1931, and the acetylene problem was not only solved, but also a source of acetylene for the local Linde cylinder-bottling plant and, later for vinyl chloride production, was established. Ironically, the ethylene-for-anesthesia market never developed, but the acetylene-removal technology developed made possible new generations of high-purity-ethylene-consuming processes in the subsequent decades.

The second acetylene-removal process developed was hydrogenation to ethylene and ethane. Though hydrogenation was subsequently used by most other ethylene producers, the ability to use the acetylene produced continues to make recovery Union Carbide's choice for removal from ethylene.

Reaction with copper was the third acetylene-removal process developed. The ethylene stream was fed to a reactor containing copper deposited on an alpha-alumina support. An acetylene polymer - cuprene - built up on the support and was subsequently burned off with air. This process was also considered for guard beds on ethylene streams requiring very high acetylene removals, but it was never used on any scale at all.

In the 1940s a process to recover butadiene for synthetic rubber was developed. Dichloroethyl ether, known as CHLOREX and formed as a byproduct in the ethylene chlorohydrin reaction, was first used as the solvent in the extractive-distillation step. The process, as do today's versions, included both extractive and straight distillations.

Through the '30s and '40s essentially all of the remaining separations used in modern-day cracking plants were developed and commerialized. Indeed, from those early days until now the primary chemical-engineering efforts have been consumed in scaleup and process optimization, rather than in the development of radically new technology for either olefin purification or production.

R&D Efforts in Support of the Petrochemicals Industry

Research and development work in support of the emerging petrochemicals industry began at the Mellon Institute as discussed previously and is closely tied to the chemical engineering that was required to achieve commercial production. As noted earlier, from the Mellon Institute, expansion of the R&D effort moved to labs at Buffalo. Around about 1922 a research lab was also set up on Long Island, which functioned for several years and eventually was consolidated with all R&D work at the South Charleston Plant.

In 1949 one of the world's most modern chemical research centers was completed at South Charleston, WV. This center was expanded in 1958 when company-wide engineering and development activities were added.

The R&D facilities in Bound Brook, NJ, primarily support the plastics industry. These facilities originally came into the Carbide family via the Bakelite Co. acquisition in the late 1930's. The mid-60's saw the construction of a new technical center in Tarrytown, NY, designed to house facilities for chemicals, consumer products, industrial gases, and basic research laboratories.

THE ALIPHATIC CHEMICAL INDUSTRY TODAY

The aliphatic chemical industry is relatively mature today, but from the foregoing discussion it is clear this industry sprang from rather humble beginnings. One of the main problems the chemical industry faces today is a stable and assured raw material source with reasonable price stability. This is a particularly thorny problem because the raw material sources upon which the chemical industry depends, natural gas and petroleum, have severe demands placed upon them by other industries for their fuel (energy) value. Hence, at present day high production rates, the chemical industry is closely coupled to the world's energy supply and therefore to a large degree dominated by the oil and gas industries who almost totally control the present raw material sources. Furthermore, direct competition by these industries is also experienced by the chemical producers because most oil and gas companies have their own chemical divisions which sell so-called value-added products. The interaction between the energy and chemical industries has been present since the beginning of the aliphatic chemical industry. As was noted earlier, Carbide even bought a small gasoline company in Clendenin in 1920. In the past, however, this raw material dependence was not very significant because the energy sources were sufficiently abundant to fill all needs. Recently, however, especially during the past 10 years as the energy supply has been tightening up this relationship has been feeling the economic strains. Resolution of the raw material supply question will take a long time to solve or at least so far as the chemical industry is concerned, arrive at a more favorable accommodation.

Carbide officials recognized the raw material supply problem from the very early days and continuously searched for alternate raw material sources over which they could have more control. As early as 1946 Dr. Curme launched a major R/D program to use coal as the basis for at least some of the chemicals already in production and perhaps to obtain many more new ones. Dr. George T. Felbeck was the man in charge of this project. This was an all-out effort which included buying up coal interests, testing underground gasification, inventing a coal mining machine called a coal mole (later sold), and running several pilot and larger-scale plants, including a commercial scale unit for coal hydrogenation at Institute, WV. This entire effort was finally abandoned in 1962 due to the availability of cheap oil from the Middle East.

The R&D and Engineering programs of today will determine the nature of the petrochemical industries of the 21st century. Ethylene used for ethylene-based derivatives today is being produced almost exclusively from both concentrates and refined petroleum feedstocks. The continued growth of ethylene utilization will, however, create serious problems of supply and production, not to mention cost. These problems are now being dealt with by various companies throughout the world. Basically, two distinct approaches are currently being considered for the future production of what has traditionally been called ethylene-based derivatives. One approach looks for alternate ways of producing ethylene making the implicit assumption that all the ethylene-based chemistry will be left intact. Here we have (i) ethylene produced directly from crude oil and (ii) the production of ethanol from biomass which can subsequently be dehydrated to ethylene. The other approach aims at making the traditional ethylene-based chemical derivatives from non-ethylene feedstocks. Again we have (i) the direct synthesis of oxygenated chemicals from synthesis gas and (ii) chemicals derived directly from biomass. In the future all these options may become feasible and actually practiced depending on the particular economic climate and various government involvements around the world.

CURRENT PETROCHEMICAL PROJECTS

The current assessment of the raw material supply situation by Union Carbide and the petrochemical industry in general is that adequate oil supplies, at a price, will carry us into the 21st century. The availability of refined naphtha for ethylene production is of course not as reliable because of the control oil companies can exert. At present both concentrates and petroleum distillates are being used as feedstocks by Carbide and the industry in general. In the future an ever larger fraction of the chemicals produced will be derived from petroleum. Today approximately 50% of the ethylene produced is oil based. Consequently, this situation will key the cost of ethylene directly to cost of oil and energy in general.

At Carbide two basic programs are currently underway which are designed to alter the raw material requirements for the Corporation. The first active program is the Union Carbide/Kureha-Chiyoda joint venture called the Advanced Cracking Reactor (ACR) (22,23), designed to crack deasphalted crude oil directly to olefins. At present this process is being proved in a demo-unit operating in Seadrift, TX. The ability to use crude oil directly instead of just its lightest fractions greatly increases the possible sources of raw material, thus tending to insure a favorable raw-material price-volume situation. The second program involves the production of oxygenated chemicals from synthesis gas. Since synthesis gas can be made from virtually any carbonaceous raw material (including coal and residuum to mention but two), it represents the epitome of precursors in terms of raw-material flexibility.

The production of oxygenated chemicals from regenerable (biomass) raw materials is felt to be another but longer-range alternative to the chemical-industry raw material dilemma. (The only exception to the longer-range time frame scenario is fermentation ethanol, which is now becoming competitive with ethylene-based ethanol.) The success of biomass conversion to other large-volume chemicals will depend to a considerable extent on the relative rate of increase in cost of biomass production compared to the cost of traditional feedstocks. Tong (24) in a recent article describes the fermentation of carbohydrates to produce a whole family of short chain organic compounds.

A recent book by Kniel et al (25) summarizes the major ethylene derivatives and end uses at the present time. Based on this end use section it is clear that, in an industrialized society, ethylene today affects everybody's life in some way.

CONCLUSIONS

We have presented a brief history of how that part of the aliphatic chemical industry based on ethylene originated, concentrating primarily on ethylene itself. Before ethylene could become a major intermediate, it was necessary for what were then four somewhat disparate companies to unite to form the Union Carbide and Carbon Corporation, bringing together the financial and directional wherewithal to mount a pioneering effort. The key elements which led to the success of the effort were:

- A mission oriented research program at Mellon Institute, brilliantly led by Dr. George Curme, to develop practical means of producing ethylene and its derivatives;

- The availability of abundant quantities of cheap raw materials;

- The acquisition and development of separation-engineering and reaction-engineering skills, which were critically important for efficient raw material recovery and process scale-up;

- The development of the concept of the technical salesman, who could work with customers to define major new uses for ethylene derivatives; and, finally,

- The preseverance of key officials in the Corporation during the early, less-than-profitable period in supporting the effort.

Ethylene, in the course of nearly 60 years of commercial production, has reached the status of a very-large-volume, mature chemical. Because of its rapidly escalating price during the '70s and the likelihood of the continuance of this trend for the foreseeable future, its possibilities for continued growth are somewhat in question. Synthesis-gas-based routes to some derivatives will undoubtedly begin to make inroads over the next 10 to 20 years, while renewable-raw-material-based fermentation processes may also come into vogue. Thus king ethylene, though now firmly ensconced on his chemical-intermediate throne, may by the beginning of the 21st century see his empire in a state of contraction, the extent of which cannot yet be predicted.

Highlights

The subject of this paper was to outline the role Union Carbide Corporation played in launching and developing the olefin-based petrochemical industry, primarily as it evolved through the commercialization of ethylene as the principal feedstock. The following is an overview of the salient features of this paper.

Formative Events Which Launched the Petrochemical Industry.
1913 - Mellon Institute Formed
1914 - Illumination Fellowship - Prest-O-Lite Co.
1915 - Organic Synthesis Fellowship - Union Carbide Co.
     - Nitrogen Fixation Fellowship - Linde Co.
1917 - Union Carbide and Carbon Corp. Formed
     - Fellowships Combined

- Objective, to Develop Ethylene Derivatives Industry
1918 - Construction of Ethylene Cryogenic Separation
       Plant - Buffalo, NY
1919 - Corporate Management Decision to Commercialize
       Diethyl Sulfate
1920 - First Pure Ethylene Shipped to Mellon
     - Move to Clendenin WV
     - Carbide and Carbon Chemicals Corp Formed
1922 - Bottled Gas - Propane, Pyrofax: Propane/Butane,
       Ruralite
     - Commercial Source of Ethylene Established
1925 - First Chemicals Sold - CELLOSIZE Solvent
       (Ethylene Glycol Monoethyl Ether) Ethylene Glycol,
       Ethylene Chlorohydrin, Ethylene Dichloride,
       Ethylene
     - Move to South Charleston
1935 - Whiting In Plant Start-Up

Major Technological Advancements.
Thermal Cracking of Ethane/Propane to Produce Ethylene
Cryogenic Distillation for Ethylene Separation and
Purification
Basic Design of the Cracking Furnace
Introduction of the Stabilizer Column for Hydrocarbon
Separation
Demonstrating the Compressability of Ethylene
Introduction of the Chemical Tank Car
Introduction of the Chemical Pipeline
Large Scale Production of Ethylene Derivatives -
Development of Separation Engineering and
Reaction Engineering Skills
The Use of Steam During the Cracking Process
Developmment of the Concept of the Technical Salesman

Basic Requirements for Establishing the Ethylene Based
Petrochemical Industry.
Process for Producing Pure Ethylene
Knowledge of the Underlying Chemistry of Ethylene
and Derivatives
Commercial Source of Feedstock for Ethylene Production
Engineering Skills for Process Scale-Up
Market Demand

ACKNOWLEDGEMENTS

Numerous people provided help to the authors during preparation of this manuscript, in particular we would like to thank Jon L. Brockwell, Norma Lawson, Richard C. Grimm,

Simon Meyer*, Paula Sandry, Hubert G. Davis, Gerard R. Kamm, Donald Gatewood, William S. Kelly, Robert G. Keister, Bertram D. Ash, and Sally A. Goedecke. A special note of thanks to Karen McClung for her expert secretarial work.

BIBLIOGRAPHY

1. S. Field, "World Petrochemical Feedstocks to 1985", Stanford Res. Inst. Report No. 121 (1978).

2. Union Carbide Products and Sales Directory, 1978.

3. Union Carbide Corporation Annual Report, 1977.

4. Moody's 1978 and 1969 Industrial Manual.

5. "Historical Highlights" and "Back-Grounder: Our History", Union Carbide Corp. Pamphlets.

6. Union Carbide and Carbon Corporation, Fortune Magazine Articles, June, July, and Sept. (1941).

7. A Research Center Directory, Gales Research Co., p. 366 (1965).

8. J. N. Compton, "Informal Personal Observations on the History of the Carbide and Carbon Chemicals Corporation", Eng. Dept., UCC, South Charleston, Oct. 4, 1937. J. N. Compton and E. T. Crawford, Jr., "The Chemical Industry of the Great Kanawha Valley, Trans. Am. Inst. of Chem. Eng. $\underline{34}$, 199 (1938), and, "Chemical Pioneering in the Kanawha Valley, Chem. and Met. Eng. $\underline{45}$, #6, 306 (1938).

9. UCC Publication, "A Marketing Revolution - Early Sales Efforts of Union Carbide Chemicals Co." (1962).

10. Abstracted from Union Carbide files.

11. A. B. Kinzel, National Academy of Science, submitted for publication 1978.

12. W. L. Nelson, "Petroleum Refinery Engineering, 4th Edition McGraw-Hill Book Co. (1958).

13. Petroleum Panorama 1859-1959, The Oil & Gas J. $\underline{57}$, No. 5 (1959).

---

*A Carbide pioneer from the Clendenin days.

14. S. A. Miller Ed., "Ethylene and its Industrial Derivatives", Ernest Benn Ltd., London (1969).

15. G. O. Curme, Jr., "Importance of the Olefin Gases and Their Derivatives I. Sources and Uses of Ethylene and Propylene". Chem. & Metal. Eng., 25, 907 (1921).

16. G. O. Curme, Jr., and H. R. Curme, "Importance of the Olefin Gases and Their Derivatives II. Diethyl Sulphate", ibid, 25, 957 (1921).

17. G. O. Curme, Jr., "...III. Ethylene Dichloride", ibid, 25, 999 (1921).

18. G. O. Curme, Jr. and E. W. Reid, "...IV. Isopropanol", ibid, 25, 1049 (1921).

19. G. O. Curme, Jr. and C. O. Young, "...V. Ethylene Chlorohydrin and Ethylene Oxide" ibid, 25, 1091 (1921).

20. F. Johnston, "A Brief History of Chemistry in the Kanawha Valley", Kanawha Valley Section, ACS (1977).

21. C & E News, p. 24, May 7, 1979.

22. T. Ishikawa and R. G. Keister, "A Petrochemical Alternative - ACR", Hydrocarbon Processing, p 109, December (1978).

23. J. D. Kearns, D. Milks and G. R. Kamm, "Development of Scaling Methods for a Crude Oil Cracking Reactor by Using Short Duration Test Techniques," Advances in Chem. Series, No. 183, Thermal Hydrocarbon Chemistry, p 108 (1979) ACS-Washington, D.C.

24. G. E. Tong, Chem. Eng. Prog., April, 1978, p. 71.

25. L. Kniel, O. Winter and K. Stork, "Ethylene, Keystone to the Petrochemical Industry", Marcel Dekker, Inc., NY (1980).

GENERAL REFERENCES

1. E. R. Kane, "Trends in U. S. Chemical Industry", Chem and Ind., June 2 (1979).

2. N. W. Haynes, "Am. Chem. Ind. - A History", D. Van Nostrand Co. NY (1945-54).

3. R. Landau, "The Chemical Industry 2000 AD" CEP P. 27, October (1978).

4. K. N. McKelvey (Ed), "Alternate Raw Materials for Organic Chemicals Feedstocks". Am. Inst. Chem. Eng. Reprint from Petroleum and Refining Exp. Houston, April 1-6 (1979).

5. P. A. Sermon, "Catalysis of the Petrochemical Economy," H. M. Stanley/BP Lecture, Chem & Ind. October 20 (1979).

6. P. Desprairies, "Worldwide Petroleum Supply Limits - Ultimate Resources and Maximum Annual Production of Conventional Petroleum: Possible Resources of Unconventional Petroleum CEER, 10, No. 2, P.7 (1978).

7. M. Klaus, "Outlook of the Petrochemical Industry in the World - Its Feedstocks, Technical Aspects and Problems", CEER 10, No. 2, P.16 (1978).

8. T. Matsuura, "Technological Outlooks for Petrochemical Industry", CEER 10, No. 2, P.21 (1978).

9. A. L. Woddams, "Chemicals from Petroleum" Ch. 8 4th Ed., John Murray, London (1978).

10. N. W. Haynes and E. L. Lardy, (Eds.), Chemical Industry's Contribution to the Nation: 1635-1935. A Suppl. to Chem Ind., May (1935), John Winthrop, Jr., NY.

11. J. A. Kent (Ed.), "Riegel's Industrial Chemistry," Reinhold Publ. Corp., NY (1962).

12. D. E. Brandt, "Long-Term Changes Are Possible In Ethylene Feedstocks" The Oil and Gas J., Feb. 5, P. 51 (1979).

13. B. D. Berger and K. E. Anderson, "Modern Petroleum - A Basic Primer of the Industry," The Petroleum Publ. Co. Tulsa, Oklahoma (1978).

14. S. Meyer, "Carbide-Recollections of the Early Years," S. Meyer, Charleston, WV, Dec (1979).

15. J. M. Winton, "Adequate Supply of Feedstocks Likely in 1980's," Chem. Week, April 16, P. 44 (1980).

SOME HISTORICAL NOTES ON THE USE OF

MATHEMATICS IN CHEMICAL ENGINEERING

        Arvind Varma

        Department of Chemical Engineering
        University of Notre Dame
        Notre Dame, Indiana 46556

    Chemical Engineering as a distinct discipline began in January 1888, when George E. Davis gave a series of twelve lectures on the subject at the Manchester Technical School. It was in 1880 that Davis had coined the word "chemical engineer," and promoted it (unsuccessfully) to found a society of chemical engineers. The first undergraduate chemical engineering curriculum was begun at MIT in Fall 1888, followed by those of Pennsylvania in 1892, Tulane in 1894, and Michigan in 1898. Fascinating accounts of early developments in the curriculum have recently been given by Aris (1977) and by Hougen (1977; the development of chemical technology was reviewed recently by Pigford (1976).

    Most early curricula had two years of college mathematics, but rarely was anything done beyond calculus. Formal chemical engineering courses until the appearance of Walker, Lewis and McAdams (1923) consisted largely of qualitative industrial chemistry. Although this influential book charted the course of chemical engineering for decades by unifying Arthur D. Little's 1915 "unit actions" concept into "unit operations," the use of mathematics in chemical engineering remained at what may now be called an elementary level.

    On a large scale, the situation did not change substantially until about the late fifties when three signal events occurred. One was the appearance of the famous book by Bird, Stewart and Lightfoot (1960), which had a tremendous impact on the curriculum by introducing quantitative mathematical treatment of chemical engineering problems. The other significant event was the general availability of computers, whereby it became feasible to carry out numerical simulation of process models to identify optimal design and operation conditions. The third event was the appearance of a person, Neal R. Amundson, then of Minnesota. Amundson had earned bachelor's and master's degrees in chemical engineering at Minnesota (his master's thesis was the first non-experimental one that had ever been accepted in that department), and obtained a Ph.D. (1945) in Mathematics also from Minnesota with a dissertation on the nonlinear diffusion equation of drying and shrinking of a gel. By the mid-50's, he had launched a thorough, sustained and systematic mathematical attack on understanding the behavior of many diverse chemical engineering systems, including separation processes and chemical reactors--which he continues to this day. Through the intelligent use of mathematical models, "No-one has made greater contributions to chemical engineering science than N. R. Amundson," as J. F. Davidson (1981) recently observed in reviewing a selection of his papers (Amundson, 1980).

On a smaller scale, however, there were many instances where 'sophisticated' mathematics was used in solving important chemical engineering problems. I shall identify some of these in the sequel. It might be worthwhile to first examine which were the books in applied mathematics available to early chemical engineers; a list is shown in Table 1.

The first, by J. W. Mellor--a distinguished inorganic chemist, who was to later author many standard reference books-- was written specifically for students of chemistry and physics. This 530-page book, along with differential and integral calculus, ordinary and partial differential equations, also contained some examples of interest to chemical engineers--such as in reaction kinetics, chemical equilibrium, and thermodynamics. In the preface, Mellor writes that "...I have sought to make clear how experimental results lend themselves to mathematical treatment. I have found by trial that it is possible to interest chemical students and to give them a working knowledge of mathematics by manipulating the results of physical or chemical observations..." He goes on to record indebtedness to his Professor of Chemistry, H. B. Dixon, whose encouraging letter is included.

It is in the topic of heat conduction and the related one of mass diffusion, that we see the first use of higher mathematics by chemical engineers. I shall discuss several specific contributions in a later section. Carslaw's 1906 book on "Fourier Series and Integrals, and the Mathematical Theory of the Conduction of Heat" was subsequently divided and published separately as two books in 1921. Among these, the one on heat conduction was particularly influential, and was the precursor of the now famous Carslaw and

Table 1. <u>Influential Early Books in Applied Mathematics</u>

1902 J. W. Mellor, "Higher Mathematics for Students of Chemistry and Physics--with Special Reference to Practical Work," Longmans, Green.

1906 H. S. Carslaw, "Fourier Series and Integrals, and the Mathematical Theory of the Conduction of Heat," Macmillan.

1913 L. R. Ingersoll and O. J. Zobel, "An Introduction to the Mathematical Theory of Heat Conduction," Ginn.

1921 H. S. Carslaw, "Fourier Series and Integrals," Macmillan.

1921 H. S. Carslaw, "Introduction to the Mathematical Theory of the Conduction of Heat in Solids," Macmillan.

1922 H. B. Phillips, "Differential Equations," John Wiley.

1934 I. S. and E. S. Sokolnikoff, "Higher Mathematics for Engineers and Physicists," McGraw-Hill.

1936 R. E. Doherty and E. G. Keller," Mathematics of Modern Engineering," Wiley, Volume I (1936); Volume II (1942).

1938 H. W. Reddick and F. H. Miller,"Advanced Mathematics for Engineers," Wiley.

1940 Th. von Karman and M. A. Biot," Mathematical Methods in Engineering," McGraw-Hill.

1941 H. S. Carslaw and J. C. Jaeger, "Operational Methods in Applied Mathematics," Oxford.

1941 R. V. Churchill, "Fourier Series and Boundary-Value Problems," McGraw-Hill.

1944 R. V. Churchill, "Modern Operational Mathematics in Engineering," McGraw-Hill.

1947 H. S. Carslaw and J. C. Jaeger, "Conduction of Heat in Solids," Oxford.

Jaeger (1947). Ingersoll and Zobel (1913) was on the same topic, and is noteworthy for it is the first in the area by American authors (both physicists at the University of Wisconsin) that influenced early chemical engineers.

The book by Phillips (1922) on differential equations is the first where a direct influence of chemical engineers on an applied mathematics book can be traced. In the preface, Phillips, a professor of mathematics at MIT, acknowledges that "...The problems have been collected from a variety of sources...the notes on Mathematics for Chemists prepared by Professors W. K. Lewis and F. L. Hitchcock."

The books of I. S. and E. S. Sokolinikoff (1934) and its later editions with Redheffer (1958), Doherty and Keller (1936,1942), Reddick and Miller (1938), von Karman and Biot (1940), and Churchill (1941, 1944) continued to be used actively till the early sixties. It is an example to modern administrators that Doherty was President of Carnegie Institute of Technology when his 1936 book appeared.

Three early books on applied mathematics for chemical engineers are shown in Table 2. The earliest is by F. L. Hitchcock—a professor of mathematics, and C. S. Robinson—a well-known professor of chemical engineering—both of MIT. It was based on a course given to chemical engineering students at that institution. Acknowledging their sources for problems, the authors write that "...a large number of them are due to Professor W. K. Lewis, who was the first to recognize the need of such a course..." They also acknowledge that "...Some of the simpler problems on rates of reaction have been adapted from 'Chemical Statics and Dynamics' by J. W. Mellor, a work to be commended as collateral reading." This is, of course, the same J. W. Mellor whose 1902 book was discussed earlier in Table 1.

The books of Sherwood and Reed (1934), and Marshall and Pigford (1947) were both also based on courses, given at MIT and the University of Delaware, respectively. Both still remain sources for many interesting and instructive problems. The former was revised in a popular second edition, prepared by Mickley (1957).

The first all-electronic digital computer (ENIAC) was built at the Moore School of Electrical Engineering, University of Pennsylvania in 1945. By 1951, a large-scale computer (UNIVAC), developed by Sperry Rand division of Remington-Rand Corporation was available (Williams, 1961). Compiling systems for easier programming (FORTRAN) were developed in 1957. The first book dealing specifically with digital computation for chemical engineering problems was that of Lapidus (1962). It came shortly after a Ford Foundation Project (Katz, 1960) on computers in engineering education was completed at the University of Michigan.

Table 2. Early Books in Applied Mathematics
for Chemical Engineers

1923 F. L. Hitchcock and C. S. Robinson, "Differential Equations in Applied Chemistry," John Wiley.

1939 T. K. Sherwood and C. E. Reed, "Applied Mathematics in Chemical Engineering," McGraw-Hill.

1947 W. R. Marshall, Jr. and R. L. Pigford, "The Application of Differential Equations to Chemical Engineering Problems," University of Delaware Press.

The first commercial analog computer (REAC) was built by the Reeves Instrument Company in 1948. Acrivos and Amundson (1953) were the first to use an analog computer to solve chemical engineering problems.

CHRONOLOGY OF COURSES IN APPLIED MATHEMATICS FOR CHEMICAL ENGINEERS

The first time a specialized applied mathematics course was taught to chemical engineering students was in the Fall semester, 1919 at MIT. The course was given as a service course by the Mathematics Group (not a separate department then) specifically to seniors in chemical engineering. It was out of this course that the book by Hitchcock and Robinson (1923) developed (see Table 2). Today, almost all undergraduate and graduate curricula have one or more courses devoted to the subject matter.

A chronology of the early graduate courses, titled along the lines of "Applied Mathematics for Chemical Engineers," is given in Table 3. The first course was that of Sherwood at MIT, first taught in Fall 1935. Amundson's course at Minnesota began in Fall 1943 as a one-quarter offering, and became a full-year sequence in 1947; he taught it continuously till he left the university in 1977. Amundson was completing his doctorate and was an Instructor when he first gave the course to a group of chemical engineering graduates enrolled in the Army Specialized Training Program (ASTP). The course taught by Marshall and Pigford at Delaware also has a World War II connection. It was first taught in an evening program when veterans were returning from the War, and many were eager to refresh their engineering background by enrolling in graduate classes. Both instructors were employed with duPont at the time, having joined T. H. Chilton's group on the same day (September 1, 1941).

Table 3. <u>Applied Mathematics in Chemical Engineering
Chronology of Graduate Courses</u>

1. MIT　　　　　Fall 1935　　T. K. Sherwood
    "Analytical Treatment of Chemical Engineering Processes"

2. Minnesota　　Fall 1943　　N. R. Amundson
    "Advanced Mathematics for Chemical Engineers"

3. Delaware　　 Fall 1945　　W. R. Marshall, Jr. and R. L. Pigford
    "Application of Differential Equations to Chemical Engineering Problems"

4. Columbia　　 1945　　　　 T. B. Drew

5. Wisconsin　　Spring 1948　W. R. Marshall, Jr.
    "Chemical Engineering Problems"

6. Michigan　　 Winter 1948　J. J. Martin
    "Advanced Mathematics for Chemical Engineers"

7. Yale　　　　 Fall 1948　　R. H. Bretton and R. Southworth
    "Applied Mathematics for Chemical Engineers"

8. Notre Dame　 Fall 1949　　M. Howerton
    "Mathematical Methods in Chemical Engineering"

It should be mentioned here that availability of textbooks (Tables 1 and 2) was a major factor in these courses getting started. Other than Neal Amundson, at least one other prominent chemical engineer was sufficiently interested in applied mathematics to take a formal degree in the subject. S. W. Churchill (1978) writes that while an undergraduate at Michigan during 1938-42, he was so influenced by his freshman mathematics teacher that he ended up getting a B.S. degree in Engineering Mathematics as well as in Chemical Engineering.

I shall now devote the rest of this paper to some specific topics, and indicate the early uses of mathematics by chemical engineers in solving and understanding important problems. I should mention here that entirely due to my own limitations, I have largely reviewed only the English literature.

HEAT CONDUCTION IN SOLIDS

In Table 4, I have presented some of the earliest works in the area of heat conduction in which chemical engineers made a unique contribution. The earliest such work that I was able to trace is that of Gurney and Lurie (1923). For one-dimensional geometries (slab, cylinder, and sphere), with or without surface resistance, they computed results in terms of dimensionless variables and presented them graphically to indicate the unaccomplished temperature-change ratio as a function of time and position within the body--initially at a uniform temperature, when a step or a linearly varying change is made in the temperature of the surrounding media. They did not report either the mathematical solutions or any references for the solution; one has to assume that the source is the books of Carslaw (1906, 1921), and Ingersoll and Zobel (1913)--given in Table 1.

Their graphical results, later appropriately called Gurney-Lurie charts in books such as McAdams (1933), were presented in form of dimensionless quantities so as to be useful for many situations. [It is worth mentioning that just a few years before the Gurney and Lurie paper, physicists Williamson and Adams (1919) had reported rather detailed solutions for various geometries, and also had noted the existence of dimensionless-position and time variables. They remark that "The numerical computations can be applied to other cases with a very little manipulation owing to the fact that the only physical constant used (κ, the so-called diffusivity constant), occurs, at least when it occurs in any complicated term, multiplied by the time. Hence the values of these terms for different substances are the same at equal values of the product of the time by the diffusivity constant." The dimensionless graphical charts of Williamson and Adams, however, are not as comprehensive as those of Gurney and Lurie; they also did not consider thermal resistance at the surface, as Gurney and Lurie did.] To illustrate use of their charts, Gurney and Lurie presented three examples dealing with heating and cooling of a slab of rubber, a steel ingot, and a slab of glass. Both authors at the time were employed with the Boston Belting Company. In the abstract, they expressed the hope that their "...graphical presentation of the formulas of heat diffusion may help to realize Fourier's wish (1812)...that his mathematics be applied to industry."

G. G. Brown and C. C. Furnas (1926) presented an ingenious graphical solution of the heat conduction equation in a cylinder, when the thermal properties are functions of temperature. They also indicated how the method could be used to evaluate heats of phase transition. Experimental use of the method was made to determine heat capacity of powdered alumina by Newman and Brown (1930), in a paper prepared from Newman's Ph.D. Thesis research.

The remaining five papers in Table 4 are of the prodigious A. B. Newman, who appears to be the first chemical engineer to solve a variety of new heat conduction problems. At the time, Newman was a Professor of Chemical Engineering at Cooper Union in New York City,

Table 4. Heat Conduction in Solids

1923 H. P. Gurney and J. Lurie, "Charts for Estimating Temperature Distributions in Heating or Cooling Solid Shapes," Ind. Eng. Chem., $\underline{15}$, 1170-72.

1926 G. G. Brown and C. C. Furnas, "Heat Conduction of Solids," Trans. AIChE, $\underline{18}$, 295-307.

1930 A. B. Newman, "Temperature Distribution in Internally Heated Cylinders," Trans. AIChE, $\underline{24}$, 44-53.

1934a A. B. Newman, "The Temperature-Time Relations Resulting From the Electrical Heating of the Face of a Slab," Trans. AIChE, $\underline{30}$, 598-612.

1934b A. B. Newman and L. Green, "The Temperature History and Rate of Heat Loss of an Electrically Heated Slab," Trans. Electrochem. Soc., $\underline{66}$, 345-58.

1935 A. B. Newman and A. H. Church, "The Temperature History During the Constant-Rate Heating of a Solid Circular Cylinder," J. Appl. Mech., $\underline{2}$, A-96--A-98.

1936 A. B. Newman, "Heating and Cooling Rectangular and Cylindrical Solids," Ind. Eng. Chem. $\underline{28}$, 545-548.

---

and was later President of AIChE in 1948. Professor C. F. Bonilla of Columbia recalls (1980) that in the early 1930s, Newman regularly held seminars on topics in applied mathematics which chemical engineers from other universities and industry in New York City attended. His biographical listing in Who's Who in Engineering--1948 reveals that he not only published on heat conduction and mass diffusion problems, but also wrote a paper titled "Restoring the Panama Mural Paintings"--a reference to which I have not been able to find.

Written discussion by C. C. Furnas at the end of Newman's paper (1930) concludes with the prescient comment that "Mr. Newman is to be commended on his execution of the work on this particular heat transfer problem, and I wish he might extend his energies to some of the other difficult problems in this field." In the discussion, Furnas also discloses that "I have recently developed a method of computation...and have reported the method and results..." for evaluating modified Bessel functions of the first kind and zero order (Furnas, 1930)--this paper may well be the first by a chemical engineer in a mathematics journal.

MASS DIFFUSION IN SOLIDS

In Table 5 are shown some of the earliest papers by chemical engineers in the area of mass diffusion in solids, an area closely related to that of heat conduction in solids. All of these early papers dealt with the problem of drying of solids. The very first is by W. K. Lewis (1921), "considered by many to have been the father of the (chemical engineering) profession in the U.S." (Pigford, 1976). Lewis states that "The drying of a solid of necessity involves two independent processes, first, the evaporation of the moisture from the surface of the solid, and, second, the diffusion of the moisture from the interior of the solid out to the surface." He goes on to write the proper diffusion equation, but admits that "the exact integration of this equation is, however, so involved that we have chosen to integrate it by approximation," treating the moisture content within the solid slab to be a straight line--although he realized that "this line will (really) not be straight." In the paper, remarkable for clarity of exposition, Lewis goes on to consider special cases in which surface evaporation or diffusion are limiting factors, provides useful formulae, and interprets a rather large amount of experimental data--also commenting at a point that "the graphical determination of slopes is always inaccurate, and it is far more satisfactory to use the integrated expression."

T. K. Sherwood took up the problem in much more detail, in a seven-part series titled "The Drying of Solids," of which the first three were from his 1928 MIT doctoral thesis. There was, however, before Sherwood's, a fascinating paper of F. Tuttle (1925) which "advanced a mathematical theory of the drying of wood, based by analogy on the Fourier heat-conduction analysis." Tuttle was a physicist at the University of Wisconsin, and was suggested the analogy by a W. Karl Loughborough of the Forest Products Laboratory, with whose cooperation "experimental data in support of the theory" were obtained. Tuttle writes the diffusion equation for mositure content as a function of position and time, and credits Ingersoll and Zobel (1913) as source for his solutions. Using a single experimental observation (moisture content at a specific location within a 2-inch Sitka spruce plank after drying had continued for a specific length of time), he computed the value of "transfusivity" (i.e., mass diffusivity), and then compared experimental moisture contents at six different locations for five different durations of exposure with results of theoretical predictions from the diffusion equation. The agreement between theory and experiment was excellent.

In his seven-part series, Sherwood analyzed various aspects related to the drying of solids, and clarified drying mechanisms. In 1929b, he explicitly stated the general mechanisms of drying as three:

"I-Evaporation of the liquid at the solid surface; resistance to internal diffusion of liquid small as compared with the resistance to removal of vapor from the surface.

II-Evaporation at the solid surface; resistance to internal diffusion of liquid great as compared with the resistance to removal of vapor from the surface.

III-Evaporation in the interior of the solid; resistance to internal diffusion of the liquid great as compared with the total resistance to removal of vapor."

Under constant ambient conditions, very wet solids lose moisture by the first mechanism, in a constant rate-of-drying period. As drying proceeds, comes the falling rate period governed by mechanisms II and III."

Table 5. Mass Diffusion in Solids

1921   W. K. Lewis, "The Rate of Drying of Solid Materials," Ind. Eng. Chem., 13, 427-432.

1929a   T. K. Sherwood, "The Drying of Solids-I," Ind. Eng. Chem., 21, 12-16.

1929b   T. K. Sherwood, "...-II," Ind. Eng. Chem., 21, 976-80.

1930   T. K. Sherwood, "...-III, Mechanism of the Drying of Pulp and Paper," Ind. Eng. Chem., 22, 132-6.

1931a   A. B. Newman, "The Drying of Porous Solids: Diffusion and Surface Emission Equations," Trans. AIChE, 27, 203-16.

1931b   A. B. Newman, "...: Diffusion Calculations," Trans. AIChE, 27, 310-33.

1932   T. K. Sherwood, "...-IV. Application of Diffusion Equations," Ind. Eng. Chem., 24, 307-10.

1933a   T. K. Sherwood and E. W. Comings, "...-V. Mechanism of Drying of Clays," Ind. Eng. Chem., 25, 311-6.

1933b   E. R. Gilliland and T. K. Sherwood, "...-VI. Diffusion Equations for the Period of Constant Drying Rate," Ind. Eng. Chem., 25, 1134-6.

1934   E. W. Comings and T. K. Sherwood, "...-VII. Moisture Movement by Capillarity in Drying Granular Materials," Ind. Eng. Chem., 26, 1096-8.

Aware of Tuttle's work, Sherwood (1929a) considered the second mechanism in detail with the diffusion equation. He felt that it was better to obtain an integrated expression for the moisture content in the entire solid as a function of time, and compare this with experimental results. This expression is not linear with dimensionless time on a semi-logarithmic paper. With typical ingenuity, "a special plotting paper was constructed, using a uniform abscissa (time) scale but so changing the ordinate scale so as to force the theoretical relation to be a straight line." He reanalyzed Tuttle's data, and compared his own experimental data for wood slabs of two different thickness with theoretical results. He also reported experiments for drying of soap, and observed that the data on the special plotting paper were no longer on a straight line but rather on a curve, which is "due to the decrease in the diffusion coefficient K as the drying proceeds...The decrease in K is undoubtedly connected with the shrinkage of the soap."

Sherwood's 1932 paper analyzed the constant rate-of-drying period, and using ingenious intuitive arguments supported by experiments, showed that "after drying an infinite slab for an appreciable length of time at a constant rate, the moisture-gradient curve approaches a parabolic form, no matter what the initial moisture gradient may have been." This argument was made rigorous in part VI of the series, with E. R. Gilliland (1933b).

The two papers of A. B. Newman shown in Table 5 developed equations governing diffusion-controlled drying for solids of various shapes. Newman's (1931b) remark "Many engineers are disinclined to accept equations on faith, and for their benefit, the derivations have been given in full in such a way as to be intelligible to anyone having a working knowledge of the ordinary differential and integral calculus" is a comment on the mathematical background of chemical engineers of the time.

HEAT TRANSFER--OTHER

Some other works in heat transfer through which chemical engineers contributed in a mathematical way to solving important problems are listed in Table 6. I have divided this topic into two categories.

The first category is that of fluid-solid heat transfer, in which the papers of Schumann (1929) and Furnas (1930) are the foremost. Both address the problem of heat transfer between a flowing fluid stream and a bed packed with solid particles. Schumann's paper is quoted to this day. He formulated and solved the simplest model possible for this case -- involving plug flow of the fluid in

Table 6. Heat Transfer-Other

A. Fluid-Solid

1929 T. E. W. Schumann, "Heat Transfer: A Liquid Flowing Through a Porous Prism," J. Franklin Inst., 208, 405-16.

1930 C. C. Furnas, "Heat Transfer from a Gas Stream to a Bed of Broken Solids," Trans. AIChE, 24, 142-86.

B. Heat Exchangers

1931 T. B. Drew, "Mathematical Attacks on Forced Convection Problems: A Review," Trans. AIChE, 26, 26-79.

1933 A. P. Colburn, "Mean Temperature Difference and Heat Transfer Coefficient in Liquid Heat Exchangers," Ind. Eng. Chem., 25, 873-7.

1934 A. P. Colburn, "Calculation of Condensation with a Portion of Condensate Layer in Turbulent Motion," Ind. Eng. Chem., 26, 432-4.

1934 A. P. Colburn and O. A. Hougen, "Design of Cooler Condensers for Mixtures of Vapors with Noncondensing Gases," Ind. Eng. Chem., 26, 1178-82.

---

the axial direction, no temperature gradient within the solid particles, and Newtonian heat transfer between the fluid and the solid. Equations describing the model, when transformed into a reference frame moving with the fluid velocity, lead to a pair of coupled first-order partial differential equations (for the fluid and solid temperatures, as a function of position and time), which were solved in terms of a series of Bessel functions.

Schumann did not compare theoretical results with experimental data, as his "main purpose being to present the mathematical treatment of the problem." Comparison with experiments was made by Furnas (1930) who noted that "It is possible that the experimental difficulties would have proved insurmountable had it not been for a very important theoretical paper recently published by Schumann."

The second category is that of heat exchangers. Drew's 1931 paper was a monumental review of the literature dealing with heat exchange between a solid surface and a fluid flowing in forced convection. It served admirably to make the chemical engineer aware, in a comprehensive way, of developments in the area by illustrious contributors, primarily from physics, fluid mechanics, and applied mathematics.

Of the three other papers that I have chosen to mention in this category, two are by A. P. Colburn and the third by Colburn and O. A. Hougen. The first (1933) provided a simple relationship for the rate of heat transfer in a countercurrent flow heat exchanger, when the overall heat transfer coefficient varies (linearly) with temperature. In the 1934 paper, Colburn provided an analytic expression for heat transfer (for condensing vapor) in a vertical condenser where the condensate layer near the lower end of the condenser becomes so thick that it is in turbulent motion. The corresponding problem for laminar flow had been treated earlier by Nusselt in 1916. The paper of Colburn and Hougen (1934) was concerned with the design of heat exchangers when the condensing vapor contains some noncondensible gases; the design procedure required a graphical integration.

Although later in time, the distinguished efforts of S. W. Churchill and colleagues (e.g., Martini and Churchill, 1960; Hellums and Churchill, 1961) in solving a variety of natural convection problems by finite-difference methods should also be mentioned here, as examples of some of the early uses of computers in solving chemical engineering problems.

FLUID FRICTION-HEAT TRANSFER-MASS TRANSFER ANALOGY

The two influential papers on the analogy of transport processes are indicated in Table 7. They deal with correlating data, and are themselves not mathematical. However, as is well-known as a result of these papers, the concept that heat transfer and mass transfer data can be correlated with fluid friction data, owes to Reynolds' analogy--which is a mathematical argument. Interesting details on historical development of the concept were given by Colburn in the Appendix of his 1933 paper. The situation in correlating heat transfer data at the time was apparently so bad that O. A. Hougen was led to remark in the written discussion following Colburn (1933) that "...it would be well for technical periodicals to restrict for publication only the experimental work of those who have had extensive experience in this type of work."

Reynolds' analogy was only between fluid friction and heat transfer. It was the highly original contribution of Chilton and Colburn (1934) to extend it to mass transfer.

DISTILLATION

The topic of distillation serves as a convenient one to discuss the mathematical contributions of chemical engineers in the general area of separation processes (see Lewis, 1959 for an interesting perspective on the evolution of unit operations). I have divided this topic in two, dealing separately with binary (Table 8) and multicomponent (Table 9) mixtures, respectively.

Binary Mixtures

The earliest contribution to distillation theory is that of Sorel (1889). In a series of papers, Sorel wrote the general mass and energy balance equations in multiplate columns, and also assumed that equilibrium was realized between the vapor and liquid streams leaving a plate. His method involved algebraic plate-by-plate calculation, successively applied starting at the top plate. The computation was, of course, tedious. The work of Hausbrand (1893) was clearer in presentation, but the basic method was the same as that of Sorel; both also dealt with alcohol-water systems, because these mixtures were the only ones for which equilibrium data were then available.

Ponchon (1921) observed that for adiabatic columns, straight operating lines would result on an enthalpy-composition diagram. His graphical method was simplified and applied extensively by Savarit (1923) to the distillation of alcohol-water mixtures. As is well-known, the method is now called the Ponchon-Savarit method, and is still useful as the most general method for binary distillation (cf. Treybal 1980).

Lewis (1922) was the first to make the constant molal overflow approximation; this when invoked, eliminates the need for enthalpy balances, and only material balances are then required from plate-to-plate. He, however, assumed that enrichment per plate was small, which permitted him to convert the plate-to-plate difference equation into a differential equation; this he graphically integrated to determine the number of plates. Because of replacing differences by differentials, the larger the number of plates, the more accurate is the method.

McCabe and Thiele (1925) realized that when the number of plates in a column are relatively small, with Lewis' assumption "the error introduced by assuming continuous for stepwise conditions is appreciable." They showed that with the constant molal overflow assumption, it is possible to graphically draw "on the same rectangular plot, the equilibrium curve for vapor and liquid compositions and straight lines representing the equations for enrichment from plate-to-plate, and passing from one to the other in a series of steps."

Table 7. Fluid Friction-Heat Transfer-Mass Transfer Analogy

1933   A. P. Colburn, "A Method for Correlating Forced Convection Heat Transfer Data and a Comparison with Fluid Friction," Trans. AIChE, $\underline{29}$, 174-210.

1934   T. H. Chilton and A. P. Colburn, "Mass Transfer (Absorption) Coefficients: Prediction from Data on Heat Transfer and Fluid Friction," Ind. Eng. Chem. $\underline{26}$, 1183-7.

Table 8. Distillation--Binary Mixtures

1889   E. Sorel, "Sur la Rectification de l'Alcool," Comp. Rend., $\underline{108}$, 1128-31, 1204-7, 1317-20.

1893   E. Hausbrand, "Die Wirkungsweise der Rectificer-und Destillir-Apparate," Springer.

1921   M. Ponchon, "Étude Graphique de la Distillation Fractioneé Industrielle," La Tech. Moderne, $\underline{13}$, 20, 55.

1922   W. K. Lewis, "The Efficiency and Design of Rectifying Columns for Binary Mixtures," Ind. Eng. Chem., $\underline{14}$, 492-7.

1923   R. Savarit, "Étude Graphique des Colonnes á Distiller les Mélanges Binaires et Ternaires," Chemie et Industrie, Special Number, May 1923; pg. 737.

1925   W. L. McCabe and E. W. Thiele, "Graphical Design of Fractionating Columns," Ind. Eng. Chem. $\underline{17}$, 605-11.

1928   K. Thormann, "Destillieren und Rektifiezieren," Leipzig.

1932   A. J. V. Underwood, "The Theory and Practice of Testing Stills," Trans. Inst. Chem. Engrs., $\underline{10}$, 112-52.

1938   E. H. Smoker, "Analytic Determination of Plates in Fractionating Columns," Trans. AIChE, $\underline{34}$, 165-72.

1944   F. M. Tiller and R. S. Tour, "Stagewise Operations: Application of the Calculus of Finite Differences to Chemical Engineering," Trans. AIChE, $\underline{40}$, 317-31.

1946   N. R. Amundson, "Application of Matrices and Finite Difference Equations to Binary Distillation," Trans. AIChE, $\underline{42}$, 939-46.

Near the ends of the composition range, the equilibrium curve may be considered a straight line. Thormann (1928) showed that with straight operating lines, an analytic expression can be obtained for the number of plates in this region.

For binary systems with constant relative volatility, Underwood (1932) developed an analytic expression for the number of theoretical plates in the case of extremely pure separations. For constant relative volatility, he also developed exact analytic expressions for minimum reflux and for minimum number of plates at total reflux; Fenske (1932) had developed the same expressions just a few months earlier.

Smoker (1938) noted that the MaCabe-Thiele method, being a graphical one, "is quite tedious and almost impractical to apply to systems involving a large number of theoretical plates." With constant molal overflow and constant relative volatility assumptions, by algebraic iteration he obtained an analytic result for the number of plates--with no restrictions on product purity, as Underwood (1932) had placed. His work is the first where the number of plates was determined explicitly, completely analytically.

Tiller and Tour (1944) were the first to solve various separation problems, including distillation, using the calculus of finite differences. They noted that "Because of the directness and generality of attack, the methods of finite differences offer distinct advantages...in solving the problems arising in stagewise processes."

N. R. Amundson (1946), in his very first published paper, was the first to use elegant matrix methods for distillation, which he systematically utilized later to explore steady state and transient behavior of other separation and reaction processes (Lapidus and Amundson 1950, Acrivos and Amundson 1955). He stated that although "The result obtained is similar to that of Smoker but the method of solution is interesting in that the use of matrices simplifies a repititive calculation considerably." In the chemical engineering literature at that time, it was necessary to "list without proofs a few of the facts about matrices." The paper, remarkable for its preciseness which is his hallmark, concludes by noting that "The application of matrices to such a problem and the occurrence of a difference equation in matrices in a physical problem is unknown also to the author although the theory of functions defined on matrices is well known in pure mathematics."

Multicomponent Mixtures

Considerations of multicomponent mixtures were motivated by the distillation of petroleum. The earliest work appears to be that

MATHEMATICS IN CHEMICAL ENGINEERING

of Lewis and Robinson (1922), who treated simple (or differential) distillation of mixtures containing "a very large number of components,...whose boiling points as function of their molecular weight may be indicated at least approximately by a curve." Assuming Raoult's law to describe vapor-liquid compositions, they utilized a combined analytic-graphical method to determine the so-called Engler curve (or boiling-point curve) which relates the percent of original mixture distilled with the vapor temperature. It is of historical interest that both this and Lewis' 1922 paper on binary mixtures were given at a 1922 ACS symposium on distillation.

Robinson was also the author of the first English language book on distillation (1922), and is the same one who coauthored "Differential Equations in Applied Chemistry" discussed in Table 2.

Most of the other papers to which I refer deal with continuous rectification of multicomponent mixtures. The basic problem with such systems, as B. D. Smith (1963) points out, is that the specification of a relatively small number of variables causes a distillation column to produce a unique set of stream flowrates, compositions, and temperatures. Calculation methods must therefore predict the unique set of conditions which exist in the column under any desired set of specified variables.

Lewis and Wilde (1928) suggested an extension of the Sorel-Hausbrand method for binary systems, by writing algebraic plate-to-plate balances combined with the use of Raoult's Law and the boiling-point curve. The method consisted of breaking up the boiling-point curve of the feed into fractions boiling within $\sim 10°F$ temperature ranges, and treating these 'cuts' as individual components. They also introduced the idea of 'key' components, which define the fractionation. Fenske (1932) defines this idea rather clearly, by the use of which he derived analytic equations for the minimum number of plates and for minimum reflux. He observed that for these extremes of column operation, "the separation of a complex mixture may be treated as the separation of a simple binary mixture of the key components."

In two papers with Matheson (1932) and Cope (1932), Lewis furthered his earlier work with Wilde; the latter extended the McCabe-Thiele graphical method for binary mixtures to multicomponent ones, by constructing a separate x-y plot for each component. The graphical method "not only shortens computation but also enables one to visualize far more clearly what is happening in the middle of the column." The papers of Fenske, Lewis and Matheson, and Cope and Lewis, were all presented at an ACS symposium on chemical engineering processes in the oil industry.

Table 9. Distillation--Multicomponent Mixtures

1922	W. K. Lewis and C. S. Robinson, "The Simple Distillation of Hydrocarbon Mixtures," Ind. Eng. Chem., 14, 481-4.
1928	W. K. Lewis and H. D. Wilde, Jr., "Plate Efficiency in Rectification of Petroleum," Trans. AIChE, 21, 99-126.
1932	M. R. Fenske, "Fractionation of Straight-Run Pennsylvania Gasoline," Ind. Eng. Chem., 24, 482-5.
1932	W. K. Lewis and G. L. Matheson, "Studies in Distillation: Design of Rectifying Columns for Natural and Refinery Gasoline," Ind. Eng. Chem., 24, 494-8.
1932	J. Q. Cope, Jr. and W. K. Lewis, "Studies in Distillation: Graphical Method for Computation for Rectifying Complex Hydrocarbon Mixtures," Ind. Eng. Chem., 24, 498-501.
1933	E. W. Thiele and R. L. Geddes, "Computation of Distillation Apparatus for Hydrocarbon Mixtures," Ind. Eng. Chem., 25, 289-95.
1934	G. G. Brown, M. Souders, et al., "Fractional Distillation, I-III," Trans. AIChE, 30, 438-56, 457-76, 477-503.
1940	E. R. Gilliland, "Multicomponent Rectification: Estimation of the Number of Theoretical Plates as a Function of the Reflux Ratio," Ind. Eng. Chem., 32, 1220-3.
1948	A. J. V. Underwood, "Fractional Distillation of Multicomponent Mixtures," Chem. Eng. Prog., 44, 603-13.
1949	J. R. Bowman, "Distillation of an Indefinite Number of Components," Ind. Eng. Chem., 41, 2004-7.
1955	A. Acrivos and N. R. Amundson, "On the Steady State Fractionation of Multicomponent and Complex Mixtures in an Ideal Cascade," Chem. Eng. Sci., parts I-VI, 4, 29-38, 68-74, 141-8, 159-66, 206-8, 249-54.
1958	N. R. Amundson and A. J. Pontinen, "Multicomponent Distillation Calculations on a Large Digital Computer," Ind. Eng. Chem., 50, 730-6.
1958	A. Rose, R. F. Sweeney and V. N. Schrodt, "Continuous Distillation Calculations by Relaxation Method," Ind. Eng. Chem., 50, 737-40.
1958	J. Greenstadt, Y. Bard and B. Morse, "Multicomponent Distillation Calculation on the IBM 704," Ind. Eng. Chem., 50, 1644-7.

Thiele and Geddes (1933) developed a plate-to-plate method, which simplified and accelerated Lewis-Matheson type computation. They also treated other cases involving infinite number of components, side-stream stripping, and multiple feeds. Brown and Souders (1934) suggested the use of an "absorption factor" method, that Kremser (1930) had then recently developed in the context of gas absorption, to distillation. In this method the column was divided into many sections over which constant vapor and liquid flowrates could be assumed. Treating each component at a time, they developed equations for the number of theoretical plates required in the section to provide the desired change in terminal conditions of an individual component.

Gilliland (1940) provided a useful correlation relating the minimum number of theoretical plates and minimum reflux to the actual number of theoretical plates and actual reflux required to produce a desired separation. The correlation was developed by performing calculations on eight different multicomponent systems, with various reflux ratios.

Underwood (1948) was the first to provide rigorous analytic equations which gave compositions on any plate in the column, under the assumptions of constant molal overflow and constant relative volatilities.

Bowman (1949), in a unified theory, treated distillation of an indefinite number of components, ranging from finite to infinite. He denoted compositions by functions of the form $x(\alpha)$, where α is the relative volatility, and the function x having the property that $x.d\alpha$ represents the amount of material with relative volatility between α and $\alpha + d\alpha$. He thus treated the case of total and minimum reflux, and showed column performance in these cases. For simple distillation, he also derived explicit expressions for batch and distillate composition as a function of fraction distilled.

In a six-part series, Acrivos and Amundson (1955) gave closed form analytic solutions for the fractionation of mixtures having finite or infinite number of components. The case treated was that of constant relative volatilities and constant reflux flow rate. For finite number of components, the method of solution was an elegant separation of variables for the partial difference equation representing composition of component i on stage n, while the solution for infinite number of components was performed by means of a novel integral transform. Formulae for minimum reflux were given in both cases. A numerical example was "considered in some detail since these calculations are not of a routine nature," solved using an IBM 602A Calculating Punch. Expensions to packed columns, and using perturbation methods to cases where small deviations from constant relative volatility and constant reflux occur, were also given.

All-in-all, the series was a frontal mathematical attack on "a well-known problem." It is worth noting that more recently, Ramkrishna and Amundson (1973) showed that the Acrivos problem, "when cast in the proper mathematical framework" utilizing an ingenious inner product, becomes self-adjoint, and hence "amenable to a more direct solution."

Noting that "The use of large digital computers for distillation calculations has not been investigated to date, although the high speed of computation seems to offer economies and present the opportunity of making calculations not otherwise possible," Amundson and Pontinen (1958) performed such calculations on a Univac 1103. The material and energy balance equations for the whole column were solved simultaneously by an iterative procedure. Greenstadt et al. (1958) used an IBM 704, while Rose et al. (1958) used lengthier relaxation calculations where the steady state result is calculated as a gradual change from initial startup. The Amundson-Pontinen and Rose papers, presented as part of a 1957 ACS symposium on Application of Machine Computation to Petroleum Research, heralded the coming era of computer-aided process design and analysis.

CHEMICAL AND CATALYTIC REACTORS

This subject is rather broad in scope, and so in the hope of keeping these notes within reasonable bounds, I have limited consideration to a relatively few topics within the subject matter. I shall discuss four specific topics here: diffusion-reaction in catalyst pellets (Table 10), empty and packed-bed tubular reactors (Table 11), chemical reactor stability and control (Table 12), and reactor optimization (Table 13). In addition, I have created a reactors-other category (Table 14) to discuss certain contributions which are too important mathematically to overlook, but which do not fit into the other categories.

With the mention of chemical and catalytic reactors, one has to, at least in the English language, immediately give credit to the pioneering textbook of Hougen and Watson (1947). It was the first organized material dealing with many aspects of the subject, and so far-reaching was its influence that the origin of reaction engineering courses at most universities can be traced to the appearance of this book; today, every undergraduate and graduate curriculum in chemical engineering contains at least one course in the subject.

Diffusion-Reaction in Catalyst Pellets

The historical development of this topic has been treated by far abler pens than mine (Thiele, 1967; Aris, 1974), and everything about the mathematical theory of the topic (including related problems in biological areas--to which Weisz, 1973 has also appealed in

Table 10. Diffusion-Reaction in Catalyst Pellets

1909	F. Jüttner, "Reaktionskinetik und Diffusion," Z. für Phys. Chem., $\underline{65}$, 595-623.
1939	E. W. Thiele, "Relation Between Catalytic Activity and Size of Particle," Ind. Eng. Chem., $\underline{31}$, 916-20.
1949	E. Wicke and W. Brötz, "Diffusion, strömung und Reaktionsgeschwindigkeit im Innern poröser Kontaktkörper," Chem. Ing. Tech., $\underline{21}$, 219-26.
1951	A. Wheeler, "Reaction Rates and Selectivity in Catalyst Pores," Adv. in Catalysis, $\underline{3}$, 249-327.
1957	R. Aris, "On Shape Factors for Irregular Particles," Chem. Eng. Sci., $\underline{6}$, 262-8.
1958	C. D. Prater, "The Temperature Produced by Heat of Reaction in the Interior of Porous Particles," Chem. Eng. Sci., $\underline{8}$, 284-6.
1961	R. E. Schilson and N. R. Amundson, "Intraparticle Diffusion and Conduction in Porous Catalysts," Chem. Eng. Sci., $\underline{13}$, 226-36, 237-44.
1962	P. B. Weisz and J. S. Hicks, "The Behavior of Porous Catalyst Particles in View of Internal Mass and Heat Diffusion Effects," Chem. Eng. Sci., $\underline{17}$, 265-75.

interesting remarks about the interdisciplinary nature of the problem) is available in Aris' two-volume treatise (1975). I shall therefore keep the discussion brief.

The basic problem of simultaneous diffusion and reaction in catalyst pellets is that the actual reaction rate attained by the pellet can be different from the intrinsic reactivity of the catalyst, by a factor commonly denoted by η--the effectiveness factor. This feature was brought to attention by E. W. Thiele (1939) who noted that "... In general, it appears to be tacitly assumed by workers in this field that the reacting fluid penetrates to the pores in the interior of the grains and maintains substantially a constant composition throughout all the pores of a single grain, which is the same as the composition of the bulk of the fluid bathing the grain at the time. ... Qualitatively, however, it is evident that the size of the grains cannot be indefinitely increased without ultimately reaching a point at which the reaction will produce products in the interior of the grain faster than diffusion can carry them away. The reaction will then tend to be confined to the outer layers of the grain, the interior being relatively inactive. As the grain size is further increased, the catalytic activity will tend to become proportional to the external surface of the grains (or lumps). There appears to be little or no published information on this point." He went on "to treat the matter mathematically," and showed that a dimensionless quantity called 'modulus' by him (and very appropriately the Thiele modulus subsequently) governs "the effect of varying grain size on activity." At about the same time, Damköhler (1937) in Germany and Zel'dovich (1939) in Russia had independently obtained similar results, although Thiele's work was more comprehensive and also included plots of the effectiveness factor as a function of the modulus.

As Thiele (1967) has noted, there was as early as 1899, a German patent with implicit reference to the possibility of diffusional limitation in catalysis. However, Wicke and Brötz (1949) were the first to make experiments on two sizes of catalyst pellets, with knowledge of the theory and intention to test it.

All along while this was going on, there had already been in the literature, a very detailed paper by the physicist Jüttner (1909) who treated the mathematical theory of diffusion and reaction in catalyst pellets in far greater depth than the three investigators in the thirties. This paper had been completely overlooked until it was discovered by R. L. Gorring of Mobil in 1971.

In a comprehensive review, Wheeler (1951) treated the influence of diffusion on reactions involving selectivity, and also on catalyst poisoning.

The issue of the catalyst pellet temperature being different from that of the surrounding bulk had by this time not been raised in a major way; Prater (1958) was the first to show by calculation that for exothermic reactions, this difference could be significant. Schilson and Amundson (1961) were the first to make calculations for the radial temperature profile within a spherical pellet, and also to report effectiveness factors greater than one--since for normal reactions, although the reduction in reactant concentration within the pellet due to diffusion reduces the actual reaction rate relative to the intrinsic one, "the high temperature within the particle produces an abnormally high rate of reaction." This same feature was also reported almost simultaneously by Carberry (1961), Tinkler and Metener (1961), and Tinkler and Pigford (1961). Only another year had passed before Weisz and Hicks (1962) reported comprehensive numerical solutions of the nonisothermal diffusion-reaction problem, and also demonstrated the existence of multiple steady states.

Aris (1957) gave a normalization whereby for first-order reactions, the effectiveness factor--Thiele modulus curves were shown to

MATHEMATICS IN CHEMICAL ENGINEERING

lie very close to each other for pellets of any shape. Almost simultaneously and independently, Petersen, Aris, and Bischoff provided the corresponding normalization for arbitrary reactions in 1965.

Empty and Packed-Bed Tubular Reactors

At least in this country, R. H. Wilhelm is credited with the first systematic studies of tubular reactors--although Damköhler (1937) had conducted studies in Germany earlier. In his 1943 paper, Wilhelm wrote and solved the differential equation for radial temperature distribution at any axial position, and using this developed charts through which temperature and concentration profiles within the reactor could be calculated by a "step-by-step analysis in the direction of gas flow from one cross section to the next with the aid of heat and material balances." A simplifying assumption that the Arrhenius temperature dependence of the rate constant could be approximated by a term linear in temperature made the analytical solution possible. Amundson was to later use this same approximation in a series of three papers (1956) to analyze steady state and transient behavior of fixed-and moving-bed reactors in a comprehensive way, using the method of generalized Finite Fourier Transforms which "has an automatic quality that the separation of variables seems to lack." Similar but simpler problems had been solved earlier by Singer and Wilhelm (1950).

Very soon after the 1943 paper of Wilhelm, Hulburt developed equations for isothermal homogeneous (1944) and heterogeneously catalyzed (1945) reactions in tubular reactors. He included axial diffusion along with convection in the 1944 paper, and for this, although he arrived at the correct boundary condition at the reactor outlet, his inlet condition was incorrect (I shall have more to say about this later). By obtaining the analytic solution for first-order reactions, he correctly concluded that "diffusion shortens the contact time and leaves more unreacted material in the outlet gas than when it is negligible." His 1945 paper contains an elegant solution in terms of a series of Bessel functions, for a first-order reaction occurring at the tube-wall.

With ingenious experiments for dispersion of a point source of solute (methylene blue or carbon dioxide) into a flowing fluid stream (water or air) conducted so as to be compared with mathematical solutions, Bernard and Wilhelm (1950) obtained radial dispersion coefficients in a packed-bed -- since "under the conditions obtaining in this investigation, it can be shown that the eddy diffusivity (E_t) in the longitudinal direction has little effect on the shape of the concentration profile."

In a detailed analysis of flows in a packed bed of spheres, Ranz (1952) showed that the radial Peclet number for mass transfer is $\cong 11$, in accord with Bernard and Wilhelm.

Table 11. <u>Empty and Packed-Bed Tubular Reactors</u>

1908	I. Langmuir, "The Velocity of Reactions in Gases Moving Through Heated Vessels and the Effect of Convection and Diffusion," J. Amer. Chem. Soc., <u>30</u>, 1742-54.
1943	R. H. Wilhelm, W. C. Johnson, and F. S. Acton, "Conduction, Convection and Heat Release in Catalytic Converters," Ind. Eng. Chem., <u>35</u>, 562-75.
1944	H. M. Hulburt, "Chemical Reactions in Continuous-Flow Systems: Reaction Kinetics," Ind. Eng. Chem; <u>36</u>, 1012-7.
1945	H. M. Hulburt, "... : Heterogeneous Reactions," Ind. Eng. Chem., <u>37</u>, 1063-9.
1950	E. Singer and R. H. Wilhelm, "Heat Transfer in Packed Beds: Analytical Solution and Design Method--Fluid Flow, Solids Flow, and Chemical Reaction," Chem. Eng. Prog., <u>46</u>, 343-57.
1953	P. V. Danckwerts, "Continuous Flow Systems: Distribution of Residence Times," Chem. Eng. Sci., <u>2</u>, 1-13.
1953	G. I. Taylor, "Dispersion of a Soluble Matter in Solvent Flowing Slowly Through a Tube," Proc. Roy. Soc., <u>A219</u>, 186-203.
1956	R. Aris, "On the Dispersion of a Solute in a Fluid Flowing Through a Tube," Proc. Roy. Soc., <u>A235</u>, 67-77.
1956	N. R. Amundson, "Solid-Fluid Interactions in Fixed and Moving Beds," Ind. Eng. Chem. <u>48</u>, 26-35, 35-43, 43-50.
1960	H. A. Deans and L. Lapidus, "A Computational Model for Predicting and Correlating the Behavior of Fixed-Bed Reactors," AIChEJl., <u>6</u>, 656-68.

Bernard and Wilhelm also acknowledged that "It is conceivable that in a packed bed, the longitudinal and radial values of E_t might be somewhat different." This was indeed the case as shown almost simultaneously by comparing experimental results with solutions of the convective-diffusion equation by McHenry and Wilhelm (1957), Carberry and Bretton (1958), and analytically by Aris and Amundson (1957), all of whom showed that the Peclet number for axial dispersion of mass at high Reynolds numbers is $\cong 2$.

Danckwerts (1953) had studied axial dispersion in packed-beds earlier, as had Lapidus and Amundson (1952) in the context of adsorption. Danckwerts' widely quoted paper dealt with the distribution of residence times in flow systems, where he also arrived at the proper boundary conditions for the axial dispersion model. He developed an analytic solution for an isothermal first-order reaction as well, and concurring with Hulburt (1944) noted that "the effect of diffusion is to decrease the fractional conversion compared to that for piston-flow."

Danckwerts is generally given credit for the correct boundary conditions for a tubular reactor with axial dispersion. However, Langmuir had arrived at the same conditions by much the same arguments in 1908, and had reported the analytic solution for the first-order reaction as well. Like Jüttner's, this paper was also discovered only recently. Wehner and Wilhelm (1956) derived these boundary conditions in a mathematically rigorous manner for first-order reactions; Bischoff (1961) later showed that they hold for reactions of arbitrary order.

Kramers and Alberda (1953) studied residence-time distributions in flow systems by frequency response methods, and were the first to arrive at a well-known formula relating the number of mixers in series which are equivalent to a tubular reactor with a specific axial dispersion.

A 'cell' model "based on a two-dimensional network of perfectly stirred tanks" useful for fixed-bed reactor calculations on a computer was first described by Deans and Lapidus (1960). Coste, Rudd, and Amundson (1961) were the first to demonstrate numerical difficulties in solving the nonisothermal axial dispersion model equations by forward integration, and showed how they could be overcome either by backward integration or by the series of mixers model. Carberry and Wendel (1963) were the first to consider nonisothermal fixed-bed reactors for reactions involving selectivity.

Taylor (1953) had considered dispersion of a solute in a fluid flowing in an empty tube, as a consequence of molecular diffusion and velocity variations over the tube cross-section. With typically ingenious arguments, he provided an approximate expression for the solute dispersion coefficient about a point moving with the mean velocity of flow. In only his second published paper, Aris (1956) gave the correct asymptotic expression for this dispersion coefficient by using the method of moments.

Chemical Reactor Stability and Control

A complete paper on this topic by R. Aris appears elsewhere in this volume, so I shall at best do an inadequate job here.

Until recently, it was thought that the area of chemical reactor multiplicity and stability was initiated by van Heerden (1953), who showed that "...autothermic processes...are characterized by a simple diagram consisting of two curves which give the production and the consumption of heat..." Using the diagram, he demonstrated the existence of three steady states in a continuous-flow stirred tank reactor (CSTR), and gave qualitative arguments to correctly identify instability of the intermediate steady state. He concluded with the observation, "It is remarkable that the elementary considerations given in this paper, which are essential for a right understanding of all autothermic processes, have not been presented before."

He was, of course, not aware that a Swedish engineer, F. G. Liljenorth, then with du Pont, had some 35 years earlier in a remarkable paper (1918) reached exactly the same conclusions about reactor stability by about the same arguments. I believe the discovery of this paper is due to D. Luss in 1973.

Using techniques of nonlinear mechanics (Minorsky, 1947), Bilous and Amundson (1955) were the first to give a rigorous formulation of the stability criteria for a CSTR. They showed that for a single exothermic reaction, there are actually two conditions required to ensure stability of the reactor to small perturbations. They also gave phase portraits of the system in the concentration-temperature phase plane, obtained by numerically integrating the governing nonlinear transient equations on an analog computer, which provided the first complete understanding of a chemical reactor as a dynamical system.

In part II of the paper, Bilous and Amundson (1956) analyzed for parametric sensitivity of empty tubular reactors. "The calculations tend to show that there are regions of operation in which the reactor effluent is very sensitive to operating conditions."

These two papers were the start of a campaign that Amundson was to wage in understanding chemical reactor behavior by analyzing their stability and control. Several of his early papers are jointly with Aris, in a famous collaboration that produced some of the most admired pieces of modern chemical engineering research. The types of problems tackled included CSTRs, empty and packed-bed tubular reactors, catalyst pellets, two-phase reacting systems, and polymerization reactors. So numerous are these contributions that they cannot be listed even by title alone, in less than many pages. In Table 12, I have indicated the first three papers in a 15-part series that appeared in Chemical Engineering Science. Part II of this series confirmed the existence of self-sustained oscillations in the reactor, which had been conjectured in the paper with Bilous (1955). Although many illustrious contributions came in between, it has been only recently that, using Hopf bifurcation theory, Uppal, Ray, and Poore (1974, 1976) provided a comprehensive picture of the various responses possible in a CSTR, even for a single exothermic reaction.

Table 12. Chemical Reactor Stability and Control

1918   F. G. Liljenroth, "Starting and Stability Phenomena of Ammonia-Oxidation and Similar Reactions," Chem. Met. Eng., 19, 287-93.

1953   C. van Heerden, "Autothermic Processes: Properties and Reactor Design," Ind. Eng. Chem., 45, 1242-7.

1955   O. Bilous and N. R. Amundson, "Chemical Reactor Stability and Sensitivity," AIChE Jl, 1, 513-21.

1956   O. Bilous and N. R. Amundson, "...-II. Effect of Parameters on Sensitivity of Empty Tubular Reactors," AIChE Jl, 2, 117-126.

1958   R. Aris and N. R. Amundson, "An Analysis of Chemical Reactor Stability and Control," Chem. Eng. Sci.; -I, 7, 121-131; -II, 7, 132-147; -III, 7, 148-155.

1958   C. van Heerden, "The Character of the Stationary State of Exothermic Processes," Chem. Eng. Sci., 8, 133-145.

1969   T. Furusawa, H. Nishimura, T. Miyauchi, "Experimental Study of a Bistable Continuous Stirred-Tank Reactor," J. Chem. Eng. Japan, 2, 95-100.

1970   S. A. Vejtasa and R. A. Schmitz, "An Experimental Study of Steady State Multiplicity in an Adiabatic Stirred Reactor," AIChE Jl., 16, 410-9.

---

    van Heerden (1958) pointed out the possibility of multiple steady states in an empty tubular reactor with axial dispersion; Raymond and Amundson (1964) were the first to actually demonstrate them, and as well to show instability of the intermediate steady state.

    Experimental demonstrations of reactor steady state multiplicity and stability came relatively later. This is another instance where mathematical theory preceded and guided experimental results. Furusawa et al. (1969), and Vejtasa and Schmitz (1970) were the first to conduct such experiments in CSTRs. R. A. Schmitz thoroughly surveyed the entire area of reactor multiplicity, stability, and sensitivity in 1975, and included a till-then complete list of experimental studies showing such features in CSTRs, single catalyst pellets, and in tubular reactors.

Optimal Design of Reactors

The idea of attaining the optimum is very appealing, and the principles involved in maximizing reaction rates offer particular challenges for designing reactors optimally. In the first chapter of his first book, Aris (1961) reviewed the early work in this area, and so we need not go into great detail here.

The first contribution appears to be that of Leitenberger (1939), who calculated optimal temperatures in a multibed adiabatic reactor for the oxidation of sulfur dioxide, by finding conditions such that all beds operated close to the maximum reaction rate. Calderbank (1953) discussed the same reaction using a different kinetic expression. He gave an example showing that a two-bed adiabatic reactor would require some 15 times the catalyst as compared to one with an optimal temperature gradient. Annable (1952) had just the previous year calculated optimal temperature gradient for a tubular reactor for the ammonia synthesis reaction.

Denbigh made many early contributions to optimal reactor design, beginning with his landmark 1944 paper which treated many basic problems in the design of stirred tank, batch, and tubular reactors.

Using the calculus of variations, Bilous and Amundson (1956) described the optimal temperature gradient in a tubular reactor for reactions involving selectivity, and showed that the problem reduced to solving a nonlinear ordinary differential equation which they solved for many cases.

Denbigh (1958) discussed a complex reaction scheme involving four reactions, and showed that remarkable improvement in yield of the desired product could be made by correctly choosing the temperatures at which a sequence of two CSTRs would operate. Noting that Denbigh's solution for two reactors in series (by directly choosing various values for their temperatures) "if more than two reactors were considered, more than ordinary insight into the behavior of the system would be required to organize such a calculation," Aris (1960) gave the optimal design for the problem, using the then-new technique of dynamic programming which Bellman (1957) and his associates had recently enunciated. Horn (1961) made a detailed study of this reaction system for the tubular reactor, and discussed variation of the optimal policy with reaction parameters.

Starting with the paper on Denbigh's reaction, Aris began an influential eleven-part series on Studies in Optimization, in which the first seminal four (shown in Table 13) were on optimal design and the rest on optimal control. His 1961 book grew out of his Ph.D. thesis that he did without a mentor, for a Ph.D. in Chemical Engineering and Mathematics from the University of London in 1960.

MATHEMATICS IN CHEMICAL ENGINEERING 381

Table 13. Optimal Design of Reactors

1939   W. Leitenberger, "Thermische Beziehungen und Verlagerung der Kontaktmasse bei Schwefelsäurekontaktkesseln," Chem. Fabrik, 12, 281-92.

1944   K. G. Denbigh, "Velocity and Yield in Continuous Reaction Systems," Trans. Faraday Soc., 40, 352-73.

1952   D. Annable, "Application of the Temkin Kinetic Equation to Ammonia Synthesis in Large-Scale Reactors," Chem. Eng. Sci., 1, 145-54.

1953   P. H. Calderbank, "Contact-Process Converter Design," Chem. Eng. Prog., 49, 585-90.

1956   O. Bilous and N. R. Amundson, "Optimum Temperature Gradients in Tubular Reactors," Chem. Eng. Sci., 5, 81-92, 115-26.

1958   K. G. Denbigh, "Optimum Temperature Sequences in Reactors," Chem. Eng. Sci., 8, 125-31.

1960   R. Aris, "On Denbigh's Optimum Temperature Sequence," Chem. Eng. Sci., 12, 56-64.

1960   R. Aris, "Studies in Optimization," Chem. Eng. Sci., 12, 243-52; 13, 18-29, 75-81, 197-206.

1961   F. Horn, "Optimale Temperatur-und Konzentrationsverläufe," Chem. Eng. Sci., 14, 77-89.

1961   R. Aris, "The Optimal Design of Chemical Reactors: A Study in Dynamic Programming," Academic Press.

In addition to the papers cited already, Aris, F. Horn, and S. Katz contributed enormously to this area in publications too numerous to mention here.

Reactors--Other

In this category, I would like to say a few words about two specific areas in which some very important contributions to the understanding of reactions and reactors were made by the use of mathematics.

The first is that of complex reaction systems, where the pioneering work of Wei and Prater (1962) on monomolecular systems, by using matrix mathematics, is the foremost example. Some observations

Table 14. Reactors--Other

A. Complex Reaction Systems

    1962   J. Wei and C. D. Prater, "The Structure and Analysis of Complex Reaction Systems," Adv. in Catalysis, 13, 203-392.

B. Gas-Liquid Reactions

    1948   D. W. van Krevelen and P. J. Hoftijzer, "Kinetics of Gas-Liquid Reactions, Part I. General Theory," Rec. Trav. Chim., 67, 563-86.

    1951   P. V. Danckwerts, "Significance of Liquid-Film Coefficients in Gas Absorption," Ind. Eng. Chem., 43, 1460-7.

---

on the practical use and impact of this work and others in its wake have recently been made by a leading industrial practitioner (Weekman, 1979).

The second area is that of gas-liquid reactions, where absorption of a gas into a liquid is accompanied by a chemical reaction--a topic of considerable practical import. The physical absorption of the gas into the liquid is enhanced due to the chemical reaction. Pioneering contributions in this area, with the use of mathematics, were made by van Krevelen and Hoftijzer (1948), and by Danckwerts (1951). van Krevelen and Hoftijzer used the film-model of Whitman (1923) to describe the liquid film over which absorption and reaction occur simultaneously, while Danckwerts used an original surface-renewal model in a so-called penetration theory. Fortunately, with equivalent definitions of the physicochemical parameters, the difference in the rate of gas absorption from the two models is not significant, and so both are used interchangeably for most purposes (Danckwerts, 1970).

POSTSCRIPT

It should be evident from these notes that the use of higher mathematics (the definition of 'higher' has, of course, changed over the years since Mellor) has played a major role in the understanding and solving of many important chemical engineering problems. That this will continue to be, of that there seems little doubt. It must, however, also be stated that this understanding came about by the intelligent use of mathematics, coupled with knowledge of the other physical and chemical sciences--particularly, of chemistry. It is certain that such progress as we see in the profession today would not have been made if there had been "an iron curtain between chemistry and engineering," as Lewis (1959) so eloquently stated. Progress would have similarly been hampered if there had been an iron curtain between mathematics and chemical engineering.

ACKNOWLEDGMENT

In writing this paper, I have benefitted from communications and discussions with colleagues from many institutions. These include N. R. Amundson, R. Aris, J. T. Banchero, C. F. Bonilla, J. B. Butt, J. J. Carberry, S. W. Churchill, C. Georgakis, J. P. Kohn, J. O. Maloney, W. R. Marshall, Jr., J. J. Martin, E. E. Petersen, R. L. Pigford, C. J. Pings, D. Ramkrishna, R. A. Schmitz, W. E. Stewart, P. S. Virk, and C. A. Walker. To all of them, I am most grateful.

REFERENCES

Acrivos, A., Amundson, N. R., 1953, Solution of transient stagewise operations on an analog computer, Ind. Eng. Chem., 45:467.

Acrivos, A., Amundson, N. R., 1955, Applications of matrix mathematics to chemical engineering problems, Ind. Eng. Chem., 47:1533.

Amundson, N. R., 1980, "The Mathematical Understanding of Chemical Engineering Systems--Selected Papers of Neal R. Amundson," Edited by R. Aris and A. Varma, Pergamon Press, Oxford.

Aris, R., 1965, A normalization for the Thiele modulus, Ind. Eng. Chem. Fundls., 4:227.

Aris, R., 1974, The theory of diffusion and reaction--a chemical engineering symphony, Chem. Eng. Edn., 8:20.

Aris, R., 1975, "The Mathematical Theory of Diffusion and Reaction in Permeable Catalysts," Volumes I and II, Clarendon Press, Oxford.

Aris, R., 1977, Academic chemical engineering is an historical perspective, Ind. Eng. Chem. Fundls., 16:1.

Aris, R., Amundson, N. R., 1957, Some remarks on longitudinal mixing or diffusion in fixed beds, AIChE Jl., 3:280.

Bellman, R., 1957, "Dynamic Programming," Princeton University Press, Princeton.

Bernard, R. A., Wilhelm, R. H., 1950, Turbulent diffusion in fixed beds of packed solids, Chem. Eng. Prog., 46:233.

Bird, R. B., Stewart, W. E., Lightfoot, E. N., 1960, "Transport Phenomena," John Wiley, New York.

Bischoff, K. B., 1961, A note on boundary conditions for flow reactors, Chem. Eng. Sci., 16:131.

Bischoff, K. B., 1965, Effectiveness factors for general reaction rate forms, AIChE Jl., 11:351.

Bonilla, C. F., 1980, Private communication.

Carberry, J. J., Bretton, R. H., 1958, Axial dispersion of mass in flow through fixed beds, AIChE Jl., 4:367.

Carberry, J. J., 1961, The catalytic effectiveness factor under nonisothermal conditions, AIChE Jl., 7:350.

Carberry, J. J., Wendel, M. M., 1963, A computer model of the fixed bed catalytic reactor: The adiabatic and quasi-adiabatic cases, AIChE Jl., 9:129.

Churchill, S. W., 1978, Private communication.

Coste J., Rudd, D., Amundson, N. R., 1961, Taylor diffusion in tubular reactors, Canad. Jl. Chem. Eng., 39:149.

Damköhler, G., 1937, Einfluss von diffusion, strömung und wärmetransport auf die ausbeute bei chemisch-technischen reaktionen, Der-Chemie-Ingenieur, 3:359.

Dauckwerts, P. V., 1970, "Gas-Liquid Reactions," McGraw-Hill, New York.

Davidson, J. F., 1981, Book review, Trans. Inst. Chem. Engrs., 59:67.

Furnas, C. C., 1930, Evaluation of the modified Bessel function of the first kind and zeroth order, Amer. Math. Monthly, 37:282.

Hellums, J. D., Churchill, S. W., 1961, Computation of natural convection by finite-difference methods, in "International Developments in Heat Transfer," ASME, pg. 985.

Hougen, O. A., 1977, Seven decades of chemical engineering, Chem. Eng. Prog., 73(1):89.

Hougen, O. A., Watson, K. M., 1947, "Chemical Process Principles, Part III: Kinetics and Catalysis," John Wiley, New York.

Katz, D. L., 1960, "Integration of Electronic Computers into the Undergraduate Engineering Educational Program," First Annual Report--Project Supported by the Ford Foundation at the University of Michigan, Ann Arbor.

Kramers, H., Alberda, G., 1953, Frequency response analysis of continuous flow systems, Chem. Eng. Sci., 2:173.

Kremser, A., 1930, Theoretical analysis of absorption processes, Natl. Petroleum News, 22(21):43.

Lapidus, L., 1962, "Digital Computation for Chemical Engineers," McGraw-Hill, New York.

Lapidus, L., Amundson, N. R., 1950, Stagewise absorption and extraction equipment: Transient and unsteady state operation, Ind. Eng. Chem., 42:1071.

Lapidus, L., Amundson, N. R., 1952, Mathematics of adsorption in beds, VI. The effect of longitudinal diffusion in ion exchange and chromatographic columns, J. Phys. Chem., 56:984.

Lewis, W. K., 1959, Evolution of the unit operations, Chem. Eng. Prog. Symp. Ser., 55(26):1.

Martini, W. R., Churchill, S. W., 1960, Natural convection inside a horizontal cylinder, AIChE Jl., 6:251.

McAdams, W. H., 1933, "Heat Transmission," McGraw-Hill, New York.

McHenry, K. W., Wilhelm, R. H., 1957, Axial mixing of binary gas mixtures flowing in a random bed of spheres, AIChE Jl., 3:83.

Mickley, H. S., Sherwood, T. K., Reed, C. E., 1957, "Applied Mathematics in Chemical Engineering," Second Edition, McGraw-Hill, New York.

Minorsky, N., 1947, "Introduction to Nonlinear Mechanics," Edwards Bros., Ann Arbor.

Newman, A. B., Brown, G. G., 1930, Minimum voltage to reduce aluminum oxide, Ind. Eng. Chem., 22:995.

Petersen, E. E., 1965, A general criterion for diffusion influenced chemical reactions in porous solids, Chem. Eng. Sci., 20:587.

Pigford, R. L., 1976, Chemical technology in the past 100 years, Chem. & Eng. News, 54(15):190.

Ramkrishna, D., Amundson, N. R., 1973, Self-adjoint operators from selected nonsymmetric matrices: application to kinetics and rectification, Chem. Eng. Sci., 28:601.

Ranz, W. E., 1952, Friction and transfer coefficients for single particles and packed beds, Chem. Eng. Prog., 48:247.

Raymond, L. R., Amundson, N. R., 1964, Some observations on tubular reactor stability, Canad. Jl. Chem. Eng., 42:173.

Robinson, C. S., 1922, "The Elements of Fractional Distillation," McGraw-Hill, New York.

Schmitz, R. A., 1975, Multiplicity, stability, and sensitivity of states in chemically reacting systems--A review, Adv. in Chem., 148:156.

Smith, B. D., 1963, "Design of Equilibrium Stage Processes," McGraw-Hill, New York, pg. 319.

Sokolinikoff, I. S., Redheffer, R. M., 1958, "Mathematics of Physics and Modern Engineering," McGraw-Hill, New York.

Thiele, E. W., 1967, The effect of grain size on catalyst performance, Amer. Sci., 55(2):176.

Tinkler, J. D., Metzner, A. B., 1961, Reaction rates in nonisothermal catalysts, Ind. Eng. Chem., 53:663.

Tinkler, J. D., Pigford, R. L., 1961, The influence of heat generation on the catalyst effectiveness factor, Chem. Eng. Sci., 15:326.

Treybal, R. E., 1980, "Mass Transfer Operations," Third Edition, McGraw-Hill, New York.

Tuttle, F., 1925, A mathematical theory of the drying of wood, J. Franklin Inst., 200:609.

Uppal, A., Ray, W. H., Poore, A. B., 1974, On the dynamic behavior of continuous stirred tank reactors, Chem. Eng. Sci., 29:967.

Uppal, A., Ray, W. H., Poore, A. B., 1976, The classification of the dynamic behavior of continuous stirred tank reactors-- Influence of reactor residence time, Chem. Eng. Sci., 31:205.

Walker, W. H., Lewis, W. K., McAdams, W. H., 1923, "Principles of Chemical Engineering," McGraw-Hill, New York.

Weekman, V. W., Jr., 1979, "Lumps, Models and Kinetics in Practice," AIChE Monograph Series No. 11, AIChE, New York.

Wehner, J. F., Wilhelm, R. H., 1956, Boundary conditions of flow reactor, Chem. Eng. Sci., 6:89.

Weisz, P. B., 1973, Diffusion and chemical transformation: An interdisciplinary excursion, Science, 179:433.

Whitman, W. G., 1923, The two-film theory of gas absorption, Chem. Met. Eng., 29:146.

Williams, T. J., 1961, "Systems Engineering for the Process Industries," McGraw-Hill, New York.

Williamson, E. D., Adams, L. H., 1919, Temperature distribution in solids during heating or cooling, Phys. Rev., Ser. II, 14:99.

Zel'dovich, Ya. B., 1939, On the theory of reactions on powders and porous substances, Acta Physicochim. URSS, 10:583.

THE DEVELOPMENT OF THE NOTIONS OF MULTIPLICITY AND STABILITY

IN THE UNDERSTANDING OF CHEMICAL REACTOR BEHAVIOR

Rutherford Aris

Department of Chemical Engineering and Materials Science
University of Minnesota
Minneapolis, Minnesota 55455

ABSTRACT

The roots of the understanding of the importance of multiplicity and stability in chemical reactor behavior lie in theory and experiments on combustion, where the phenomenon of ignition takes a dramatic form. For more conventional chemical reactors they go back to a paper by Liljenroth and rise through the work of Denbigh, van Heerden and Amundson. Since the time of these early developments it has become increasingly clear that multiplicity is a pervasive phenomenon to be found in almost all types of chemical reactor. Some of these will be mentioned in outline to illustrate the developments of later work, but the emphasis will be on the evolution of the notion of stability itself.

INTRODUCTORY DEFINITIONS

The phenomena of multiple steady states and the disposition of their stabilities constitute so important an area in the understanding of chemical processes that it is worthwhile tracing a little of the history of the development of its understanding. Before doing so, however, it would be well to review the definitions of the various terms we shall use. Thus, multiplicity of steady states applies to any system which is capable of time-dependent operation in more than one condition. Its opposite is uniqueness of steady state, and both these properties are properties of the system rather than of the steady state or states themselves. (It could be held that uniqueness is a property of the steady state itself, but the contrary property would be non-uniqueness, and this is a clumsy usage.) On the other hand, stability is a property of the individual steady state itself and not of the system as a whole. The engineer commonly uses the term stability in the mathematician's sense of <u>asymptotic stability</u>. This implies that any sufficiently small disturbance decays to zero as time goes on and the disturbed system therefore returns to its original steady state. This return is usually of the nature of an asymptotically exponential decay of the disturbance so that, more often than not, the engineeringly stable steady state is in fact <u>exponentially</u> <u>asymptotically</u> <u>stable</u>. Stability is usually therefore associated with having eigenvalues entirely in the left half plane and its contrary, instability, with at least one eigenvalue being in the right half plane. If the eigenvalue of largest real part has its real part equal to zero, we can speak of marginal stability (this is included in the mathematician's definition of stable, though not asymptotically stable). If the steady state is unique and stable then it may also be <u>globally stable</u>. By this we mean that not merely a sufficiently small deviation but any deviation whatsoever will eventually die out and the system return to the unique steady state. It is clear that in a system with multiple steady states all of them cannot be globally stable and the most common pattern is that the stable steady states are associated with open regions of attraction such that deviations within the one region of attraction will decay to zero but those which take the state into the region of attraction of another steady state will not. The unstable steady states then lie in the interfaces between these regions of attraction. (See for example, the discussion in Aris, [9]).

As well as steady states, there may also be periodic solutions, represented by closed curves in the state space. Such a closed curve is invariant as a whole, though the point representint the state at any particular time moves along the curve. A periodic solution is often called a limit cycle if it is stable in the sense that a small perturbation away from it will eventually die down to zero and the limiting path be asymptotically assumed again. A periodic solution can be unstable if a deviation from it does not die away to zero. The stability of a periodic solution can be discussed in terms of the Poincaré map which maps a point in some transverse space into the point in that space to which it returns.

In recent years much interest has been aroused by a class of phenomena known as "chaos". These are solutions of differential equations which are neither periodic or asymptotic to a solution of any period. It is believed that such phenomena exist in chemical systems (see for example Schmitz and Sheintuch [13]). We shall not discuss this phenomenon here.

An associated question which must be distinguished is that of sensitivity. Amundson and Bilous [3] coined the phrase "parameteric sensitivity" for the situation when the two steady

CHEMICAL REACTOR BEHAVIOR

states of a system for two quite close values of one of its parameters were themselves markedly different. Thus, the dependence of the hot spot temperature on the inlet temperature or on the heat transfer to the cooling jacket in a tubular reactor shows a very sensitive behavior in certain regions.

Early French Work on Combustion

It is now well known that the question of multiplicity can often be decided from the intertwining of curves of heat generation and rejection. Though the early work using these curves dealt rather with the dynamics of combustion, it is worth turning aside for a moment to consider this, for a remarkable series of papers has recently been called attention to by Gray [23]. In 1913 the French scientists Taffanel and Le Floch studied the delay in the combustion of gaseous mixtures [32-35]. Their work was done at the testing station at Liévin in connection with their study of the dangers of the combustion of coal dust in the mines. They used the technique devised in 1880 by Mallard and Le Chatelier for studying the length of time lag before the explosion of a mixture of methane and air. The instant of explosion was taken to be the time of the very rapid rise in pressure which took place in one or two-hundredths of a second and was followed by a subsequent slower relaxation period. They attributed the duration of this lag to the length of time spent at temperatures where the curve of heat generation was close to that of heat loss. In fact, they showed the curve (D) of heat loss as a straight line dependent on the temperature difference between the gas mixture and the wall,

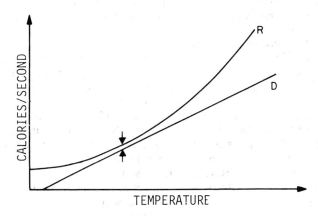

Fig. 1   Curves of heat generation and removal in the combustion of gaseous mixtures: after Taffanel and Le Floch, 1913.

whereas that of heat generation (R) was a curve which is convex downwards. As such, it was possible that there should be a considerable duration in which the distance between the curves, which was a measure of the available heat for temperature increase, would be quite small (cf. Fig. 1). Their conclusion was that "les différentes durées de retard que manifestent, dans un même récipient, divers mélanges gazeux sont simplement la consequence de différences dans la position et la forme de la courbe R."

These results were communicated in the Comptes Rendues for the 19th May 1913 and later that year they extended their work to measure the temperature of ignition (15 September 1913). Here they regarded the straight line D to be tangent to the heat of generation curve R and took the point of tangency to be the temperature of ignition. In their next communication (13 October 1913), they measured the rate of reaction and found it to be fitted with a positive exponential dependence on temperature.

The Work of Lilijenroth

If the first use of the heat generation and rejection curves was to show the meaning of the distance between them rather than their intersection, the work of Lilijenroth [26] in 1918 clearly illustrates the importance of the intersections of these curves as defining the steady states. The practice of equating the heat produced by the reaction to the heat removal in various forms of cooling must have been as old as the first attempts to design a reactor of any kind, but the recognition that the forms of those curves might give rise to a situation where there were multiple steady states was a distinct step forward and, as far as we can judge, first taken by F. G. Liljenroth (the reference to this paper was, I believe, uncovered by R. Gorring some years ago). Liljenroth was at the time an employee of Dupont in Wilmington. He was a Swede by nationality and seems to have returned to Sweden a few years later and left no further trace on the literature. His paper in the September 15, 1918 issue of Chemical and Metallurgical Engineering on "Starting and Stability Phenomena of Ammonia Oxidation and Similar Reactions" is however, a truly seminal one. It may be noted in passing that it seems to have entirely escaped notice of those who worked on this question some thirty years later, for the work of Van Heerden [39] and others was quite independent of this early work. Liljenroth started out by showing that the curve of combustion versus temperature for an ammonia-oxidation burner changes from convex to concave at a temperature of about 550° and goes through maximum in the region of 750°. At low temperatures very little nitric oxide is formed but the conversion increases very rapidly at temperatures of about 500 because of the highly temperature dependent reaction rate. Apparently he took the curve shown in his figures (and here in Fig. 2) from observed data and did not attempt to link it with

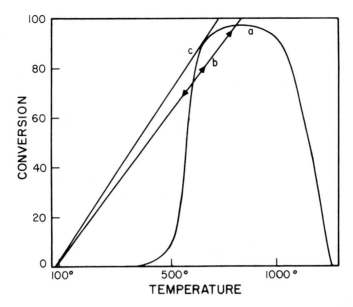

Fig. 2   Conversion curve: after Liljenroth, 1918.

anything which might be known of the thermodynamics and kinetics of reaction. To calculate the slope of the straight line of heat removal he allows a 5% heat loss by radiation and hence reduced the heat of reaction to 53 kilocals per gram-mol of nitric oxide formed. Allowing 9 moles of air for each one of ammonia and taking an average molecular heat capacity as 7.4, Liljenroth gives an adiabatic temperature rise for complete conversion of 675°. With gas entering at 25° the temperature of reaction at the gauze will be approximately 700°. This is the temperature at the peak of his conversion curve. It is represented by the point $\underline{a}$ of his figure (Fig.2). The low temperature intersection can be ignored but he calls attention to the intersection of the line at $\underline{b}$ with the very steep part of the conversion curve. He points out that such a state cannot be stable using the (now familiar) argument that a slight temperature increase would result in a greater heat generation that removal and hence to further increases of temperature, and vice-versa for a slight temperature decrease.

He applies the same argument point $\underline{a}$ at which a slight temperature increase would tend to lead to a net heat removal while a decrease of temperature would lead to a net generation of heat. This argument as we shall see in discussing the work of Amundson and Bilous, is not conclusive but provides only a necessary condition for the stability of a steady state. He uses this diagram to draw practical conclusions for the guidance of operators, pointing out that unless the initial temperature is above that of the point b (about 470°) there is little hope of reaching the reactive steady state a. On the other hand he shows that if the slope of the line were to be changed to be just tangential to the conversion curve then there could be no stability at such a tangential point and the reaction would be snuffed out. This leads him to a discussion of the influence of catalytic poisons and the gas velocity and "strength" - by "strength" he means the percentage of ammonia. "It would be beyond the scope of this article to go into these questions, but I hope the above examples have been sufficient to demonstrate the usefulness of the method of investigation set forth. It not only explains in a very simple manner the starting and stability phenomena and the influence of catalytic poisons on stability, but it might also be useful for the further study and proving of ammonia oxidation and similar processes. It is therefore desirable that pure raw material conversion curves be determined for different gas strengths and gas velocities, both for ammonia-air and ammonia-oxygen mixtures. If such curves were available, it would be possible to adjust or design a burner for maximum commercial efficiency giving proper consideration to conversion efficiency, etc. This applies also to other processes." The article finishes with the usual disclaimer that the shapes of the curves are more than qualitative and the few inches at the end of the concluding page are filled with a discussion for the treatment of caustic soda burns. It is, altogether, a remarkable paper, innocent even of algebra and containing only a few diagrams and numerical ratios, but pregnant with the principal notions of multiplicity and big with one of the key arguments concerning stability.

Wagner's Treatment of the Catalyst Particle

Diagrams of this kind and the corresponding arguments occur again in the article by Carl Wagner "Über die Temperatur einstellung an Höchstleistungskatalysatoren" in 1945 [41]. The article is dated as having been written in August of 1944. In contrast to Liljenroth's feeling that the curve of heat generation has to be determined experimentally, Wagner uses the equations of mass and heat balance with the assumption of a first-order reaction. Thus his first equation is a mass balance in which diffusion from a gas stream of concentration c_G takes place through a boundary layer of thickness δ to the surface. c_K denotes the concentration on the catalyst. The resulting

$$kc_K = \frac{D}{\delta}(c_G - c_K) \qquad (1)$$

can be solved for c_K as a function of c_G and the temperature, and then substituted in the heat balance, Wagner's Eq. 2.

$$QKc_K = \frac{\lambda_K}{\delta}(T_K - T_G) \qquad (2)$$

$$\dot{Q}_{II} = \frac{\lambda_G}{\delta}(T_K - T_G) = Qc_G \frac{(D/\delta)A\,e^{-q/RT_K}}{(D/\delta) + A\,e^{-q/RT_K}} = \dot{Q}_I \qquad (3)$$

This equation can yield one or three solutions for T_K, namely $T_K^{(1)}$, $T_K^{(2)}$ and $T_K^{(3)}$. In discussing the stability of the two extreme steady states Wagner identifies the two sides of Eq. 3 as rates of heat removal or generation and incorporates the expressions obtained for the rates of heat removal and generation in a heat balance for the distributed system. In this equation x is a coordinate in the pellet and T_K is a function of x and T.

$$c_{pK}\rho_K \frac{\partial T_K}{\partial t} = \lambda_K \frac{\partial^2 T_K}{\partial x^2} + \frac{2}{\Delta}[\dot{Q}_I(T_K) - \dot{Q}_{II}(T_K)] \qquad (4)$$

If the initial conditions

$$\left.\begin{array}{l} T_K = T_K^{(1)}, \quad x < 0 \\ \\ T_K = T_K^{(3)}, \quad x > 0 \end{array}\right\} t = 0 \qquad (5)$$

are imposed on this equation and the generation term at the end is ignored, the solution can be expressed in term of error function by Eq. 6 follows that a point of temperature T_K will move with the speed given by Eq. 7

$$T_K(x,t) = \frac{1}{2}[T_K^{(1)} + T_K^{(3)}] + \frac{1}{2}[T_K^{(3)} - T_K^{(1)}]\,\Phi\left[\frac{x}{2}\sqrt{\frac{c_{pK}\rho_K}{t\lambda_K}}\right] \qquad (6)$$

$$\left(\frac{dx}{dt}\right)_{T_K} = -\left(\frac{\partial T_K}{\partial t}\right)/\left(\frac{\partial T_K}{\partial x}\right) = -\frac{\lambda_K}{c_{pK}\rho_K}\left(\frac{\partial^2 T_K}{\partial x^2}\right)/\left(\frac{\partial T_K}{\partial x}\right) \qquad (7)$$

From the shape of the solution given by Eq. 6 he deduces that $T_K^{(2)}$ will increase if it is greater than the mean of $T_K^{(1)}$ and $T_K^{(3)}$. Wagner explicity realizes that there will be a difference in the response to small and large perturbations and that when there are three steady states the extreme states can only be stable to mild perturbation. However, he suggests a condition for the stability of the high temperature steady state is that the intermediate temperature be less than the mean of the high and the low. He then points out that the high temperature will be near the maximum possible temperature which is the gas temperature, T_G, plus the adiabatic temperature rise. Similarly the low temperature steady state will be close to the feed temperature of the gas T_G. Moreover the intermediate steady state temperature will be in the steeply rising part of the curve which is in the neighborhood of the point at which the rate constant is just equal to D/δ. Thus as a practical condition for the stability of the high temperature steady state he suggests that

$$T_{K=D/\delta} = q/R\ln[A\delta/D] < T_G + Qc_G D/2\lambda_G \qquad (8)$$

This goes on to discuss the ignition phenomenon and the effects of varying the different parameters.

van Heerden on Autothermic Processes

It is clear that Wagner has the general picture completely in mind and that his condition [8], though hardly rigorous, is a very practically sensible one. However the concept of a catalyst particle as a dynamical system is only alluded to and not exploited. No doubt these notions were current among a few chemical engineers, but they do not seem to have gained general currency until the very valuable exposition of autothermic processes by van Heerden in a 1953 article in Industrial and Engineering Chemistry [39]. It does not review the literature but begins by a simple discussion of the autothermic reactor. Without specifying the nature of the reactor he points out that the heat generated will generally depend on the reaction temperature in a sigmoidal manner. Assuming the reactor to be adiabatic, the heat removal is proportional to the difference between the reaction temperature and the feed temperature, the constant of proportionality being the heat

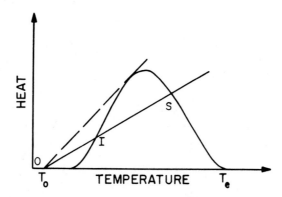

Fig. 3 van Heerden's diagram of heat generation and removal in an equilibrium process.

capacity of the stream. He suggests that the point I of his
figure (Fig. 3) would correspond to the intermediate and unstable
steady state may be taken to be the ignition temperature. The
point 0 is the initial feed temperature or low temperature steady
state whereas S is the steady state achieved after ignition. The
perfectly correct argument showing the instability of the intermediate steady state is given and he avoids giving the incomplete
converse argument for the stability of the high temperature steady
state. However he does seem to assume that the upper steady state
will be stable and discussed the variation of this with the feed
rate and other factors. To illustrate how valuable this kind of
presentation is, van Heerden gives a detail calculation for an
ammonia synthesis converter with Temkin kinetics. The good agreement between known practice and the calculation gives confidence
both in the usefulness of the approach and in the accuracy of the
kinetic and other constants which he has used. In his conclusion
he says "it is remarkable that the elementary considerations given
in this paper, which are essential for a right understanding of
all autothermic processes, have not been presented before." This
clearly indicates that he was unaware of the work of Liljenroth
and Wagner.

Denbigh on Open Systems

Let us turn now to the specific consideration of the stirred
tank which is the model in which progress was now really made.
It is in a certain sense implicit in the earlier work being
identical in form to the equations that Wagner used and implied
though not stated by the work of Liljenroth and van Heerden. K.
G. Denbigh had been concerned with the velocity and yield in
continuous reaction systems and published a paper on this subject
in the Transactions of the Faraday Society [16-17] in 1944. He
took up the calculation of the steady state and the comparison of
the performance of the stirred tank with the tubular reactor or
batch system and, in particular, he raised with the question of
optimality in a pioneer work that lies outwith our present scope.
It was in a later paper with Hicks and Page [18] that he discussed
something of the dynamics of the reactor and relates it to the
biological considerations of cell physiology. The recognition of
the global stability of a steady state and consequently of the
fact that the final steady state would be reached independently of
the earlier states had been considered earlier by Burton [15],
who applied the term "equifinality" to this notion. Denbigh and
his colleagues extend the work of Burton on the single first-order
reaction. They show, not only a complete solution for transition
between steady states with conditions for overshoot or a false
start, but also extending the calculation to the sequence of reversible reactions A→B→C. They also report some experimental results
which illustrate the theory. Piret and Mason also integrated the
transient equations for isothermal first-order systems.

Amundson and Bilous on Conditions for Stability

This is the background of the work of Amundson and Bilous, who appear to be the first to recognize the essentially dynamical character of stability questions [3]. It must be remembered that during the second world war a lot of mathematical activity had been going on and there had been particular activity on the part of a group associated with Minorsky, concerned with nonlinear dynamical systems. The most available source of understanding of this at the time was the book of Minorsky called "Nonlinear Mechanics" published by Edwards of Ann Arbor in 1947. Bilous was a French student who had come to Minnesota to do his Ph.D. with Piret, which indeed he did; his thesis was on the relationship between batch and stirred tank reaction paths and was published with Piret in several papers. However Bilous also discussed problems with Amundson and together they achieved some remarkable results with the nonisothermal stirred tank behavior. At the time the University of Minnesota had an antique REA analogue computer with which Bilous had remarkable empathy. In the dead of night he was known to be able to tune this rather erratic instrument to give excellent phase planes of the behavior of the stirred tank with a single reaction.

Their paper in 1955 deals not only with the steady state as Heerden had done but since then the equations for the stirred tank in which a first-order irreversible reaction is taking place. Their equations appear in the form

$$V \frac{dA}{d\theta} = qA_o - qA - pV e^{-E/RT} A \qquad (9)$$

$$Vc\rho \frac{dT}{d\theta} = qc\rho (T_o - T) - pV e^{-E/RT} A(\Delta H) - U(T-T') \qquad (10)$$

where θ is the time, V the volume, A the concentration of the reactant, q the flowrate, and the rate constant is $k = p \exp -E/RT$. In the second equation c is the heat of capacity, and ρ the density of the reacting mixture, T the temperature, T_o the feed temperature, T' the coolant temperature and U a heat transfer coefficient. By linearizing about the steady state A_s and T_s and applying the conditions that the real parts of the characteristic equation should be negative for stability they showed that there were two conditions necessary and sufficient for stability which they gave in the form

$$\frac{2q}{V} + k_s + \frac{U}{Vc\rho} > k_s (\frac{-\Delta H}{\rho c}) A_s \frac{E}{RT_s^2} \qquad (11)$$

$$\frac{q}{V} + k_s + \frac{U}{Vc\rho} + k_s \frac{U}{qc\rho} > k_s \left(\frac{-\Delta H}{\rho c}\right) A_s \frac{E}{RT_s^2} \qquad (12)$$

Their paper is not confined to the single first-order reaction but considers also a general set-up of N chemical species and, in particular, gives an example of a simultaneous second-order reversible reaction A + B = C + F with a first-order irreversible side reaction C → E. The multiplicity and disposition of the steady

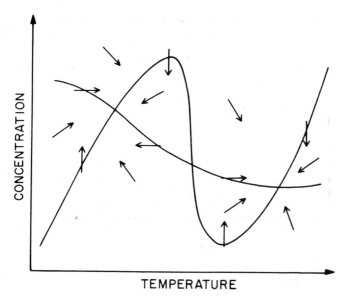

Fig. 4 Diagram used by Amundson and Bilous to show the flow of the solution in the case of three steady states.

states for the sequential reaction A → B → C is also considered. What is even more remarkable is the introduction of the phase plane. This clearly is taken over from the nonlinear mechanics that had grown up during the war and it serves to show in compact form the totality of solutions to equations. In particular, before even integrating the equations, the qualitative plot, showing the trends of the reaction paths (as in their Fig. 4) shows clearly that there is a tendency to swing away from the intermediate steady state. This qualitative picture is completed by some direct

calculations. The single stable steady state is clearly shown in
Fig. 5 and all the reaction paths lead to it. However they swing
to considerably higher temperatures on those paths which come
from high concentration and moderate temperatures. The further
system of a reactor with a recycle stream is considered and is
shown that the same principles of linearization can be applied to
give a set of criteria for stability.

This paper seems to have marked the real introduction into
the Chemical Engineering literature of the dynamical system
approach to the stability of reactors. It clearly spells out the

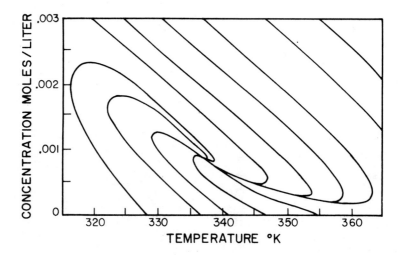

Fig. 5 Amundson and Bilous' illustration of a globally stable
steady state.

limitations of the linearization technique and shows how useful
the phase plane is in getting a complete picture of the system.

Some Later Developments

To go beyond this point would take us into the vast jungle of
literature which has flowed from it. It might be mentioned that
the direction that Amundson took was that of control and in a
series of papers studied certain questions of "chemical reactor
stability and control". Though tradition has it that once Bilous
coaxed a limit cycle out of the old analog computer at the dead of
night, the first well-established limit cycles were obtained in
connection with the control problem and the partial stabilization
of a steady state [2]. Amundson's work then took him into a number

of special areas such as that on two-phase reactors with Schmitz
[6] and polymerization with Goldstein and Liu [13] and later to
the tubular reactor with Varma [7]. He also espoused the direct
method of Liapounov in the study of stability, as in the work with
Warden and this was further taken up by Perlmutter [29] and his
coworkers. The full complexity of behavior of the system however,
even without control, was not fully brought out until the fascinating work of Uppal, Ray and Poore in 1974 and 1976 [35,36]. Most
recently Golubitsky and Keyfitz (22) have been able to bring the
stirred tank reactor within the purview of singularity theory.

In a general way a great deal of credit belongs to George
Gavalas [21] whose monograph on the "nonlinear differential
equations of chemically reacting systems" in 1968 did much to draw
the attention of mathematicians to the rich field of problems that
was to be found in chemical engineering. He also recognized that
the behavior of the catalyst particle was analogous to that of the
stirred tank. This theme has been taken up in many ways and
catalyst particle studies of great interest have been made by Luss
[27,28] and his colleagues, Villadsen [39,40] and others. The
references will give a good idea of the variety of work that has
been done and the titles truly reflect the contents [1,14,20,25,
27,30,37]. It was the work of Humphrey in microbiology that drew
attention to the existence of a continuous line of steady states
[12] . An infinite multiplicity of this kind is related to the
marginal or neutral stability of the resulting states since only
isolated critical points can have non-zero eigenvalues.

REFERENCES

1. N. R. Amundson, "Nonlinear Problems in Chemical Reactor Theory," SIAM-AMS Proc. 8, 59 (1974).
2. N. R. Amundson and R. Aris, "An Analysis of Chemical Reactor Stability and Control," Chem. Engng. Sci. 7, 121 (1958).
3. N. R. Amundson and O. Bilous, "Chemical Reactor Stability and Sensitivity" AIChE Journal 1, 513 (1955).
4. N. R. Amundson and L. Lapidus (eds.), "Chemical Reactor Theory," Prentice-Hall, Englewood Cliffs, 1978.
5. N. R. Amundson and L. R. Raymond, "Some Observations on Tubular Reactor Stability," Canadian J. Chem. Eng. 42,173 (1964).
6. N. R. Amundson and R. A. Schmitz, "An analysis of Chemical Reactor Stability and Control: Two-Phase Systems," Chem. Engng. Sci. 18, 265, 391, 415 (1963).
7. N. R. Amundson and A. Varma, "Local Stability of Tubular Reactors," AIChE Journal 19, 395 (1973).
8. N. R. Amundson and A. Varma, "Some Observations on Uniqueness and Multiplicity of Steady States in Nonadiabatic Chemically Reacting Systems," Canadian J. Chem. Eng. 51, 206 (1973).
9. R. Aris, "A Note on the Structure of the Transient Behavior of Chemical Reactors," Chem. Eng. J. 2, 140, (1971).

10. R. Aris, "On Stability Criteria of Chemical Reaction Engineering," Chem. Engng. Sci. 24, 149 (1969).
11. R. Aris, "The Mathematical Theory of Diffusion Reaction in Permeable Catalysts," 2 vols., Clarendon Press, Oxford, 1975.
12. R. Aris and A. E. Humphrey, "Dynamics of Chemostat in Which Two Organisms Compete for a Common Substrate," Biotech. and Bioeng. 19, 1375 (1977).
13. R. Aris and A. Varma (eds.), "The Mathematical Understanding of Chemical Engineering Systems: Selected papers of N. R. Amundson," Pergamon Press, Oxford, 1980.
14. B. Bunow and C. K. Colton, "Substrate Inhibition Kinetics in Assembleges of Cells," Biosystems 7, 160 (1975).
15. A. C. Burton, "The Properties of the Steady State Compared to Those of Equilibrium as Shown in Characteristic Biological Behavior," J. Cellular Comparitive Physiology 14, 327 (1939).
16. K. G. Denbigh, "Velocity and Yield in Continuous Reaction Systems," Trans. Farad. Soc. 40, 352 (1944).
17. K. G. Denbigh, "Continuous Reactions. The Kinetics of Steady State Polymerization," Trans. Farad. Soc. 43, 648 (1947).
18. K. G. Denbigh, M. Hicks and F. M. Page, "The Kinetics of Open Reaction Systems," Trans. Farad. Soc. 44, 479 (1948).
19. M. M. Denn, "Stability of Reaction in Transport Processes," Prentice-Hall, Englewood Cliffs, 1975.
20. M. P. Dudukovič, "Micromixing Effects on Multiple Steady States in Iso-Thermal Chemical Reactors," Chem. Engng. Sci. 32, 985 (1977).
21. G. R. Gavalas, "Nonlinear Differential Equations of Chemically Reacting Systems," Springer Verlag, Heidelberg, 1968.
22. M. Golubitzky and B. L. Keyfitz, "A Qualitative Study of the Steady-State Solutions for a Continuous Flow Stirred Tank Chemical Reactor," (to appear).
23. P. Gray, "Thermokinetic Oscillations in Gaseous Systems," Ber. der Bunsenges. für Phys. Chemie, 84, 309 (1980).
24. O. A. Hougen, "Reaction Kinetics in Chemical Engineering," Chem. Eng. Prog. Monograph Series 47 (1), 1 (1951).
25. V. Hlavácek and J. Votruba, "Hysteresis and Periodic Behavior in Catalytic Chemical Reaction Systems," Adv. in Cat. 27, 59 (1978).
26. F. J. Liljenroth, "Starting and Stability Phenomena of Ammonia-Oxidation and Similar Reactions," Chem. Met. Eng. 19, 287 (1918).
27. D. Luss, J.S.Y. Ding and S. Sharma, "Steady-State Multiplicity Control of the Chlorination of n-decane in an Adiabatic Continuously Stirred Tank Reactor," Ind. Eng. Chem. Fund. 13, 76 (1974).
28. D. Luss and C.A. Pikios, "Steady-State Multiplicity of Lumped-Parameter Systems in Which Two Consecutive or Two Parallel Irreversible First-Order Reactions Occur," Chem.

Engng. Sci. 34, 919 (1979).
29. D.D. Perlmutter, "Stability of Chemical Reactors," Prentice-Hall, Englewood Cliffs, 1972.
30. R. A. Schmitz and M. Sheintuch, "Oscillations in Catalytic Reactions," Cat. Revs. 17, (1977).
31. Taffanel, "Sur la combustion des melanges gazeux et les vitesse de réaction," Comptes Rendues. 157 , 714 (1913).
32. Taffanel and Le Floch, "Sur la combustion des mélanges gazeux et les retards à l'inflammation." Comptes Rendues 156, 1544, (1913).
33. Taffanel and Le Floch, "Sur la combustion des mélanges gazeux," 157, 595, (1913).
34. Taffanel and Le Floch, "Sur la combustion des mélanges gazeux et les températures d'inflammation" 157, 469 (1913).
35. A. Uppal, W. H. Ray and A. B. Poore, "On the Dynamical Behavior of Continuous Stirred Tank Reactors," Chem. Engng. Sci. 29, 967 (1974).
36. A. Uppal, W. H. Ray and A. B. Poore, "The Classification of the Dynamic Behavior of Continuous Stirred Tank Reactors - Influence of Reactor Residence Time," Chem. Engng. Sci. 31, 205 (1976).
37. D. A. Vaganov and N. G. Samoilenko and V. G. Abramov, "Periodic Regimes of Continuous Stirred Tank Reactors," Chem. Engng. Sci. 33, 1133 (1978).
38. C. van Heerden, "Autothermic Processes. Properties and Reactor Design," Ind. Eng. Chem. 45, 1242 (1953).
39. J. Villadsen and M. L. Michelsen, "Diffusion and Reaction on Spherical Catalyst Pellets: Steady State and Local Stability Analysis," Chem. Engng. Sci. 27, 751 (1972).
40. J. Villadsen and W. E. Stewart, "Graphical Calculation of Multiple Steady States and Effectiveness factors for Porous Catalysts," AIChE Journal 15, 28 (1969).
41. C. Wagner, "Über die Temperatureinstellung an Höchsteistungskatalyssatoren," die Chemische Technik, Vol. 18,28, 1945.

HISTORICAL DEVELOPMENT OF POLYMER BLENDS AND INTERPENETRATING POLYMER NETWORKS

L. H. Sperling

Materials Research Center
Coxe Laboratory #32
Lehigh University
Bethlehem, Pennsylvania 18015

INTRODUCTION

The concepts of mixing and reacting two or more different kinds of polymer molecules may be condiered on several levels. These include the mechanics of blending, reaction engineering, or from the point of view of molecular architecture. By way of defining several structures of interest, Figure 1 illustrates mixtures of two polymer species characterized by blending (no polymer I-polymer II bonds) or by various combinations of graft, block, or crosslink formation. As will be explored below, work on mixing two polymers dates back at least to 1927.

The concept of an interpenetrating polymer network, IPN, arose early in the history of polymer science and engineering, dating back to at least 1941. Apparently discovered and rediscovered independently by many investigators, the characteristic arrangement of the chains in space in an IPN has titillated chemist's minds for a generation. Now, after a slow and sometimes faltering start, many people in different parts of the globe are investigating IPN's. For some, the purpose is to gain a better understanding of polymer structure and behavior. For others, based on the patent literature, IPN's have a diversity of practical applications. This paper briefly reviews the history of IPN's in context with polymer blend ideas being developed during the same time period.

The author wishes to acknowledge the support of the National Science Foundation through Grant No. DMR-8015802, Polymers Program.

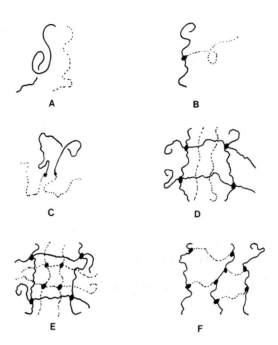

Fig. 1. Simple two-polymer combinations. A illustrates a polymer blend, without covalent linkages. B is a graft copolymer. C is a block copolymer, D illustrates a semi-I IPN, E a full IPN, and F designates an AB crosslinked Polymer. Structure F requires two polymers to make one network, while E has two independent networks.

Definition of an Interpenetrating Polymer Network

In its broadest definition, an IPN is any material containing two polymers, each in network form (1-10). A practical restriction requires that the two polymers have been synthesized and/or crosslinked in the immediate presence of each other. Two types of IPN's are illustrated in Figure 2. The sequential IPN begins with the synthesis of a crosslinked polymer 1. Monomer II, plus its own crosslinker and initiator, are swollen into polymer I, and polymerized in situ. Simultaneous interpenetrating networks, SIN's, begin with a mutual solution of both monomers and their respective crosslinkers, which are then polymerized by noninterfering modes, such as stepwise and chain polymerizations. Both syntheses produce structure E in Figure 1.

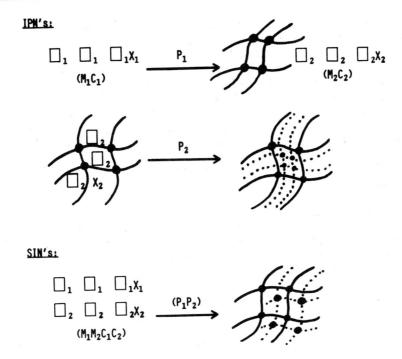

Fig. 2. A schematic comparison of IPN and SIN synthesis.
(1) R. E. Touhsaent, D. A. Thomas and L. H. Sperling
J. Polym. Sci., 46, 175 (1974).

Another mode of IPN synthesis takes two latexes of linear polymers, mixes and coagulates them, and crosslinks both components simultaneously. The product is called an interpenetrating elastomeric network, IEN. There are, in fact, many different ways that an IPN can be prepared; each yields a distinctive topology.

The term IPN implies an interpenetration of the two polymer networks, and was coined before the full consequences of phase separation were realized. Molecular interpenetration only occurs in the case of total mutual solubility; however, most IPN's phase separate to a greater or lesser extent. Thus, molecular interpenetration may be restricted or shared with supermolecular levels of interpenetration. In some cases, true molecular interpenetration is thought to take place only at the phase boundaries.

Given that the synthetic mode yields two networks, the extent of continuity of each network needs to be examined. If both networks are continuous throughout the sample, and the material is phase separated, the phases must interpenetrate in some way. Thus some IPN compositions are thought to contain two continuous phases

(11). The extent of molecular mixing at the phase boundaries has aroused significant interest lately (12).

When only one of the polymers is crosslinked, the product is called a semi-IPN, see Figure 1D. If the polymerizations are sequential in time, four semi-IPN's may be distinguished. If polymer I is crosslinked and polymer II is linear, the product is called a semi-IPN of the first kind, or semi-1. If polymer I is linear and polymer II is crosslinked, a semi-2 results. The remaining two compositions are materialized by inverting the order of polymerization. For simultaneous polymerizations, of course, only two semi-SIN's may be distinguished.

Historical Aspects

Like many other areas of scientific and engineering endeavor, it is difficult to pinpoint an exact time of origin for the ideas leading to IPN's. Clearly the discovery of a synthetic means of crosslinking polymers belongs in this chain of events. Since no other logical time of origin appears better, Table 1 begins with Goodyear's work on vulcanization, or crosslinking, of rubber.

Since the chain-like characteristics of polymeric materials were not understood until H. Staudinger's work beginning in 1920, systematic research on polymer topology could hardly have preceded his efforts. However, within the decade Ostromislensky's patent in 1927 clearly shows an understanding of graft copolymer structure. During the period of the 1940's and 1950's, great advances were made with polymer blends, grafts, and block copolymers. Through the leadership of Amos and others, polymer blends and grafts found uses as rubber-toughened plastics, known today as high-impact polystyrene (HiPS) and acrylonitrile-butadiene-styrene (ABS) plastics. As further illustrated in Table 1, block copolymers containing a water soluble block and an oil soluble block became important as surfactants through the work of Lunsted, while other block copolymers, composed of elastomer and plastic blocks, were useful as thermoplastic elastomers.

J. J. P. Staudinger, H. Staudinger's son, began his efforts during this time, applying for the first IPN patent in 1941, which was issued in the U.S. ten years later on the manufacture of smooth-surfaced, transparent plastics. The first use of the term "interpenetrating polymer network" was by Millar in 1960, who also made the first serious scientific study of IPN's. Millar used polystyrene/polystyrene IPN's as models for conventional gel ion-exchange materials (13-15).

Dr. Millar was kind enough to amplify on the early history of IPN's (16a). He wrote that the original name for IPN's was Re-swollen Polymer Networks, but that Interpenetrating Polymer Networks

TABLE 1
HISTORY OF IPN'S AND RELATED MATERIALS

Event	First Investigator	Year	Ref.
Vulcanization of Rubber	Goodyear	1844	(1)
Polymer Structure Elucidated	Staudinger	1920	(2)
Graft Copolymers	Ostromislensky	1927	(3)
Interpenetrating Polymer Networks	Staudinger and Hutchinson	1951	(4)
Block Copolymers	Dunn and Melville	1952	(5)
HiPS and ABS	Amos, McCurdy and McIntire	1954	(6)
Block Copolymer Surfactants	Lunsted	1954	(7)
Homo-IPN's	Millar	1960	(8)
AB-Crosslinked Copolymers	Bamford, Dyson and Eastmond	1967	(9)
Sequential IPN's	Sperling and Friedman	1969	(10)
Latex IEN's	Frisch, Klemper and Frisch	1969	(11)
Simultaneous Interpenetrating Networks	Sperling and Arnts	1971	(12)
IPN Nomenclature	Sperling	1974	(13)

1. C. Goodyear, U.S. 3,633 (1844).
2. H. Staudinger, Ber. dtsch. Chem. Ges., $\underline{53}$, 1073 (1920).
3. I. Ostromislensky, U.S. 1,613,673 (1927).
4. J. J. P. Staudinger and H. M. Hutchinson, U.S. 2,539,377 (1951).
5. A. S. Dunn and H. W. Melville, Nature, $\underline{169}$, 699 (1952).
6. J. L. Amos, J. L. McCurdy, and O. R. McIntire, U.S. 2,694,692 (1954).
7. L. G. Lunsted, U.S. 2,674,619 (1954).
8. J. R. Millar, J. Chem. Soc., 1311 (1960).
9. C. H. Bamford, R. W. Dyson, and G. C. Eastmond, J. Polym. Sci., $\underline{16C}$, 2425 (1967).
10. L. H. Sperling and D. W. Friedman, J. Polym. Sci., $\underline{A-2}$, $\underline{7}$, 425 (1969).
11. H. L. Frisch, D. Klempner, and K. C. Frisch, Polymer Letters, $\underline{7}$, 775 (1969).
12. L. H. Sperling and R. R. Arnts, J. Appl. Polym. Sci., $\underline{15}$, 2317 (1971).
13. L. H. Sperling in "Recent Advances in Polymer Blends, Grafts, and Blocks," L. H. Sperling, Ed., Plenum, 1974.

was preferred for publication because it was descriptive of the product, not the process. Millar wrote, "Incidently, I knew J. J. P. Staudinger quite well, as he lived and worked down the road from me in Barry (South Wales), and we were at one time coworkers of the Royal Institute of Chemistry local section. I don't recall discussing IPN's with him, as I believe by that time he'd moved to Epsom with Distillers Central Lab. George Solt, a colleague of mine at the time, came up with the idea of using re-swelling to increase bead size and got a patent on it." The Solt patent referred to by Millar was by G. S. Solt, Br. Pat. 728,508 (1955) on anionic/cationic IPN's for ion-exchange resins (14). Millar continues (16b) "George Solt was a process engineer and concerned with ways of reusing the 'fines' from our S/DVB bead production. We discussed the technique of re-swelling beads in the monomers one day at lunch. He was excited about the possibility of increasing the size of the fines to enable them to be put back into production, while I was being cautious about their showing the same properties, in particular their swelling. My feeling was that they were going to be structurally different, and might show unique behavior, which I wanted to determine in case there was some unexpected and patentable advantage."

Unlike most other areas of science and technology, apparently the IPN topology was discovered and rediscovered several times. While Millar was aware of much of the earlier work, the Sperling and Frisch teams some nine years later arrived at the idea independently but at nearly the same time.

The Frisch team was originally composed of two brothers, Harry at SUNY, Albany, NY, and Kurt, at the University of Detroit. Daniel Klempner was at that time Harry Frisch's student. Later he became a faculty member at Detroit and a full-fledged team member. H. Frisch had long been interested in catenanes, which consist of interlocking ring structures, physically bound together as illustrated in Figure 3. In an earlier work, Frisch and Wasserman (17,18) described the syntheses of the catenanes by directed ring closure. The Frisch team conceived of the IPN's as the macromolecular analog of the catenanes. It should be pointed out that K. Frisch is one of the world's top polyurethane scientists. Because of Kurt's interest in polyurethanes, one component of the Frisch team IPN's nearly always consists of a polyurethane.

Sperling was originally interested in producing finely divided polyblends without the need for heavy mechanical equipment. The dual network idea was fancied as a means of suppressing gross phase separation. Shortly after Sperling began his IPN studies, he was joined by Dr. David Thomas, a metallurgist. Dr. Thomas brought considerable electron microscopy and mechanical behavior understanding into the program, while Sperling contributed viscoelasticity and the vagaries of network topology and phase structures.

While all of the original thought patterns cannot be described here, two more people must be mentioned. Dr. Yury Lipatov of the Academy of Sciences of the Ukranian SSR, Kiev, USSR became interested in the interphase mixing regions of IPN's caused by thermodynamic incompatibility (2,12). Lipatov et al. also explored the actual locus of polymerization of monomer II in polymer-filler composites.

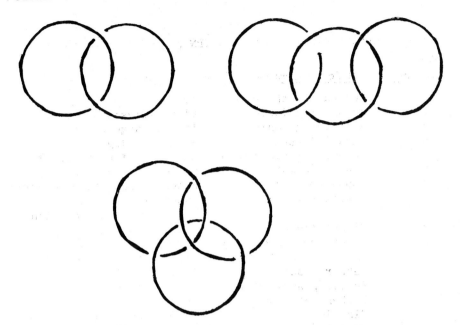

Fig. 3. Small molecule catenanes.

Dr. Guy Meyer, Louis Pasteur University, Strasbourg, France, came to IPN's through a consideration of semi-IPN's as adhesives (19). In addition to many scientific papers on IPN's and related materials contributed by numerous authors, the burgeoning patent literature describes quite different approaches to IPN chemistry and technology (20), as described in Table 2. The proposed or actual uses range from tough plastics to medical applications.

Toughening of Plastics with Rubber

Two other trains of thought need to be included for completeness. The first is the theories of rubber toughening and the second is the thermodynamics of mixing.

It is of significant interest to review the history of the theories of toughening of plastics by the addition of rubber because many IPN's are, in fact, rubber-toughened plastics. The concept of energy absorption by the rubber was first treated in 1956 by Merz et al. (21), who took the position that rubber particles, by anchoring themselves in both sides of a widening crack and by

TABLE 2
ACTUAL OR ANTICIPATED USES FOR IPN AND RELATED MATERIALS

Reference Numbers	Mode of Combination	Application
1	Natural leather/rubber	Improved leather
2	Anionic/cationic	Piezodialysis membranes
3	Anionic/cationic	Ion-exchange resin
4	Plastic/rubber	Noise damping
5	Rubber/plastic semi-IPN	Impact-resistant plastic
6	Plastic/plastic	Optically smooth surfaces
7	Rubber/rubber	Pressure-sensitive adhesive
8	Plastic/plastic	Compression-molding composition
9	Plastic/rubber	Tough plastic
10	Water-swellable/water swellable	Soft contact lenses
11	Rubber/plastic	Impact modifier
12	Rubber/plastic SIN's	Adhesive coatings
13	Rubber/water-swellable	Arterial-venous shunts
14	Plastic/plastic	Denture base materials
15	Plastic/plastic semi-II	Dental fillings

1. S. H. Feairheller, A. H. Korn, E. H. Harris, E. M. Filachione, and M. M. Taylor, U.S. Pat. 3,843,320 (1974).
2. L. H. Sperling, V. A. Forlenza, and J. A. Manson, J. Polym. Sci., Polym. Lett. Ed., $\underline{13}$, 713 (1975).
3. G. S. Solt, Br. Pat. 728, 508 (1955).
4. L. H. Sperling and D. A. Thomas, U.S. Pat. 3,833,404 (1974).
5. B. Vollmert, U.S. Pat. 3,055,859 (1962).
6. J. J. P. Staudinger and H. M. Hutchinson, U.S. Pat. 2,539,377 (1951).
7. H. A. Clark, U.S. Pat. 3,527,842 (1970).
8. Anonymous (Ciba, Ltd.) Br. Pat. 1,223,338 (1971).
9. K. C. Frisch, H. L. Frisch, and D. Klempner, Gr. Pat. 2,153,987 (1972).
10. J. J. Falcetta, G. D. Friends, and G. C. C. Niu, Ger. Offen. Pat. 2,518,904 (1975).
11. C. F. Ryan and R. J. Crochowski, U.S. Pat. 3,426,101 (1969).
12. H. L. Frisch, J. Cifaratti, R. Palma, R. Schwartz, R. Forman, H. Yoon, D. Klempner, and K. C. Frisch, in "Polymer Alloys," D. Klempner and K. C. Frisch, Eds., Plenum, 1977.
13. P. Predecki, J. Biomed. Materials Res., $\underline{8}$, 487 (1974).
14. B. E. Causton, J. Dental Res., $\underline{53(3)}$, 1074 (1974).
15. L. N. Johnson, J. Biomed. Mater. Res. Symposium, $\underline{1}$, 207 (1971).

subsequent extensive elongation, absorbed large amounts of energy. In 1965, Newman and Strella (22) asserted that the presence of the dispersed elastomer particle served to trigger yielding in the continuous phase. The enhanced toughness was attributed to cold drawing of the matrix and could be quantitatively assessed by measuring the area under the stress-strain curve.

Later, Kambour (23) and Bucknall and Smith (24) identified stress-whitening with crazing. Crazes, as distinct from open cracks, had oriented fibrils of polymer running from roof to floor, absorbing large amounts of energy in their formation. Rubber particles initiated crazes and craze-splitting, leading to great absorption of mechanical energy by small volumes of polymer. Most recently, Bucknall and Street (25) studied shear-yielding and shear-band formation as an energy-absorption mechanism. Such shear bands also function as crack- and craze-stoppers.

While the IPN work has yet to contribute any theory to rubber toughening, its unique dual phase characteristics provide a novel mode of energy absorption, as recently reviewed by Paul (26).

Thermodynamics of Mixing

Polymer blends, blocks and grafts have long been observed to be incompatible; i.e., not mutually soluble in the thermodynamic sense (27). Frequently this was true even in the presence of a mutual solvent. For example, Dobry and Boyer-Kawenoki (28) in 1947 investigated the phase relationships existing in ternary systems: polymer I-polymer II-mutual solvent III. They prepared separate solutions of the several polymers in common solvents, and then mixed the two solutions of interest. All of the polymer pairs investigated by this team underwent phase separation at a level of only 5-10% combined polymer concentration. For instance, cellulose acetate and polystyrene were immiscible at 5% concentration in toluene. These investigators concluded that incompatibility of the two polymers was the normal circumstance.

Shortly thereafter, Scott (29) offered a theoretical explanation based on the free energy of mixing. Using a quasi-lattice model, Scott showed that long polymer chains exhibited a very low entropy of mixing. This arises because one mer in a chain cannot be interchanged with another mer, and hence the number of distinguishable conformations that the mixture of chains can assume in space is much smaller than is true for an equivalent mass of small molecules.

Since the heat of mixing for non-interacting materials is usually positive, a small entropy term led to a positive free energy of mixing. Hence phase separation will occur. Experiments by Slonimskii (30) and Tompa (31) confirmed these results.

These early works must be viewed in the context of the time during which they were performed. While both the theory and subsequent experiments themselves were correct, their combined conclusions tended to discourage efforts to find and study compatible and semi-compatible polymer pairs.

In more recent studies, Krause (32) and Meier (33) showed that the presence of block-copolymer formation in part overcame the entropic problem. Similar results are true for graft copolymers and these two methods of joining polymer I to polymer II have been a principle route to controlling both thermodynamic miscibility and phase domain size.

In the case of sequential IPN's, Donatelli et al. (34) showed that the crosslink density of polymer I controls the phase domain size of polymer II. Other factors were interfacial tension and overall composition. The main point, however, is that crosslinks in an IPN replace block and graft formation in influencing miscibility considerations.

The problem in IPN's has been carried further. Even homo-IPN's (both polymers identical) are not completely uniform, as shown by Siegfried et al. (35). Separately, Thiele and Cohen (36) have provided an equation, based on the studies of Flory and Rehner (37), which govern the swelling of sequential IPN's.

Anatomy of a Research Program

When the term IPN was first conceived, as mentioned above, the full consequences of phase separation were not understood. This was certainly the case when Sperling undertook his first study with Friedman (38), who synthesized his IPN's as part of his undergraduate senior research project in Chemical Engineering.

During that time period, as now, some of the most important research in the field of polymer blends, grafts, and blocks was being done in Japan. Polymer blends, grafts, blocks, and IPN's are all multicomponent polymer systems, differing by their spatial molecular arrangements (10), Fig. 1. The works of three people must be mentioned in the present context: Prof. M. Takayanagi, an expert in the viscoelastic behavior of polymer blends who developed the widely used Takayanagi models (39); and two great polymer morphologists: Dr. K. Kato (40) and Dr. M. Matsuo (41). (Dr. Matsuo had been Dr. Takayanagi's student.) Dr. Kato, it should be pointed out, had developed the osmium tetroxide staining technique, which all polyblend scientists have employed to great advantage. Because of these three researchers, polymer scientists and engineers have a much clearer idea of how a rubber-toughened plastic works and how thermoplastic elastomers should be formulated. In particular,

Sperling was influenced by these men and set out to do a similar study on IPN's.

Sperling's first Ph.D. candidate in the field of IPN's, Volker Huelck, displayed great talent at the electron microscope (42). By thought habit, he was a geometrician, with unusual capability of imagining three dimensional structures. (It should be pointed out that the majority of scientists and engineers are algebraists by thought habit or training.)

At the time that Huelck's Ph.D. committee was organized, Sperling learned that Matsuo had come to Bell Laboratories as a Senior Research Associate. He was invited to be a committee member and contributed enormously to Lehigh's IPN program.

As it was, Matsuo also did key electron microscopy for the Frisch team during this time period (43), so that he stands out as a major contributor to both programs. (Serving as a friend of the court, Matsuo later helped Huelck to go to Japan as a postdoctoral research associate where he studied under Prof. Takayanagi. Dr. Huelck is now associated with Hoechst in West Germany.)

Current Status of Polymer Blends and IPN's

Recently, researchers in the field of polymer blends took a new interest in the thermodynamics of polymer-polymer miscibility. McMaster (44, 45) showed that polymer-blend phase diagrams exhibit a lower critical solution temperature, LCST, rather than the usual upper critical solution temperature, UCST.

This phenomenon was explained in terms of Flory's equation of state thermodynamics (46, 47). In this case, both the enthalpic and entropic mixing terms are near zero, and phase separation is dominated by the differences in the coefficients of thermal expansion, normally a minor consideration. This research has been reviewed by Olabisi, et. al. (48).

A very new engineering application in the field of multiphase polymer materials has involved the development of reaction injection molding, RIM. The RIM process utilizes segmented polyurethanes, which are multiblock copolymers of the type similar to that shown in Figure 1, except each chain has a fairly large number of blocks (49). The RIM process provides a very rapid method of molding large plastic or elastomeric materials suitable for automotive bumpers and fascia.

In the field of IPN's, major interests at present include the thermoplastic IPN's (50-53) and novel ion exchange resins (54-57). The thermoplastic IPN's are crosslinked via block-copolymer, ionic, or semicrystalline bonds, each of which loosen and permit flow at elevated temperatures.

The IPN ion-exchange resins have anionic groups on one network and cationic groups on the other. The new advances (54-57) relate to a controlled collapse of network I to make a swiss-cheese structure, network II fills up the pores. The controlled collapse is brought about by polymerizing monomer I in a solvent for the monomer, but a precipitant for the polymer. A critical condition includes the gelation of the network at an early stage of the polymerization, before precipitation.

It is appropriate to conclude this review by mentioning that two books on IPN's have been written, one by Lipatov and Sergeeva (58), and one by the present author (59).

In summary, it was through the pioneering efforts of many people that the IPN program emerged. Concepts of a two-phased morphology, thermodynamics, and multicomponent polymer structure each played its part. At this time of writing, the field is growing enormously. For example, over half of the 180 papers and 80 patents identified as IPN works were published since 1976.

REFERENCES

1. K. Shibayama and Y. Suzuki, Rubber Chem. Tech., 40, 476 (1967).
2. Y. S. Lipatov and L. M. Sergeeva, Russian Chem. Revs., 45(1), 63 (1976).
3. D. Klempner, Agnew. Chem., 90, 104 (1978).
4. L. H. Sperling, Encyl. Polym. Sci., Technol. Suppl. 1, 288 (1976).
5. L. H. Sperling, J. Polym. Sci., Macromol. Revs., 12, 141 (1977).
6. D. A. Thomas and L. H. Sperling, in "Polymer Blends," Vol. 2, D. R. Paul and S. Newman, Eds. Academic Press, 1978.
7. H. A. J. Battaerd, J. Polym. Sci., 49C, 149 (1975).
8. D. S. Kaplan, J. Appl. Polym. Sci., 20, 2615 (1976).
9. H. L. Frisch, K. C. Frisch, and D. Klempner, Modern Plastics, Apr. P. 76, May, p. 84 (1977).
10. J. A. Manson and L. H. Sperling, "Polymer Blends and Composites", Plenum, 1976, Ch. 8.
11. L. H. Sperling, Polym. Eng. Sci., 16, 87 (1976).
12. Yu. S. Lipatov, L. M. Sergeeva, L. V. Karabanova, A. Ye. Nesterov and T. D. Ignatova, Vysokomol. Soyed., A18(5), 1025 (1976).
13. J. R. Millar, J. Chem. Sco., 1311 (1960).
14. G. S. Solt, Br. Pat. 728, 508 (1955).
15. J. R. Millar. D. G. Smith, and W. E. Marr, J. Chem. Soc., 1789 (1962).
16. (a) J. R. Millar, private communication, December 5, 1978
    (b) January 11, 1979.
17. H. L. Frisch and E. Wasserman, J. Am. Chem. Soc., 83, 3789 (1961).
18. H. L. Frisch and D. Klempner, Adv. Macromol. Chem., 2, 149 (1970).
19. G. C. Meyer and P. Y. Mehrenberger, European Polym. J., 13, 383 (1977).

20. L. H. Sperling, K. B. Ferguson, J. A. Manson, E. M. Corwin and D. L. Siegfried, Macromolecules, 9, 743 (1976).
21. E. H. Merz, G. C. Claver and M. J. Baer, J. Polym. Sci., 22, 325 (1956).
22. S. Newman and S. Strella, J. Appl. Polym. Sci., 9, 2297 (1965).
23. R. P. Kambour, Macromol. Rev., 7, 134 (1971).
24. C. B. Bucknall and R. R. Smith, Polymer, 6, 437 (1965).
25. C. B. Bucknall and D. G. Street, SCI Monograph No. 26, p. 272 (1967).
26. D. R. Paul, Ch. 1 in "Polymer Blends," D. R. Paul and S. Newman Eds., Academic Press, 1978.
27. I. Ostromislensky, U.S. 1,613,673 (1927).
28. A. Dobry and F. Boyer-Kawenoki, J. Polym. Sci., 2, 90 (1947).
29. R. L. Scott, J. Chem. Phys., 17, 279 (1949).
30. G. L. Slonimskii, J. Polym. Sci., 30, 625 (1958).
31. S. Krause, Macromolecules, 3, 84 (1970).
33. D. J. Meier, J. Polym. Sci., 26C, 81 (1969).
34. A. A. Donatelli, L. H. Sperling and D. A. Thomas, J. Appl. Polym. Sci., 21, 1189 (1977).
35. D. L. Siegfried, J. A. Manson and L. H. Sperling, J. Polym. Sci., Polym. Phys. Ed., 16, 583 (1978).
36. J. L. Thiele and R. E. Cohen, Polymer Preprints, 19(1), 137 (1978).
37. P. J. Fory and J. Rehner, J. Chem. Phys., 11, 521 (1943).
38. L. H. Sperling and D. W. Friedman, J. Polym. Sci., A-2, 7, 425 (1969).
39. M. Takayanagi, Mem. Fac. Eng., Kyushu Univ., 23, 11 (1963).
40. K. Kato, Japan Plastics, 2, (April) 6, (1968).
41. M. Matsuo, Japan Plastics, 2, (July) 6, (1968).
42. V. Huelck, D. A. Thomas and L. H. Sperling, Macromolecules, 5, 340, 348 (1972).
43. M. Matsuo, T. K. Kwei, D. Klempner and H. L. Frisch, Polym. Eng. Sci., 10, 327 (1970).
44. L. P. McMaster, Macromolecules, 6, 760 (1973).
45. L. P. McMaster, in "Copolymers, Polyblends, and Composites," N.A.J. Platzer, Ed., Adv. Chem. Series 142, American Chemical Society, Washington, D.C., 1975.
46. P. J. Flory, R. A. Orowall, and A. Vry, J. Amer. Chem. Soc., 86, 3515 (1964).
47. P. J. Flory, J. Amer. Chem. Soc., 87 1833 (1965).
48. O. Olabisi, L. M. Robeson, and M. T. Shaw, "Polymer-Polymer Miscibility," Academic Press, New York, 1979.
49. W. E. Becker, Ed., "Reaction Injection Molding", Van Nostrand-Reinbold, N. Y., 1979.
50. F. G. Hutchinson, R. G. C. Henbest, and M. K. Leggett, U. S. 4,062,826 (1977).
51. W. K. Fischer, U.S. 3,862,106 (1975).
52. W. P. Gergen and S. Davison, U.S. 4,101,605 (1978).

53. D. L. Siegfried, D. A. Thomas, and L. H. Sperling, Polymer Preprints, 21(1), 186 (1980).
54. J. H. Barrett and D. H. Clemens, U.S. 4,152,496 (1979).
55. B. N. Kolarz, J. Polym. Sci., Symposium No. 47, 197 (1974).
56. W. Trochimczuk, presented at the IUPAC sponsored conference, "Modification of Polymers," Bratislava, Czechoslovakia, July, 1979.
57. M. J. Hatch, U.S. 3,957,698 (1976).
58. Yu. S. Lipatov and L. M. Sergeeva, "Interpenetrating Polymeric Networks," Naukova Dumka, Kiev, 1979.
59. L. H. Sperling, "Interpenetrating Polymer Networks and Related Materials," Plenum Press, N. Y., publication expected March, 1981.

THE ORIGINS AND GROWTH OF NAPHTHA REFORMING

H. B. Kendall

Department of Chemical Engineering
Ohio University
Athens, Ohio 45701

Naphtha reforming is a very complex process involving a variety of reactions: exothermic and endothermic, fast and slow, etc.[1] But the rewards for operating successful reforming processes are great, both from the standpoint of improved octane ratings and in terms of improved aromatic yields in the product streams. So it is not surprising to discover that much effort was directed toward the development of reforming processes, especially in the 1940's. And efforts to improve these processes continue to this day.

In an earlier paper, the author suggested that catalytic naphtha reforming had roots in several other petroleum conversion processes, perhaps most notably cracking (thermal and catalytic), thermal reforming, and oil hydrogenation.[2] In this presentation, some of these early contributions to catalytic reforming will be summarized briefly and included as part of the total series of events that lead to the introduction of modern reforming processes in the late 1940's and early 1950's.

The early work in thermal cracking by Dr. William M. Burton and his co-workers at Standard Oil of Indiana, at the beginning of the second decade of this century, broke the gasoline "bottleneck" of that time by providing a method for doubling the quantity of gasoline that could be obtained from crude oil. Subsequent improvements in thermal cracking techniques, fashioned by many persons and companies, were incorporated into refinery operations in the next twenty years. The development of catalytic cracking processes in the 1920's and 30's represented another step forward in the evolution of motor fuels. These techniques brought better control of the cracking reactions than was possible with thermal techniques, with the result that higher yields of desirable molecular types could be formed in the reactors. Catalytic cracking technology continued to develop rapidly through World War II, and the introduction of processes for the continuous regeneration of the catalyst ("moving bed" and "fluidized bed" processes) brought the art to a high degree of refinement by the middle of the century. This story has been well told elsewhere (e.g., by Enos[3]).

However, cracking wasn't the only innovation that lead to better fuels. Catalytic hydrogenation, polymerization, and alkylation processes, for example, also contributed greatly toward the improvements in gasoline. But even more important than these was the significant role played by catalytic reforming.

Reforming reactions differ from cracking reactions in that the latter involve, for the most part, the decomposition of large molecules into smaller ones; whereas the former involve rearrangements of molecules into others of about the same size. Thus, while cracking processes have been based upon decomposing heavy oil molecules into gasoline (and smaller) species, reforming processes have been directed toward converting poor grades of gasoline (or "naphthas") into gasolines with more desirable properties. Thus reforming reactions include such things as:

1. Conversion of straight-chain hydrocarbons into branched ("isomerization") or cyclic ("cyclization") molecules;

2. Dehydrogenation of cyclic molecules ("aromatization");

3. Addition of hydrogen to unsaturated molecules ("hydrogenation");

and so forth.[4]

Once the complexity of reforming reactions is suggested, it is not surprising to find that the concept of reforming developed rather slowly between 1920 and 1940, and that catalytic reforming processes grew from several stems, mentioned earlier.

The connection between cracking processes and reforming lies in the fact that certain "reforming-type" reactions are known to occur in cracking reactors. In other words, the presence of a complicated mixture of hydrocarbon molecules - both reactants and products - in a reactor, at elevated temperatures, creates a very complex reaction situation. Under these circumstances, it is not unusual to find that, in addition to the desired decomposition reactions, many other kinds of reactions also occur: polymerizations, hydrogenations, isomerizations, etc.

It probably didn't take refinery operators long to realize that if reforming reactions could occur in a cracking unit, it should be possible to set up thermal reactors where gasoline or naphtha molecules could be exposed to conditions that would be favorable for their rearrangements (reforming) into "better" gasoline molecules. Thermal reforming processes grew, perhaps, from this realization.

Another input to reforming technology at this time was provided by the catalytic cracking process developed over a period of years by Eugene Houdry.[5] Houdry had begun his studies on catalytic cracking in France, in the mid-1920's. Unable to interest French refiners in his work, he moved to the United States in 1930 at the invitation of the Vacuum Oil Company and set up shop at Vacuum's refinery in Paulsboro, N.J. By 1938, the Houdry Process Company, now controlled jointly by the Socony-Vacuum Oil Co., the Sun Oil Company, and Houdry, was licensing catalytic cracking plants. Construction was handled by an engineering firm, E.B. Badger and Sons Co.[6] These plants were easily converted to reforming operations by modifying operating conditions and by changing the feedstock from gas oil to naphtha. The product was a superior, high anti-knock gasoline. Hence the claim was made that this process made available "important increases in the potential supply of high-octane gasoline due to requiring greatly decreased amounts of iso-octanes for a given product as compared to straight-run aviation gasoline."[7] This was an important point just prior to World War II.

In many ways, the most intriguing development leading to modern reforming methods was that of oil hydrogenation. The fact that hydrogen could be added to, or reacted with, hydrocarbons in the presence of a metal catalyst was well established by Sabatier at the beginning of the century.[8] In addition, at about the same time, Ipatieff demonstrated the use of metals and oxide catalysts in hydrogenating hydrocarbons at superatmospheric pressures, with little decomposition of reactants or products.[9] By combining this knowledge with LeChatelier's concepts of gas-phase reaction equilibria, and recent advances in the methods for the fabrication of high-pressure laboratory equipment, Fritz Haber and his students at Karlsruhe demonstrated the feasibility of synthesizing ammonia from nitrogen and hydrogen in 1909.[10] The commercial process for accomplishing this synthesis was perfected by Carl Bosch, of the Badische Anilin und Soda Fabrik (BASF), in 1913.[11] These events not only provided Germany with a supply of nitrates during World War I, but also marked the begnining of high-pressure hydrogenation technology in that country. The importance of this special technology in terms of the "momentum" it provided for continuing research and development in Germany is pointed out by Hughes.[12]

Whatever the cause, the next milestone in the use of hydrogenation by the petroleum industry was due to the work of Friedrich Bergius, who had come into contact with Haber's work during a short stay at Karlsruhe.[13] In 1910, at Hanover, he began his research

in hydrogenation, working at pressures up to 300 atmpspheres.[14]
In 1914, Bergius began his work on an "enlarged scale", using a
40 liter bomb (reactor) for the hydrogenation of gas oils at
430°C and 120 atmospheres pressure. Yields showing the conversion
of fifty percent of the feed to gasoline, with no appreciable coke
formation and with the sulfur in the oil removed as hydrogen sulfide,
were most encouraging. During the first World War, German emphasis
was shifted to the hydrogenation of lignite and coal, which provided
the experience needed for the design of the larger scale hydrogenation equipment installed at Mannheim-Rheinau between the end
of the war and 1924. With this equipment, using cylindrical
reactors 8 meters long and 80 centimeters in diameter, both coal
and oil could be hydrogenated. High-sulfur residual oils, fed to
the reactors at a rate of 30 tons per day, provided yields of 25%
to 35% gasoline and 40% to 50% gas oil. After proving the technical
feasibility of the process, Bergius concluded his participation in
this field in 1927. However, BASF, later I.G. Farbenindustrie
(IG), continued the work on hydrogenation techniques, including
the development of successful catalysts used in subsequent processes.
It is interesting to note here that Bergius evidently never undertook a systematic search for catalysts for his reactions, although
his work in 1920-21 shows that he appreciated the usefulness of
these substances.

In 1926, officials of the Standard Oil Development Company
(SOD), the research arm of the Standard Oil Company of New Jersey
(Jersey), became aware of the German work in hydrogenation and in
other areas. Jersey's interest led to agreements, in 1927, whereby
SOD joined in the commercial development of the process.[15] This
joint effort led to the formation of a new organization, the
Standard-I.G. Co., to manage the patents that Jersey had obtained
from IG.[16,17] SOD immediately set up special research and
development facilities for hydrogenation at the Baton Rouge
refinery of the Standard Oil Co. of Louisiana, a Jersey affiliate;
and within two or three years, two commercial petroleum hydrogenation plants, each rated at 5,000 barrels per day, were on
stream. It was noted that these plants, which operated at
pressures of 3000 to 4000 psi and at temperatures of 650° - 950°F,
could have considerable value in upgrading petroleum fractions in
many ways, for example:

    The manufacture of high-octane aviation gasoline from
       kerosine and gas oil;

    The obtaining of motor gasoline from gas oil;

    The manufacture of aviation blending stock by
       saturation of branch-chain olefinic polymers
         (dimers of butylene);

The "Hydrofining" of gasoline to remove sulfur;

The preparation of high-flash, high-octane safety fuels.[18]

Truesdell, quoting from The Lamp, a Jersey publication, indicated the optimism and enthusiasm with which this new process was greeted in the Jersey organization:

> The importance of this development...can hardly be over-emphasized...it removes the fear of a permanent shortage of petroleum products and permits the development of our crude oil supplies on a rational basis.

and:

> As everyone knows, it is impossible to meet the present day demand for gasoline without at the same time making a quantity of fuel oil and coke that is out of all proportion to the size of the market. Hence these two products must be disposed of at a loss.

The article went on to express the hope that the hydrogenation process would make it possible to convert most residuals and fuel oil to gasoline. Finally:

> It is confidently believed that this process makes a real forward step in the art of petroleum refining. For the first time the petroleum refining industry has available to it a process for the constructive upbuilding of its products without a corresponding degradation of a portion thereof.[19]

At the same time, there was recognition that the economics of petroleum refining would play an important part in determining the extent to which the process would be used. The increased yield of desired products had to be balanced against the expenses of high-pressure equipment, hydrogen, and drastic operating conditions. One "neutral" authority is quoted:

> (The process)...is unnecessary at the present time due to the fact that there is an ample supply of gasoline to take care of needs. The chief importance...is in the fuel conservation program...(it) permits conversion of practically all crude oil into gasoline...and leaves the field of heavy fuel to coal to which it rightfully belongs.[20]

By 1940 Jersey officials noted that other new catalytic processes had been developed for manufacturing aviation gasoline, and that:

"The relative economics of these different methods depends on the particular situation involved."[21]

With the successful commercialization of its hydrogenation process, Jersey set about trying to interest other companies in it. To do this, it organized in 1930 the Hydro Patents Co. to take over the U.S. Patent rights to the process from the Standard-IG Co. Stock in the new company was offered to other major refineries, and soon seventeen companies representing 80% of the refining capacity in the U. S. had joined in the venture.[22] As it turned out, high pressure hydrogenation did not become a popular refining technique in America after its early promise. Other methods - more economical - did appear, some based on the patents held by the Hydro Patents Company. But there can be no questions as to the important part that the high-pressure hydrogenation process played in pointing out to the American oil industry the feasibility of upgrading its products through the use of catalytic conversion processes.

It is interesting to note that the Germans also continued to develop hydrogenation processes. Birkenfeld traces the growth of this industry from the construction in 1926 of the "Leuna-Werk" plant, until 1939, when Germany had seven hydrogenation plants, providing 1,200,000 tons per year of fuel.[23] In the early 1930's the process had proven to be uneconomical in competition with imported gasoline, but national policy, for a time, overruled economic considerations. In 1940, when the German leaders were confident that the war would be short, they did not push for additional hydrogenation capacity. Later, when Allied air strikes had put several hydrogenation plants out of commission, the resulting fuel shortage paralyzed the German air defenses.

Before continuing on into discussions of other rather specific developments in catalytic reforming, it is interesting to take a broader look at the variety of work that was being performed between the World Wars in attempts to increase the quality and quantity of gasoline. While high-pressure hydrogenation was perhaps the most spectacular of these developments, it was certainly not alone in its impact on the refining industry. One has only to survey the patent literature of the 1920's and 1930's, both in the United States and elsewhere, to get a feel for the activity of the times. While much of this activity was directed toward improving the thermal processes already in use, considerable effort was expended in searching for new catalysts and for processes that would make use of these catalysts.

The driving forces behind these activities were: first, the development of improved engines, with higher compression ratios,

NAPTHA REFORMING

requiring higher octane fuels; and second, the rapid increase in the number of motor vehicles. The cracking processes helped fill the second of these needs, and - indirectly at least - the first; but improved quality had to depend upon other kinds of processes. as suggested earlier.

A real knowledge of what components were desirable, in high quality fuels, developed rather gradually. For example, benzene - an aromatic - was considered a useful "anti-knock" fuel during the first World War; but by 1928 engineers had decided that it wasn't good in hot engines.[24] It probably wasn't until a suitable octane rating system had been accepted that engine designers and refiners began to have some specific ideas as to what they were seeking in gasoline components. By this time the value of tetraethyl lead as an additive for increasing the octane rating of gasoline had been well established; as had been the fact that not all kinds of gasoline were equally as "susceptible" to "lead". It was generally recognized that branched-chain paraffins had very high octane ratings, whereas straight chain ("normal") paraffins were very poor in this regard. Unsaturated compounds (olefins) had higher octane ratings than normal paraffins, but tended to form "gums" (polymerize) and had rather poor lead susceptibility. Cyclic compounds, both saturated (naphthenes) and aromatic were known to have good anti-knock properties. So the major efforts of gasoline manufacturers began shifting towards optimizing the output of branched paraffins and cyclics.

One way of producing high quality gasoline was through the careful separation (fractionation) of the desired components from straight-run gasoline and naphtha fractions. But the total quantity of high-test components from this source was limited. More important were the developments, in the late 1930's, of catalytic polymerization and alkylation processes.

"Polymerization", as used here, refers to the reaction between two unsaturated molecules - of three or four carbon atoms each - to form a larger unsaturated molecule, which must then be saturated by hydrogenation. In general, these unsaturated reactant molecules were available as products from thermal cracking, so that gasoline manufacturing processes could be set up by combining cracking, polymerization, and hydrogenation reactors in series or in some other combination.[25]

"Alkylation" techniques, which involve the reactions between saturated and unsaturated molecules in the presence of sulfuric or hydrofluoric acid catalysts, appeared after polymerization processes. The advantage of these is that the reaction products are saturated molecules, so that subsequent - or concurrent - hydrogenation steps are not required.

In the late '30's, catalytic reforming processes began to appear, with the Houdry process, mentioned earlier, probably the first of these. Several companies were developing catalysts and processes that could be used for the complex reforming reactions,[26] so that by 1940, Gustav Egloff could describe several kinds of catalytic processes for manufacturing quality gasolines, utilizing reforming, cyclization, and isomerization techniques.[27] Sachanan, too, described the state of the petroleum conversion art at this time, including the work of Russian scientists.[28,29]

While this flurry of activity produced many excellent results that proved to be essential in the aviation gasoline program during the second World War, one particular process that appeared shortly before 1940, "Hydroforming," seemed to stand out. Of the importance of this process in 1939, Cooke said:

> The one restraining factor in the...installation of isoctane and alkylation plants is the fear of a newcomer...catalytic reforming, (which) if conducted in the presence of hydrogen is called hydroforming.[30]

The development of this process can be traced to Jersey's work in hydrogenation that had led to the development of their high pressure hydrogenation plants, and to their continued search for better catalysts in the 1930's.[31] Of this development, Popple states:

> A modification of hydrogenation known as hydroforming was evolved by Jersey and others in this country. This eventually led to the development by Jersey of a method for quantity production of nitration-grade toluene from petroleum fractions.[32]

Paul Giddens, in his history of the Standard Oil Company (Indiana) also claims a share for that company in the development of this process. He notes that in the late 1930's, Indiana, Jersey and the M. W. Kellog Company, "were working cooperatively on a catalytic refining process." The cooperative venture led to fluidized bed catalytic cracking and, no doubt, to hydroforming.[33] As did Jersey, Indiana discovered that the new process could provide large quantities of toluene and other aromatics as well as aviation gasoline.

The first commercial Hydroformer, a 7500 barrel per day plant, was located at the Texas City refinery of the Pan American Refining Corporation, an Indiana subsidiary.[34] The plant operated at temperatures from 900° to 1100°F and at pressures in the range of 200 to 300 psi. Naphtha feed and excess hydrogen, from the process itself, were passed over the catalyst of silica and molybdena co-precipitated on alumina, with the result that the following principal reactions took place:

1. Dehydrogenation and cyclization of straight-chain hydrocarbons;

2. Dehydrogenation of naphthenes;

3. Conversion of olefins to paraffins or cyclic compounds;

4. Controlled production of selected aromatics;

5. Desulfurization.

Because of carbon deposition, the catalyst had to be regenerated every four to eight hours. To keep the process continuous, four reactors were used in order to make it possible to regenerate the catalyst in one while the others were "on stream".

The process was successful in producing both high octane gaoline blending stocks and toluene, the latter being used in the wartime manufacture of trinitrotoluene. Based upon data obtained from the operation of the Texas City plant, several Hydroformers were designed and installed at other locations around the country.

World War II was a time for cooperation among refiners, through the Petroleum Administratio for War.[35] By 1941 several kinds of catalytic processes for producing high-octane gasoline and other valuable petroleum chemicals had been developed and were in operation, as mentioned earlier. It was a time of patent-pooling, cross-licensing, and generally unified effort that contributed immeasureably to the Allied victory. The increase in production of 100 octane gasoline through the war years has been called "Without question, one of the great industrial accomplishments in the history of warfare."[36]

The end of hostilities brought the end of government-regulated inter-company cooperation. Refiners were able once again to go back to their own research and development projects and to initiate investigations for new and better conversion processes. Because of their versatility and potential as a source for both high-quality gasoline and chemical raw materials, catalytic reforming schemes received considerable attention.[37] And, of all the reforming schemes that emerged from the period immediately following the war, probably the most significant was Universal Oil Product Company's "Platforming" process.

Platforming was announced by UOP in March, 1949, and the first commercial installation of the process was completed in October of that year at the Old Dutch Refining Company in Muskegon, Michigan. It was the most quickly adopted of the "new" peacetime processes.[38,39]

Principal credit for the development of the Platforming catalyst goes to Vladimir Haensel, a research chemist with UOP. He has described the objectives of his research as being directed toward the development of a reforming catalyst that would:

1. Make the process as free as possible from thermal effects;

2. Be effective with the wide variety of reactions involved in reforming: dehydrogenation of naphthenes, hydrocracking, isomerization of paraffins, desulfurization, dehydrocyclization, etc;

3. Produce a high percentage conversion of feed to desired products, thus avoiding costly product separation steps;

4. Eliminate the need for frequent catalyst regeneration typical of reforming operations up to that time.[40]

That he was successful in his quest is quite evident. Platinum had been known as an excellent catalyst for hydrogenation-dehydrogenation reactions since the time of Sabatier, at least. What made Haensel's catalyst valuable was the fact that "exceptionally good catalysts could be prepared to contain very low concentrations of platinum," thus overcoming the high cost of the metal.[41] The patent description of the catalyst and its preparation included the claim for combining a halogen (such as fluorine) with alumina and platinum (the latter .01 to 1.0 percent by weight).[42] The patent also indicated the success that was achieved in developing a catalyst with a long active life, and subsequent experience with commercial units showed that catalyst regeneration was required only after eighteen months or so of operation (a function, of course, of the feed used and the operating conditions employed).

In the Platforming operation, naphtha feed and hydrogen were heated to a temperature of about 900°F, and passed through the catalyst beds at a total pressure in excess of 500 pounds per square inch (although more recently developed catalysts have made operation at substantially lower pressures possible). In order to provide the necessary contact time between the reactant and the catalyst particles, and to facilitate the addition of energy needed for the reforming reactions, the catalyst was placed in three or four reactors through which the reacting stream passed in series. Heat energy could then be added to the reaction mixture, as needed, between reactors.[43] Regeneration of the catalyst was usually accomplished in the reaction chambers, with a carefully controlled air-stream mixture (containing about 1% oxygen) employed for burning the "coke" from the catalyst pellets. Fresh catalyst, as needed, was added at this time.

   The development of efficient reforming processes has given the petroleum industry considerable leverage for the production of several kinds of products. For example, in a Platforming operation it is possible to maximize the production of aromatic hydrocarbons, or to emphasize the manufacture of high octane gasoline merely by modifying operating conditions. So it is no wonder that refiners have developed so many varieties of this refining concept, as noted earlier. Bell has listed more than a dozen of these that had been operated by the late 1950's.[44]

   Reforming technology has continued to improve, especially in the development of more efficient catalysts. Gates points out that the introduction of bimetallic and multimetallic catalysts has made it possible to operate reformers at lower pressures than was possible earlier, and that the new catalysts have "... enhanced resistance to deactivation by coking, which allows relatively long runs..."[45]. Reformer capacity has also grown in the last ten years or so, after levelling off during the 1960's. Increased demands for higher octane fuels, as limits were placed upon the use of lead additives for octane boosting, certainly has been a major factor in this renewed interest in the process. And new reforming schemes are appearing, such as one for continuous Platforming announced by UOP in 1971[46].

   Catalytic naphtha reforming has been, and continues to be, a major force in the production of better motor fuels and petrochemical feedstocks. It arose from many roots and as a result of the work of many people. It certainly is deserving of a place of honor in the history of chemical engineering.

FOOTNOTES

   [1]Bruce C. Gates, James R. Katzer, and G.C.A. Schuit, Chemistry of Catalytic Processes (New York: McGraw-Hill Book Co., 1979) Chapter 3, "Reforming." Many aspects of catalytic naphtha reforming are covered in this excellent new book. A general overview of reactions, including thermodynamics, kinetics, catalysis, reactor operating conditions, and reactor design is presented on p. 184-193. Detailed discussions of these topics follow in the chapter.

   [2]H. B. Kendall, "Catalytic Reforming: The Search for 'Higher Octanes'" (paper presented at the 68th National Meeting of the American Institute of Chemical Engineers, Houston, Texas, March 4, 1971).

[3] John L. Enos, Petroleum Progress and Profits (Cambridge, Mass.: M.I.T. Press, 1962).

[4] Gates, et al. p. 184-186.

[5] A. G. Peterkin, J. R. Bates, and H. P. Brown, "Catalytic Reforming for the Production of Aviation Gasoline," Refiner and Natural Gasoline Manufacturer, 18, No. 11 (Nov. 1939) 126-130. This article describes the commercial process, which was an adaptation of the successful Houdry Catalytic Cracking process, to reforming. In fact, a point of emphasis in the article is that the cracking equipment was well designed and versatile, so that it could be switched over quite readily from cracking gas oil stocks to reforming gasolines or naphthas. The same catalyst was used ("a highly active hydrosilicate of alumina") in the fixed-bed reactors, and frequent regeneration of the catalyst was necessary.

[6] Robert Schlaifer and S. D. Heron, Development of Aircraft Engines and Fuels (Boston, Harvard: 1950) 615.

[7] Peterkin, et al. p. 130.

[8] See, for example, translations of Sabatier's works by Reid in: Paul H. Emmett, Paul Sabatier, and E. Emmett Reid, Catalysis Then and Now (Englewood, N. J., Franklin Publishing Company, Inc., 1965). Chapts. VIII-XII, XIV.

[9] V. N. Ipatieff, "Earlier Work on Hydrogenation at High Temperatures and Pressures," in, The Science of Petroleum, Vol. III, eds. A. E. Dunstan, A. W. Nash, B. T. Brooks and H. Tizard (Oxford, 1938) 2133-2138.

[10] Many biographies of Fritz Haber are available. See, for example: Bonhoeffer, K. F., "Fritz Haber", in Great Chemists, ed. by Eduard Farber (New York, Interscience, 1961) 1299-1311; also: Eduard Farber, Nobel Prize Winners in Chemistry (New York, Henry Schuman, 1953) 71-75. This reference contains a brief summary of Haber's description of the process as outlined by him in his Nobel Prize address in 1919.

[11] Eduard Farber, Nobel Prize Winners in Chemistry, (New York, Henry Schuman, 1953) 123-124; 126-128.

[12] Thomas Parke Hughes, "Technological Momentum in History: Hydrogenation in Germnay 1898-1933." Past and Present, No. 44 (August, 1969) 106-132. Hughes suggests that the success of the Haber-Bosch process provided BASF with special expertise in ".. a complex new technology: high-pressure, high-temperature, catalytic hydrogenation." This opened the way for subsequent significant developments such as the methanol synthesis, the Fischer-Tropsch synthesis and the work by Berguis on the hydrogenation of coal and oil.

[13] Farber, Nobel Prize Winners, p. 124-126; 128-129.

[14] At least two accounts, essentially identical, in English, are available wherein Bergius has described his pioneer work: Friedrich Bergius, "A Historical Account of Hydrogenation," Petroleum Times, 30 (August 12, 1933) 253-256, 268; and Friedrick Bergius, "The Historical Development of Hydrogenation," in the Science of Petroleum, Vol. III, eds. A. E. Durstan, et al. (Oxford, 1938) 2130-2132. In these accounts Bergius points out that cracking was known when he began his work, but the losses in coke and gases were also known to be very large. He also knew of the constantly increasing need for motor fuel. He concluded that "...defects of the cracking process...inferior...unsaturated gasoline...could be remedied only by...replacing the hydrogen removed (during cracking)." His early work showed him that hydrogen was taken up by oils, that coke didn't form, and that the products formed were light and saturated.

[15] Several articles have appeared describing the cooperation between IG and Jersey in the late 1920's, and in outlining the subsequent commercialization of the process. Included are: R.T. Haslam & R.P. Russell, "Hydrogenation of Petroleum," Industrial and Engineering Chemistry, 22, No. 10 (October 1930) 1030-1040; approximately the same article with the same authors in: Refiner and Natural Gasoline Manufacturer, 9, No. 10 (October 1930) 116-144; and a later article: R.P. Ru-sell, "The Hydrogenation of Petroleum" in, The Science of Petroleum, Vol. III, eds. A.E. Dunstan et al. (Oxford, 1938) 2139-2148. The authors stress the importance of IG's development of "extremely rugged" sulfur-resistant catalysts, and also express their respect and admiration for (among others) Dr. Carl Bosch, who was by this time President of IG.

[16] Charles S. Popple, Standard Oil Company (New Jersey) in World War II, (New York, Standard Oil Co. (N.J.) 1952) p. 9.

[17] Joseph Borkin, The Crime and Punishment of I. G. Farben, (New York, The Free Press, 1978). Relationships between IG and Jersey in the 1920's, 1930's, and until the entrance of the USA in World War II, are examined by Borkin. Agreements involving the formation of several joint companies for the sharing and commercializing of new processes and products, involving such things as hydrogenation, synthetic rubber, lubricating oils, etc., are presented. The interested reader is referred to this book and to sources such as the records of the Anti-Trust Division of the Department of Justice and of the Truman Committee, in 1941 and 1942, for a better understanding of the cooperation between these two companies at this time in history.

[18] E.V. Murphree, C.L. Brown, and E.J. Gohr, "Hydrogenation of Petroleum", Industrial and Engineering Chemistry, 32, No. 9 (Sept. 1940) 1203-1212.

[19] Paul Truesdell, "Hydrogenation to be Used First in Conversion of Heavy Fuel Oil", <u>National Petroleum News</u>, 21, No. 28 (July 10, 1929) 27-29.

[20] Charles E. Kern, "Hydrogenation Will Assure Consumer of Adequate Supplies", <u>Oil and Gas Journal</u>, (July 24, 1930) 138.

[21] Murphree; et al., <u>op. cit.</u>

[22] Kern, <u>op. cit.</u> p. 138; and Popple, <u>op. cit.</u> p. 11.

[23] Otto Mayr, review of Wolfgang Birkenfeld, <u>Der Synthetische Treibstoff 1933-1945</u>, (Gottinger, 1964), in <u>Technology and Culture</u>, 8, No. 3, (July 1967) 425-429.

[24] Schlaifer and Heron, <u>Development of Aircraft Engines and Fuels</u>, (Boston, Harvard 1950) p. 592.

[25] "Hydrogenation-Polymerization Combined in California Plant", <u>Oil and Gas Journal</u>, 37 (July 21, 1938) p. 18; M.B. Cooke, "Demand for Aviation Fuel Coincides with Era of Catalytic Cracking", <u>World Petroleum</u>, 10, No. 11 (Annual Refinery Issue, Oct. 31, 1939) p. 72; Paul Ostergaard and Eugene R. Smoley, "Gulf Plants Produce High Gasoline Yield by <u>Polyform</u> Process", <u>Oil and Gas Journal</u>, 39, (Sept. 12, 1940) 52-59, 62; Gustav Egloff, E.F. Nelson, and J.C. Morrell, "Motor Fuel From Oil Cracking. Production by the Catalytic Water Gas Reaction." <u>Industrial and Engineering Chemistry</u>, 29, No. 5 (May, 1937) 555-559; R.P. Mase and N.C. Turner, "Thermal Reforming Plus Catalytic Polymerization", <u>Oil and Gas Journal</u>, 38 (April 11, 1940) 87-88.

[26] The list of patents dealing with reforming seems almost endless. To cite a few (by company):

a.) University Oil Products Company (U.O.P.)
- Brit. 484 368   May 4, 1938 (Feldspar and Zeolite catalysts)
- Brit. 498 247   Jan. 5, 1939 (Alumina catalyst)
- U.S. 2 194 335   March 19, 1940 (Magnetic catalyst)
- U.S. 2 289 375   July 14, 1942 (Fluorides of Al & Mg as cat.)
- U.S. 2 295 197   Sept. 8, 1942 (Alumina carrier & metal oxides or sulfides)
- U.S. 2 349 812   May 30, 1944 (Dehydrog. plus cracking cat.)

b.) Standard Oil Development Company (S.O.D.)
- Brit. 542 925   Feb. 3, 1942 (Fluidized bed of Alumina & metal oxide catalyst)

NAPTHA REFORMING

        U.S. 2 367 365   Jan. 16, 1945 (Reforming & Hydroforming; alumina plus metal oxide or sulfide)

c.) M.W. Kellog Company
     Brit. 544 155   March 30, 1942 (Reforming naphtha with hydrogen and "dehydrogenating aromatization" catalyst)

d.) Union Oil of California
     U.S. 2 279 198   April 7, 1942 (Zinc oxide catalyst)
     U.S. 2 279 199   April 7, 1942 (Uranium & Vanadium Oxide catalyst)
     U.S. 2 393 288   Jan. 22, 1946 (Cobalt Molybdate cat.)

e.) Atlantic Refining Company
     U.S. 2 285 294   June 2, 1942 (Alumina & Alkali metal borate catalyst)
     U.S. 2 289 530   July 14, 1942 (Fullers earth plus CS_2 catalyst)

This listing is by no means complete, but suggests the kinds of experimentation in reforming catalysts and processes that were being conducted just prior to World War II (when many of these patent applications were first filed) and early in the War.

[27] Gustav Egloff, "Improved Processes in the Manufacture of Motor Fuels from Petroleum", Petroleum Engineer, 11, No. 10 (1940), 21-24, 26.

[28] A.N. Sachanan, Conversion of Petroleum (New York, Reinhold, 1940).

[29] Ibid., p. 36; See also: V.I. Karzhev, M.G. Sever'yanova, and A.N. Siova, "High Octane Gasoline Obtained by Catalytic Dehydrogenation", Oil and Gas Journal, 37 (June 9, 1938) 50-53. The authors report the use of chromium oxide catalysts, at $500°-550°C$, in the dehydrogenation of naphthenes and naphthalenes to aromatics; with evidence of converting paraffins to aromatics simultaneously.

[30] M.B. Cooke, "Demand for Aviation Fuel", World Petroleum, 10, No. 11 (October 31, 1939) p. 75.

[31] John L. Enos, Petroleum Progress and Profits, (Cambridge Mass., M.I.T. Press, 1962) p. 192.

[32] Charles S. Popple, Standard Oil Company (New Jersey) in World War II. (New York: Standard Oil Co., (N.J.), 1952) p. 11.

[33] Paul H. Giddens, Standard Oil Company (Indiana), (New York, Appleton-Century-Crofts, Inc., 1955) 593-599; See also D.J. Smith and L.W. Moore, "The Hydroforming Process", News Edition, American Chemical Society, 19, No. 7 (April 10, 1941) p.432.

[34] This plant is described in many articles, among them the following by D.J. Smith and L.W. Moore: "The Hydroforming Process", News Edition, American Chemical Society, 19, No. 7 (April 10, 1941) 428, 432; "Hydroforming Is New Dehydrogenation Process", Petroleum Engineer, 12 (April 1941) 23, 24; "New Catalytic Reforming Process Easy to Convert to War Needs", Nat'l Petroleum News, 33, No. 14 (April 2, 1941) R98-R102. In addition: J.V. Hightower, "First Hydroformer Unit Put on Stream", Refiner and Natural Gasoline Manufacturer, 20, No. 5 (May 1941) 63-66; L.R. Hill, G.A. Vincent, and E.F. Everett, Jr., "Hydroforming", Nat'l. Petroleum News, 38, No. 23 (June 5, 1946) R456-R465 (This is an excellent article, written by M.W. Kellog Co. engineers); and George Armisted, Jr., "The Hydroforming Process", Oil and Gas Journal, (August 31, 1946) 85-87, 100.

[35] John W. Frey and H. Chandler Ide, eds., A History of the Petroleum Administration for War 1941-1945. (Washington; U.S. Gov't. Printing Office, 1946) 191-208.

[36] Ibid. p. 191. Quoted from the Army-Navy Petroleum Board of the Joint Chiefs of Staff.

[37] Some of these are: E.O. Saegebarth, "Catalytic Reforming Process of Standard Oil Co. of California", Petroleum Engineer, 17, No. 8 (May, 1946) 95-100. This process, using a molybdena-alumina catalyst, was in use before the end of the war. V.W. Daniels and M.W. Conn, "Perco Cycloversion for the Small Refiner", Petroleum Processing, 2, (May, 1947) 391-397. The authors are with Phillips Petroleum. Clyde Berg, "Hyperforming - Newest Development in Reforming", Petroleum Refiner, 31, No. 12 (Dec. 1952) 131-136. Union Oil of California developed this moving-bed process, using a cobalt molybdate catalyst. J.W. Teter, B.T. Borgerson and L.H. Beckberger, "New Catalytic Reforming Process", Oil and Gas Journal, 52, No. 23 (Oct. 12, 1953) 118-121, 140, 144. Development of a platinum reforming catalyst by Sinclair Research Labs and Baker and Co., Inc. H. Heinemann, et al., "Houdriforming Reactions", Industrial and Engineering Chemistry, 45, No. 1 (Jan. 1953) 130-134. More on the Houdry process. So a continued interest in reforming is indicated, well into the 1950's.

[38] "Petroleum Panorama" Oil and Gas Journal, 57, No. 5 (January 28, 1959) F15, F17.

[39] Harold F. Williamson, et al., *The American Petroleum Industry. The Age of Energy 1899-1959*. (Evanston, Northwestern University Press, 1963) p. 809.

[40] Vladimir Haensel, "Platforming", *Petroleum Refiner*, 29, No. 4 (April 1950) 131, 132. Essentially this same information, in articles also authored by Haensel, appeared in other petroleum industry journals: *Oil and Gas Journal*, 48, No. 47 (March 30, 1950) 82-85, 114, 119; *Petroleum Processing*, 5 (April, 1950) 356, 357; *Petroleum Engineer*, 22, No. 4 (April, 1950) C9-C14. See also: V. Haensel and G.R. Donaldson, "Platforming of Pure Hydrocarbons", *Industrial and Engineering Chemistry*, 43 (Sept. 1951) 2102-2104, which describes the laboratory work on the chemistry involved in the process.

[41] V. Haensel, U.S. 2 479 100 (August 16, 1949) assigned to U.O.P. The patent describes the "Process of Reforming a Gasoline with an Alumina-Platinum-Halogen Catalyst."

[42] V. Haensel, U.S. 2 479 109 (August 16, 1949) assigned to U.O.P.

[43] Descriptions of commercial Platformer operations, as found in the Old Dutch refinery, can be found in: V. Haensel, *Oil and Gas Journal*, 48, No. 47 (March 30, 1950) 85 et seq.; Merritt L. Kastens and Robert Sutherland, *Industrial and Engineering Chemistry*, 42, (April 1950) 582-593; William F. Bland, *Petroleum Processing*, 5, (April 1950) 351-355.

[44] H.S. Bell, *American Petroleum Refining*, 4th Ed. (Princeton, D. Van Nostrand, 1959) p. 287.

[45] Gates, et al. p. 290.

[46] E. A. Sutton, A. R. Greenwood, and F. H. Adams, "A New Processing Concept - Continuous Platforming" (Paper presented at the Annual Meeting of the National Petroleum Refiners Association, San Antonio, Texas, March 26-28, 1972).

CHEMICAL ENGINEERING AND FOOD PROCESSING

C. Judson King

Department of Chemical Engineering
University of California
Berkeley, CA 94720

ABSTRACT

Many important problems in food processing and preservation have a substantial chemical engineering character. However, until recently the contributions of chemical engineers themselves have been relatively small. Chemical engineering principles provide an effective a posteriori interpretation of several classical advances in canning, freezing and concentration. This contrasts with the serendipitous fashion in which these advances were actually made.

The development of freeze-dried coffee is traced as a recent chemical-engineering accomplishment, stemming directly from the crash World War II programs for manufacture of blood plasma and penicillin.

The U. S. food industry has developed along different lines in deploying chemical engineers. Larger companies started wholesale utilization of chemical engineers at quite different points in time; smaller companies and some large ones still have few chemical engineers, if any. Some companies with a strong chemical engineering tradition have branched into the food industry. Several important innovations have come from equipment vendors, from companies marketing proprietary processes, and from universities. The uneven use of chemical engineers leads to an important impact from consultants. Unique to the food industry is the strong applied research service provided by the federal government.

Important classes of problems rooted in chemical-engineering principles are outlined, with an indication of likely avenues for future developments.

INTRODUCTION

The people of the world consume over 2000 Tg (4.4 billion pounds) of food annually [extrapolated from (1)]. Foods are grown seasonally and often at some distance from the point of consumption. Also, most foods are rapidly perishable. Consequently, much of the food produced in the world must be preserved somehow--such as by heat sterilization and canning, by freezing, or by concentration or dehydration. Other processing operations change the form of foods to make them more suitable or more convenient for consumption.

As we shall see, many food processing and preservation steps depend upon chemical engineering principles for effective and efficient operation. Energetics, phase and chemical equilibria, reaction kinetics, transport and separations are all centrally involved. Yet, until recently, chemical engineers themselves have been involved to a relatively small degree in the discovery, development and improvement of processing and preservation methods for foods. This situation resulted from a number of factors: Needs for food preservation far predate the emergence of Chemical Engineering as a discipline. Chemical Engineering grew up in the seemingly unrelated contexts of petroleum refining and chemicals manufacture. Food Science, Food Technology and Agricultural Engineering became established as separate disciplines in many universities; the curricula in these fields have a very small or non-existent chemical engineering component. Food research and development were for the most part carried out by food chemists and food scientists. Also, research and development activities have always assumed a much lesser role in the food industry than in the petroleum and chemical industries. Equipment has most often been bought from vendors with a base largely in mechanical engineering, rather than being developed within the food industry. Microbiological and biochemical reactions and processes are important in the food industry, and until recently chemical engineers had little familiarity with those areas. Finally, a very significant recent deterrent to even greater growth of the number of chemical engineers in the food industry is the fact that salaries for chemical engineers are appreciably greater than those for food scientists and chemists.

THE DEVELOPMENT OF FOOD PRESERVATION TECHNIQUES

It is interesting to trace some of the developments in methods of food preservation, so as to observe the relevance of chemical engineering concepts as well as the rather serendipitous fashion in which these developments came about.

Salting

Salting is one of the oldest forms of preservation, going back to prehistoric times. It is interesting to note that there was a definite link in early Egypt between the use of salt for embalming the dead and its use for preserving foods for the living (2). Fish and some meats were salted, with the growth of Christianity and fish-on-Friday being a principal spur to the industry. In the 14th and 15th centuries the salted-herring industry in Europe was very large, and salted cod for export back to England was the first industrial product of the Massachusetts Bay Colony (3).

Two methods were used--dry salting, in which the product was buried in a bed of dry salt, and brine curing, where the product was immersed in a strong brine.

The action of salt was not understood at the time and is still not fully understood, but there seem to be at least two effects. First, the salt serves as an osmotic agent providing a water-activity driving force, causing water to pass out of the product into the salt bed or solution. This accomplishes some preservation through dewatering. Such a process has re-emerged in recent years in developmental form under the name osmotic dehydration (4), with the solute increasing the osmolarity of the surrounding medium being sugar or any of various other highly soluble substances. These current processes result from a recognition of osmotically driven water transport across cell membranes, whereas salting, of course, did not.

The second effect of salt is to penetrate the food and alter the solute composition of the intracellular solution. This inhibits or selectively controls the growth of micro-organisms. Selective control of fermentation reactions through salt concentration is used to advantage in pickling (1). Microbial and biochemical reactions are quite sensitive to water activity. A semi-quantitative understanding of mechanisms and rates of spoilage reactions at various levels of water activity is leading to expanded use of intermediate-moisture foods, which have already had a major influence on pet foods. Here water activity is reduced and controlled by adding humectants such as propylene glycol and sorbitol, which are food-compatible and do not alter taste in the way that salt does. Pemmican, invented by pre-Columbian American Indians, was an early product of this sort. The process for making it involved partial drying, as well as the addition of humectants and fat.

Dehydration and Concentration

Drying as a method of preservation was probably discovered accidentally when it was found that certain foods left out in the sun for a sufficient time were stable much longer than was the case otherwise. Drying of foods so that they could be stored for winter consumption has been traced back to ancient Mesopotamia and to "chuño", prepared by the Incas 2000 to 3000 years ago (5). Potatoes were left out to freeze overnight in the high Andes, were repeatedly trampled to squeeze out juices, and the resulting thin mash was left to dry in the sun. In hindsight, the freezing weakened cell walls, and the trampling removed enough water to give stability during drying and a thin enough material to dry before spoiling.

The use of fire to accelerate drying dates back to unrecorded times. The benefits of artificial heat sources, thin pieces to be dried, and dry air to remove moisture were all found without any systematic knowledge of drying mechanisms, although Aristotle, around 330 B.C., recognized that heat was an essential factor in other applications of drying (6). Beef jerky, cut in thin slices and sun dried, was a staple in the early Western U. S.; ox jerky, made from pack animals, saved more than one party of desperate pioneers [see, e.g., (7)]. Wars have spurred the development of drying processes for foods, since dried foods are easily stored and transported. Marco Polo reported in his diary that the Mongols made a sun-dried milk powder which was used on military expeditions (8). Horsford, a Harvard Chemistry professor, promoted the use of dried vegetables for Union troops during the Civil War. Further improvements in drying technology occurred during World Wars I and II (6,8). In the U. S., the Army remains one of the foremost sponsors of research on food dehydration.

Before the 1930's, selection and design of dryers for foods was done in the absence of quantitative mechanistic understanding of the underlying heat and mass transfer processes. Sherwood, Hougen and other chemical engineers then identified the transport principles underlying drying processes in general, and these concepts found their way into the design and analysis of dryers for foods. The subsequent work of Marshall in the late 1940's and 1950's had a considerable impact on the design of spray dryers for milk, coffee and other liquid foods.

Concentration of liquid foods (partial water removal) has been recognized for some time as a desirable way of achieving some stability and reducing weight and volume for transport. Borden in 1856 invented his process for vacuum evaporation of milk (9). Important factors for stability in that case were heat sterilization, lowering of water activity, sanitary handling, and canning to exclude the atmosphere. Evaporation is used on a very large scale today for the production of orange concentrate, for concentration of coffee extract before spray drying, and in the manufacture of sugar.

Dried foods tend to have a lower quality than those preserved by other means. Overcoming this drawback requires an understanding of the kinetics of various degradation reactions for nutrients, color, flavor, etc., in comparison to rates of heat and mass transfer involved in drying. Retention of volatile flavor and aroma has been improved through recent mechanistic studies in universities of mass-transfer factors governing the loss of these compounds in drying and evaporation. Freeze drying and various puffing methods have led to better rehydration characteristics and texture retention. These problem areas all involve chemical engineering centrally.

Evaporators for foods present special complexities due to fouling, quality loss from excessive residence time, vacuum processing to avoid high temperatures, and loss of volatile flavor components (10). These problems are also amenable to creative chemical engineering.

Heat Sterilization and Canning

The history of preservation by thermal processing has been traced by Goldblith (11,12). The initial breakthrough was that of Nicholas Appert (13), made in the 1790's in response to a reward of 12,000 francs offered by Napoleon to anyone inventing a useful method of food preservation. Once again war spurred the development of a food-preservation technique. Appert's process consisted of enclosing the food in carefully corked bottles, submitting the bottles to heat by immersion in boiling water for lengths of time depending upon the product, removing the bottles and cooling them. Subsequently Durand in England introduced sealed cans, rather than corked bottles. The chemist, Gay-Lussac, explained the success of the process in terms of a somewhat fanciful mechanism whereby oxygen was absorbed at high temperatures, producing a heat-solidified substance which precludes fermentation or putrefaction. It was not until Pasteur that it was recognized that canning achieves stability through heat destruction of microorganisms.

Subsequently, salting the boiling water and pressurization came into use as ways of accelerating the process through higher temperatures. Only much later was it realized that different microorganisms would have different kinetics of destruction and hence some might require higher temperatures in order to be killed during retorting. Quantitative schedules required for sufficient destruction of organisms were developed long before chemical engineers entered the picture; however, in recent years chemical engineers have placed formerly empirical relationships on a firmer kinetic basis, and have begun to explore ways of optimizing thermal processing so as to achieve the necessary heat transfer in ways that minimize loss of desirable food-quality properties, such nutrients and color.

Freezing

Freezing has been known as a preservation method since early times. The first commercial freezing of food in the U. S. began about 1863, as mixtures of salt and ice were used to freeze fish in the Great Lakes region (14). However, large-scale use of freezing for preservation awaited an engineering development of a different sort, ammonia mechanical refrigeration cycles, achieved by Linde in the 1870's. Starting in the 1920's, Clarence Birdseye established frozen foods in the U. S. as a major commercial entity through a company which was a forerunner of today's General Foods.

Freezing usually gives excellent retention of nutritional values, color and flavor. Texture presents a problem, since ice crystals disrupt cells. A recent important advance has been the use of liquid nitrogen for freezing, since this more rapid freezing can produce much smaller ice crystals and thereby give less cell disruption. Ice crystal sizes and locations are governed by interactions of mass and heat transport processes with nucleation rates.

FREEZE-DRIED COFFEE

Freeze-dried coffee is a comparatively recent development for which chemical engineers were principal actors. The use of freeze drying for foods is a direct outgrowth of crash World War II programs for the stabilization of blood plasma and penicillin, both of which were freeze dried in very large quantities (15). Heat and mass transfer studies to improve freeze drying of penicillin were carried out during the war by Sherwood at M.I.T. (16). It was natural for those who had worked with blood plasma and penicillin to experiment with foods, and they found that foods of very high quality could be produced by freeze drying. Freeze drying avoids shrinkage and thereby gives excellent rehydration and good texture. Because of the low temperature, it arrests most deterioration reactions, and, for reasons that have been identified only recently, it gives excellent retention of volatile flavor and aroma compounds (17).

Early developmental work on freeze drying for foods was promoted by the U. S. Army Quartermaster Food and Container Institute and by the British government at Aberdeen, Scotland, in the 1950's. In the period 1960-1968, General Foods, Inc., engaged in a large engineering program to commercialize freeze-dried coffee as a beverage superior in taste and aroma to conventional spray-dried coffee. Nestlé and other companies followed suit.

FOOD PROCESSING

Numerous problems were encountered. Freeze drying is inherently slow, complicated and expensive, because of the needs for prior freezing, vacuum, batch operation, and condensation of water as ice. Engineering development and heat and mass transfer studies were needed to increase the drying rate without damaging product quality significantly. Design extensions have eventually led to innovative, continuous freeze dryers. The bulk density of the product was another problem, since the freeze-dried product tended to give a higher bulk density, upsetting the time-tested recipe of one teaspoonful to a cup. Approaches to this problem included advertising ["Use less, it's concentrated!"] and foaming of the feed before freezing to give a lower bulk density. Color was another difficulty; freeze-dried coffee particles tended to be lighter than either spray-dried or roast-and-ground coffee. The color problem was linked to the rate of freezing, the resultant ice-crystal size and the number of reflective pore surfaces generated (18). The desired darker color could be achieved by slow freezing.

Because of the expense of freeze drying, it is desirable to preconcentrate the coffee extract by some other means. The conventional method--multi-effect evaporation--gave excessive loss of volatile flavor and aroma substances, which would offset the quality advantages of freeze drying. This led to the use of a new technique for preconcentration--freeze concentration, in which the extract is only partly frozen and the resulting ice crystals are separated from the concentrate.

Another characteristic of the food industry, worth noting here, is the need for market development. Advertising associated with building up freeze-dried coffee considerably exceeded the cost of what was, by food-industry standards, a very large engineering effort.

ROLES OF CHEMICAL ENGINEERING

Chemical engineers have entered the food industry in a number of different ways. These will be explored briefly, with some examples of advances that have come in each way.

Major Food Companies

Some major food companies have utilized chemical engineers extensively for 15 years or more. Others are just now substantially increasing employment of chemical engineers, and still others have hardly any chemical engineers. Utilization of chemical engineers is very uneven, at present, in the industry. In addition to freeze-dried coffee, some of the major advances that have come recently through chemical engineering within major food companies are synthetic meats made from spun soy fibers, and production of fructose as a less caloric sweetener than sucrose.

Smaller Food Companies

The smaller food companies employ very few chemical engineers, in part because they operate with little or no research and development effort. Process engineering needs are generally met through purchase or contract.

Expansion of Companies into the Food Industry

In a few notable cases, companies with primary interests elsewhere have chosen to enter the food industry, through acquisitions or through newly developed activities. Some of these companies already had a strong tradition of utilizing chemical engineers through their prior activities. In these companies, chemical engineers can be transferred in and out of the food operations, and cross-fertilization between different applications thereby occurs more readily. Examples of innovative products that have come about in this way include stackable synthetic potato chips made from mash, and flaked roast-and-ground coffee, which can yield a high degree of extraction in home drip machines by virtue of the shorter internal diffusion path resulting from the flaking.

Other Companies and Universities

Still other advances rooted in chemical engineering have come from equipment vendors, from non-food companies dealing in proprietary processes, and from university research. Examples here include various types of dryers and freezers; mechanistic understanding of factors controlling loss of volatile flavors and aromas; fine-pore filtration for beer, allowing marketing of "draught" beer in a can; and the production of dry whey solids obtained through ultrafiltration and reverse osmosis. Some of the developments relating to the latter two processes occurred within food companies, as well. Food-processing interests are still very under-represented in academic chemical engineering departments.

Consultants

The uneven usage of chemical engineers within the food industry has led to extensive use of and significant inputs from consultants. The Florida citrus-processing industry is based heavily upon evaporation, but has utilized relatively few chemical engineers. The most successful evaporator system currently used is the TASTE evaporator (Temperature Accelerated Short Time Evaporator), designed by an independent consultant (10). This gives the benefits of high-temperature, short-time evaporation in a multi-effect system that provides pasteurization and is amenable to cleaning and good

FOOD PROCESSING

sanitation. High-temperature, short-time operation reduces most degradation reactions, which have relatively low activation energies.

Government Research

The food industry is unique in that much research and development is carried out by the federal government and state agricultural agencies. The U. S. Department of Agriculture maintains four large Regional Research Centers, as well as several other smaller ones. Chemical engineers have been prominent in these laboratories for 25 years and more; this was one of the first ways in which chemical engineering entered the industry. Engineering research is devoted to various products and to processes which might improve or expand their utilization. Two examples of important innovations that have entered the industry through this route are reduction of water usage and pollution in the potato-processing and fruit-processing industries through "dry caustic" peeling, and the use of distillation for essence recovery from various fruit juices during concentration and jelly manufacture.

THE FUTURE

Utilization of chemical engineers in the food industry is increasing greatly at present, and should continue to increase. The catalysts for this trend are the needs to become more quantitative in process design and analysis and to become more conscious of processing costs. Several current trends have generated these needs (19):

1. Costs of energy and of materials are increasing rapidly.

2. New regulations are requiring more quantitative control and determination of product quality factors, such as nutritional content and shelf life.

3. Regulations and process economics are leading to a need for better identification, control and removal or recovery of substances in waste streams, as well as more efficient use of by-products.

4. There is a rapidly developing need for synthesizing low-cost foods from new biological, and even non-biological sources.

Becoming more quantitative and cost conscious means, in turn, that it is important to learn mechanisms of underlying processes and perform effective process modeling. This requires understanding and analyzing transport processes, phase equilibrium, and rates of chemical and biochemical reactions. Some problems involving competing transport mechanisms are selective leaching or extraction (e.g., removal of caffeine, gossypol, lactose, etc.); control of ice crystal location and size in freezing; and selective dewatering, without loss of flavor or nutritional values. Other problems, rooted in competing transport and chemical reaction, are enzyme deactivation (blanching) and sterilization, without loss of nutritional values, flavor, color, etc.; and packaging so as to provide controlled respiration during storage. Problems based on competing reactions are sterilization by thermal processing or radiation and formulation and processing of intermediate-moisture foods, where stabilization reactions are to be favored over quality-loss reactions (19).

There is an opportunity for substantially increased participation by chemical engineers in the food industry and in universities. The surface of quantitative mechanistic understanding has barely been scratched, and there remains much for creative chemical engineering to do.

REFERENCES

1. Desrosier, N. W. 1970, "The Technology of Food Preservation," 3rd ed., AVI Publishing Co., Westport, CT.

2. Tannahill, Reay, 1973, "Food in History," Stein and Day, New York.

3. Sloan, A. E., 1976, 200 years of food - a historical perspective, Food Technol., 30(6):30.

4. Karel, M., 1973, Recent research and development in the field of low-moisture and intermediate-moisture foods, CRC Critical Reviews in Food Technol., 4:329.

5. Salaman, R. N., 1940, The potato as food, Chem. Ind., 18:735.

6. Van Arsdel, W. B., 1973, Chapter 1, in : "Food Dehydration," Van Arsdel, W. B., Copley, M. J., Morgan, A. I., eds., Vol. 1, 2nd ed., AVI Publishing Co., Westport CT.

7. Manly, W. L., 1894, "Death Valley in '49," Pacific Tree and Vine Co., San Jose, CA.

8. Labuza, T. P., 1976, Drying food: Technology improves on the sun, Food Technol., 30(6):37.

9. Frantz, J. B., 1951, "Gail Borden, Dairyman to a Nation," Univ. of Oklahoma Press, Norman.

10. Veldhuis, M. K., 1971, Orange and Tangerine Juices, in: "Fruit and Vegetable Juice Processing Technology," Tressler, D. K., Joslyn, M. A., eds., 2nd ed., AVI Publishing Co., Westport, CT.

11. Goldblith, S. A., 1971, The science and technology of thermal processing, Part 1, Food Technol., 25:1256.

12. Goldblith, S. A., 1972, The science and technology of thermal processing, Part 2, Food Technol., 26(1):64.

13. Appert, Nicholas, 1810, "Le Livre de Tous les Ménages ou L'Art de Conserver, Pendant Plusiers Années, Toutes les Substances Animales et Vegetables," Chez Patris et Cie., Paris.

14. Fennema, Owen, 1976, The U. S. frozen food industry: 1776-1976, Food Technol., 30(6):56.

15. Brockmann, M. C., 1970, Freeze-drying, Chem. Eng. Prog. Symp. Ser., 66 (100):53.

16. Flosdorf, E. W., 1949, "Freeze-Drying," Reinhold, New York.

17. King, C. J., 1971, "Freeze Drying of Foods," CRC Press, Chemical Rubber Co., Cleveland, Ohio.

18. Barnett, S., 1973, Freezing of coffee extract to produce a dark-colored freeze-dried product, AIChE Symp. Ser., 69(132):26.

19. King, C. J., 1976, Contribution of chemical engineering to the development of food processing, "Chemical Engineering in a Changing World," W. T. Kotsier, ed., Elsevier Sci. Publ. Co., Amsterdam.

A HISTORY OF ULTRAFILTRATION SEPARATIONS

Anthony R. Cooper

Lockheed Palo Alto Research Laboratory

Lockheed Missiles & Space Company, Inc.
3251 Hanover St., Palo Alto, CA 94304

SYNOPSIS

The use of ultrafiltration as an industrial scale unit operation is increasing rapidly both in scope and size. A history of the various stages in the development and production of ultrafiltration membranes and their introduction into industrial processing are reviewed. The major current applications of ultrafiltration are summarized, together with an analysis of the factors influencing their selection.

HISTORICAL DEVELOPMENT

The use of membranes as a separation medium dates back many centuries. The use of animal membranes to concentrate beer has been recorded in Europe[1] and Bhutan[2]. The use of membranes for dialysis was first recorded by Fick[3] in 1855. The membranes were made from ceramic thimbles dipped into nitrocellulose (colloidon) solution in ether. Two ultrafiltration experiments were recorded in 1856 both involving animal membranes. One reported the ultrafiltration of proteins[4], the other gum arabic[5]. The preparation of colloidon sacs for dialysis was described by Schumacher[6]. In 1861 Graham[1] reported diffusion experiments and the preparation of parchment membranes. Several authors reported the use of unglazed ceramic materials to form ultrafiltration membranes with various impregnating agents. Traube is reported[7] to have used cupric ferrocyanide for this in 1870; the same material was also used by Pfeffer[8].

Sanarelli in 1896[9] introduced the use of colloidon sacs for biological applications including the ultrafiltration of blood plasma in vivo. In 1896, Martin[10] used unglazed porcelain filter candles impregnated with either gelatin or silicic acid to separate colloidal materials. Malfitano used colloidon sacs to ultrafilter colloids[11] in 1904, as did Manea[12]. The following year, Levy[13] showed that ultrafiltration and dialysis produced different results with enzymes.

These early years of membranes and applications developed a wide variety of different types of membranes and a considerable ingenuity in biological and physico-chemical applications. In the first decade of the twentieth century, attention was centered on the production and characterization of membranes, and this continued for 30 years. Bechold in 1906 termed the process "ultrafilgration" and made graded membranes from filter paper and acetic acid colloidon. He showed that pore size, which he was able to estimate, was inversely proportional to the polymer concentration used. Flat membranes were prepared by Bigelow and Gemberling in 1907[17] using ether-alcohol colloidon solutions. Graded filters were produced by a patented process[18] using these solutions. The colloidon solution was cast in a glass plate and evaporation of solvent under conditions of controlled humidity produced membranes of controlled porosity. These membranes became commercially available[19] in 1927. Manegold and his coworkers reported extensively on the structure of colloidon membranes.[20-22] In 1930 Elford produced a series of graded porosity membranes[23] for biological particle size determinations. They were produced by the addition of amyl alcohol or acetone to colloidon solution.

1930-1960 THE DORMANT PERIOD

In spite of the progress and ingenuity shown by the early researchers, very little significant progress was made for almost 30 years after 1930.

The limited availability of commercial membranes was one of the major factors responsible. Their flux was generally low, reproducibility was not satisfactory, and there was the inevitable clogging when a material is retained by a membrane. The mechanism of formation and the characterization of the porous structure was in a very early stage of development. In this period there was also a complete absence of engineering design. For some of the analytical techniques for which ultrafiltration was well suited, newer more direct techniques were being developed, notably chromatography, ultracentrifugation (1936), light scattering, and electron microscopy (1940's). On a preparative scale, alternative methods such as chromatography, preparative ultracentrifuge, electrophoresis, electrodialysis, and precipitation methods were being developed.

MODERN ULTRAFILTRATION

In 1953, Reid proposed to use membranes for the reverse osmosis desalination of sea water.[24] The first demonstration of this was achieved by the same author in 1959 using cellulose diacetate membranes.[25,26] The flux was very low and it was not until Loeb and Sourirajan's momentous discovery of anisotropy in membranes[27] that further progress became possible. These authors found that Schleicher and Schuell cellulose diacetate membranes had a rough and smooth side. After heat treatment the membrane would retain salt only if the rough side was on the high pressure side. This discovery led to the development of a highly selective, high flux anisotropic membrane based on cellulose diacetate.[28-32]

Two other significant developments in membranes occurred about this time. One was the development of polyelectrolyte complex membranes by Michaels.[33] The other involved the use of hydrous oxides or hydrolyzable salts to form a salt filtering layer on porous materials with pores as large as 5 μm.[34] Typical materials that have been employed to form these dynamic membranes are hydrous zirconium oxide, organic polyelectrolytes, and humic acid.

In 1971 Michaels[35] obtained a patent for the preparation of anisotropic noncellulosic membranes which was important in the application of UF to more applications. These exhibited increased pH stability, high temperature operation, and resistance toward sterilizing solutions. In 1973, Romicon introduced wide diameter anisotropic (noncellulosic) hollow fibers. These are manufactured by a continuous process and lead to very high surface area modules in a very compact structure.

COMMERCIAL APPLICATIONS OF ULTRAFILTRATION

In 1968, a cooperative effort was started between Amicon and Dorr-Oliver for a secondary sludge treatment process. The project was an outstanding success. The water produced contained no suspended solids, no coliform bacteria, and less than 5 ppm BOD. Several of these systems are in operation, one on Pike's Peak in Colorado with 3000/ft^2 of membrane area and the permeate is used for toilet flushing. The cost is usually only justified under special circumstances.

Laboratory scale UF membranes and equipment became available in 1965. Since that time an intense R&D effort has ensued. In the U.S.A., the Office of Saline Water funded much of the research directed toward water desalination. In Europe the development was supported by industry with the approach that UF was an exploitable unit operation for industry. Madsen summarized[36] the activity in this period as "... showing that almost everything was

possible (by UF) but very few applications were really successful ..."
In 1966 the application of membranes to hemofiltration was initiated.

The market for ultrafiltration in 1980 was estimated to be $50 million/yr[37] with a growth rate of + 20 to + 25%. Earlier estimates had sized the market at $30 to $50 million by the mid 1970's. In terms of equipment size, at present day standards, an output of 10-20 m^3/hr of permeate is considered large and 50 m^3/hr is considered maximum.

Ultrafiltration has been used where its superiority is clear-cut. In the electrocoat paint industry it was a combination of the imposition of effluent regulations, the recovery of paint for reuse, and the solids-to-conditions ratio being adjustable at will, which led to its economic acceptance. In the dairy industry, a combination of the need to reduce effluent BOD and the fact that the whey product was not denatured and was saleable promoted the success of ultrafiltration. In the enzyme concentration and purification field, the retention of activity was a major factor contributing to its successful economic applications. There are many uses of ultrafiltration where its economic performance is marginal. Each case must therefore be considered for membrane selectivity, permeate flux, and membrane lifetimes.

Ultrafiltration in the Dairy Industry

Ultrafiltration has two main applications in the dairy field, the earliest application was for the treatment of whey to reduce pollution problems and produce a saleable product from this step to favor economics. More recently the ultrafiltration of milk has been exploited in order to obtain higher yields and more uniform cheese products. It is estimated at the current time that 8,000 tons of whey per day are treated (3% of word production) from the manufacture of 60,000 tons of cheese. Table 1 lists the details supplied by various manufacturers. Matthews in 1979 described a whey treatment plant containing 702 m^2 of membrane area.[38] Maubois reported[39] that in 1979, 30,000 tons of soft cheese were made from ultrafiltered milk. New processes for ricotta, cream, and St. Paulin cheeses were developed. A spray-dried percheese was developed. Maubois also predicted that milk will be ultrafiltered at the farm to reduce transportation costs. An enzyme ultrafiltration reactor was described which could produce 3 to 7 amino acid peptides from whey.

Ultrafiltration in the Electrocoat Paint Industry

The first installations were activated in 1969. These units were first used for anodic paints. In 1975 over 300 installations

Table 1

THE APPLICATION OF ULTRAFILTRATION IN THE DAIRY INDUSTRY

MANUFACTURER	DATE	APPLICATION/SIZE
ABCOR	1971	600,000 lb/day Whey
	1976	1.4×10^6 lb/day Whey
	1976	>500 UF + RO Units in 35 Countries
DDS (Danish Sugar Co)	1980	471 Modules 150,000 ft^2 3×10^6 gal/day (Cf 300 Modules 80,000 ft^2 in all Other Applications)
DORR-OLIVER	1971	Whey Concentration
	1975	Milk Concentration
OSMONICS	1974	Whey Concentration 50,000 lb/hr
ROMICON	—	200,000 - 1×10^6 lb/day Whey

were in operation; in this application, the units tend to be small, of the order of 1-100 GPM permeate rates. With the switch to cathodic paints, the original membranes experienced severe flux declines. Several manufacturers produced modified membranes to overcome this problem. In this application there are no obvious alternatives.

Ultrafiltration in the Pharmaceutical Industry

Early penetration into this market was for enzyme purification and isolation. DDS reported this in 1969 and Dorr-Oliver in 1971. The alternative processes include a combination of dialysis, ion exchange, lyophilization, and selective precipitation. These are costly labor intensive processes so that ultrafiltration was readily accepted. DDS also reported the production of iron dextran, insulin, gums, and gelatin by ultrafiltration in 1970. At the same time, the use of UF in plasma processing was described by DDS and Millipore. In 1971 Millipore produced a high molecular weight cutoff (1×10^6 MW) polyelectrolyte membrane. This was used for virus concentration and vaccine production. In 1973 plasma processing units up to 500 ft^2 were reported. In 1979 Dorr-Oliver described the application of UF to fermentation broth clarification with 8,000 ft^2 unit.

Ultrafiltration in Effluent Treatment

Ultrafiltration has been employed in many successful and important but small applications in effluent treatment. For oily-water wastes, in 1977 Abcor reported 90 to 100 plants with capacities of 100 to 50,000 GPD Romicon reported units for 1,000 to 10,000 G batch processes at 1 to 50 GPM fluxes.

The textile industry has also used ultrafiltration to purify its waste streams. Dorr-Oliver (1976) reported the application with 2000/ft^2 of membrane area to wool/yarn effluent streams. Abcor (1975) introduced a membrane which was unaffected by 100% methyl-ethyl ketone (MEK) and this was used for latex wastes. Since the stream was severely fouling, the use of MEK in the cleaning cycle allowed restoration of flux. In 1979, Drioli reported the application to hide wastes.[40] These are a widespread source of pollution since the operations tend to be small. Dye wastes have been treated by ultrafiltration, Dorr-Oliver reported a 2000 ft^2 unit to recover indigo dye in 1979. Ultrafiltration has also been applied to the purification of slaughterhouse wastes.

Its use in the pulp and paper industries notably by the Danish Sugar Corporation for the production of lignins and lignosulfonates and for decolorizing bleaching liquors has been reported.

Ultrapure Water Production

Ultrapure water is required in large volumes by the electronics, pharmaceutical, and chemical industries. In 1979 Millipore characterized pyrogen and introduced spiral UF modules to produce pyrogen-free water. In 1979 Millipore and Continental merged to produce units for ultrapure water production having thousands of square feet of membrane area. Romicon's hollow fiber modules are also extensively used for pure water production.[41]

SUMMARY

Ultrafiltration has had a long history with exciting discoveries but also long periods with little progress. The technique is now accepted as a unit operation in several major industries and applications continue to expand. A wide variety of membrane types and engineering configurations are available. The units can be made economically in a small scale and be used by non-technical personnel. On the other hand, extremely large units can be made with very sophisticated controls to accommodate the most demanding applications. This decade promises to be extremely active for the development of new ultrafiltration hardware and applications.

BIBLIOGRAPHY

J. D. Ferry, Ultrafilter Membranes and Ultrafiltration, Chem. Rev., 18, 373 (1936)

C. Gelman, Microporous Membrane Technology, Anal. Chem., $\underline{37}$(6), 29A (1965)

H. Z. Friedlander and R. N. Rickles, Membrane Technology, Anal. Chem., $\underline{37}$(8) 27A (1965)

A. S. Michaels, Polyelectrolyte Complexes, Ind. and Eng. Chem., $\underline{57}$, 32 (1965)

H. Z. Friedlander and R. N. Rickles, Membrane Separation Processes, Parts 1-6, Chem. Eng., Feb 28 p 111, March 28 p 121, April 25 p 163, May 23 p 153, June 6 p 145, June 20 p 217, 1966

A. S. Michaels, Ultrafiltration in "Progress in Separation and Purification, Vol I, E. S. Perry, Ed., Wiley-Interscience 1968, p 297

L. C. Craig, H. C. Chen and E. J. Hornfeist, Separatations Based on Size and Conformation in "Modern Separation Methods of Macromolecules and Particles," T. Gerritsen Ed., Wiley-Interscience 1969 p 219

C. J. Van Oss, Ultrafiltration Membranes in "Progress in Separation and Purification," Vol 3, E. S. Perry and C. J. Van Oss, Eds., Wiley-Interscience 1970 p 97

C. J. Van Oss, Ultrafiltration in "Tech. Surface Colloid Chem. Phys.," $\underline{1}$, 89 (1972)

H. Goodall, Reverse Osmosis and Ultrafiltration and Their Practical Applications. The British Food Manufacturing Industries Research Association Scientific and Technical Surveys Number 77 May 1972, Surrey, England

S. Jacobs, Ultrafilter Membranes in Biochemistry in "Methods of Biochemical Analysis," Vol 22, 1974 p 307

Anon, Literature References to the Use of Amicon Ultrafiltration Systems. Pub. No. 428E, 1977 Amicon Corporation Lexington MA 02173

H. P. Gregor and C. D. Gregor, Synthetic Membrane Technology Scientific American 239(1), 112(1978)

A. R. Cooper Ed., Ultrafiltration Membranes and Applications, Plenum Press 1980

A. F. Turbak Ed., Synthetic Membranes, Volume I Desalination, Volume II Hyper and Ultrafiltration Uses, ACS Symposium Series 153, 1981

A. R. Cooper Ed., Polymeric Separation Media, Plenum Press, In Press

REFERENCES

1. T. Graham, Phil. Trans. Roy. Soc., 151, 183 (1861)
2. G. Gorer, Himalayan Village, M. Joseph 1938
3. A. Fick, Pogg. Ann. 94, 59 (1855)
4. W. Schmidt, Ann. Phys. Chem 99, 337 (1856)
5. F. Hoppe-Seyler, Virchow's Arch. 9, 245 (1856)
6. W. Schumacher, Ann. Phys. Chem., 110, 337 (1860)
7. H. P. Gregor and C. D. Gregor, Scientific American, 239 (1) July 1978, p 112
8. W. Pfeffer, Osmotische Untersuchung. Engleman, Leipzig, 1877
9. G. Sanarelli, Cent. Bakt. Parasitenk, 467 (1891)
10. C. J. Martin, J. Physiol 20, 367 (1896)
11. G. Malfitano, Compt. Rend., 139, 1221 (1904)
12. M. Manea, Compt. Rend. Soc. Biol., 57 317 (1904)
13. D. J. Levy, J. Infect Diseas. 2, 1 (1905)
14. H. Bechold, Z. Physik Chem., 60, 257 (1907)
15. H. Bechold, Biochem Z., 6, 379 (1907)
16. H. Bechold, Z. Physik Chem., 64, 328 (1908)
17. S. L. Bigelow and A. Gemberling, J. Amer. Chem. Soc., 29, 1675 (1907)
18. R. Zsigmondy and W. Bachman DP 329117, 1916, Z. Anorg. Allgem. Chem., 103, 119 (1918)
19. Membranefilter Gesellschaft, Goettingen, Germany
20. E. Manegold Kolloid Z. 49, 372 (1929)
21. E. Manegold and R. Hofmann, Kolloid Z., 50, 22 (1930)
22. E. Manegold, R. Hofman and K. Solf., Kolloid Z., 56, 142 (1931)
23. W. J. Elford, J. Path. Bact., 34, 505 (1931)
24. Development of Synthetic Osmotic Membranes for Use in De-Salting Saline Waters. Research Proposals for the U.S. Department of the Interior, February 19, 1953
25. C. E. Reid and E. J. Breton, J. Appl. Polym. Sci., 1, 133 (1959)
26. C. E. Reid and J. R. Kuppers, J. Appl. Polym. Sci., 2, 264 (1959)
27. S. Loeb and S. Sourirajan, American Chemical Society Metting, March 27, 1962
28. S. Loeb and S. Sourirajan, U.C.L.A. Dept of Engineering Report 60-60, July 1960
29. S. Loeb, S. Sourirajan and S. T. Uuster Chem. Eng. News, 411 Nov. 1960 p 64
30. S. Loeb, U.C.L.A. Engineering Report 61-42 (1961)
31. S. Loeb and S. Sourirajan, Adv. Chem. Ser., 80, 117 (1963)
32. S. Loeb and S. Sourirajan, U.S. Patent 3,133,132 (1964)
33. A. S. Michaels, Ind. Eng. Chem., 57, 32 (1965)
34. A. E. Marcinkowsky, K. A. Krauss, M. O. Phillips, J. S. Johnson Jr. and A. J. Shor, J. Amer. Chem. Soc., 88, 5744 (1966)

35. A. S. Michaels, U. S. Patent 711026, 1971, High Flow Membrane.
36. R. F. Madsen and W. K. Nielsen in "Ultrafiltration Membranes and Applications." A. R. Cooper Ed., p 423, Plenum Press 1980
37. Chem Week, Jan 25, 1980
38. M. E. Matthews, Personal Communication, 1979
39. J. L. Maubois in "Ultrafiltration Membranes and Applications," A. R. Cooper Ed., p 305-318, Plenum Press, 1980
40. E. Drioli in "Ultrafiltration Membranes and Applications," A. R. Cooper Ed., p 292-295, Plenum Press, 1980
41. B. R. Breslau, A. J. Testa, B. A. Milnes and G. Medjanis in "Ultrafiltration Membranes and Applications," A. R. Cooper Ed., p 122-126

LIST OF CONTRIBUTORS

Appl, M., 29
Aris, R., 389
Cooper, A.R., 449
Daub, E.E., 159
Furter, W.F., v
Hawke, F., 275
Hixson, A.N., 127
Katz, D.L., 139
Keller, G.E., II, 293
Kendall, H.B., 419
King, C.J., 437
Maloney, J.O., 211
Marcinkowsky, A.E., 293

Mohtadi, F., 243
Morton, F., 19
Myers, A.L., 127
Othmer, D.F., 259
Skolnik, H., 225
Sperling, L.H., 405
Tailby, S.R., 65
Thépot, A., 55
Trescott, M.M., 1
Varma, A., 353
Volckman, O.B., 275
Wilkes, J.O., 139
Yoshida, F., 195
Young, E.H., 139

INDEX

Acid industry, 15, 19, 31,
    55-64, 66-68, 128-30,
    236
Alkylation, 425
American Chemical Society, v,
    131, 147-8, 180,
    235-6, 240, 305, 369
  centennial tribute to chemical
    engineering, v
  History of Chemical Engineering,
    v-vi
American Institute of Chemical
    Engineers, v, 131, 147-8,
    235, 239-40, 305, 360
Ammonia industry, see Haber-Bosch
    process

BASF, 15-16, 29-54, 421-2
Battersea Polytechnic, see Surrey,
    University of
Bergius, F., 421-2
Bosch, C., 29-54 (see also Haber,
    F.)

Calcium carbide, 11, 293, 299-304
Cambridge University, 25, 70,
    73-4, 90
Chemical Abstracts, 235-242
Chemical Engineering
  curricula, see Chemical
      Engineering education
  education
    in Britain, 19-27, 65-126
    in Germany, 20, 73
    in Japan, 195-209
    in South Africa, 280-2, 288-9
    in US, 127-194, 211-223, 353

  emergence as a profession,
      v, 19-21, 26-27
  literature, 225-242
  origins, v-vi, 4, 10, 22, 66,
      231, 353
    in England, 4, 19-27
    in France, 55-64
    in Germany, 4, 9-10, 16-17,
      20-21, 29-33, 48-52
    in physical chemistry, 3-4,
      8
    in South Africa, 275-292
    in US, 2, 4, 50-51
  roots, v
Chemical reactors, see Reactors
    (Chemical)
Chlor-alkali industry, 30, 282
City and Guilds College London,
    65, 80-81, 87
Cracking
  of petroleum, 293-4, 419-21
  of ethane, 293, 312-16,
    335-42
Curme, G.O., Jr., 293, 304-5

Davis, G.E., 20, 23, 26-7, 66
  lecture series, v, 4, 21-3,
      31, 66, 77, 353
  *Handbook of Chemical
      Engineering*, v, 4-6,
      20, 22, 66, 78, 234
  and Society of Chemical
      Engineers (see Society
      of Chemical Engineers)
Definition of chemical engi-
    neering, 4, 6, 8, 21,
    26-7, 48-50, 81, 353

461

Desalination, 451
Diffusion
 pioneers, 243-57
Distillation, 128
 columns, 267-9
 development, 259-74, 366-72
 processes, 269-71
Doctoral theses, 211-23
Dow Chemical, 11, 14
Du Pont, 128-9, 357
Dye industry, 19, 22, 30, 68-9, 236

Eastman Kodak, 11
Edinburgh University, 65, 90-1, 94-5
Electrochemical industry, 1-2, 12-19, 30, 299-304
Ethylene chemicals, 293-352

Fick, A., 243-57
Food
 preservation, 438-43
 processing industry, 437-47
Freeze drying, 437, 442-3
Furter, W.F., v-vi, 270 (see also *History of Chemical Engineering*)

Glasgow, University of, 65, 70, 91-4, 118-121
Glass industry, 19, 55-64, 321
Graham, T., 243-57

Haber, F., 2, 9, 304, 431, 421-2 (see also Haber-Bosch process)
Haber-Bosch process, 29-54, 77, 421-2
Heriot-Watt University, 70, 91
History of Chemical Engineering, see American Chemical Society
Hougen, O., 4, 8-9, 27, 188, 205, 214, 353
Hydroforming, see Naptha reforming

Imperial College London, 23, 25, 65, 78-87, 104-11
Industrial revolution, 55-126
Institution of Chemical Engineers, 23-4, 26, 81, 84, 87-8, 90, 289-90

Kings College London, 24, 65, 74, 89-90, 116-18
Koto gakko, 195-209
Koza system, 195-209
Kyoto University, 195-209

Lewis, W.K., 7-9, 11, 66, 199, 356, 361, 366, 369-70 (see also *Principles of Chemical Engineering*)
Little, A.D., 6-8, 17, 236, 353
London, University of, 26, 74, 80, 84, 89, 90

Manchester College of Technology, 23-5, 70
Manchester Technical School, v, 4, 21-23, 31, 66, 353
Manhattan project, 135, 305
Mathematics
 in chemical engineering, 353-87
McAdams, W.H., 7, 11, 199 (see also *Principles of Chemical Engineering*)
Membrane separations, 449-58
Michigan, University of, 139-58, 211, 215-6, 353, 358
MIT, 6-7, 78, 199, 211, 215-6, 236, 353, 356-8, 362-3 (see also *Principles of Chemical Engineering*)

Naptha reforming, 419-36
Nuclear industry
 in Britain, 25
 in US, 135-148

INDEX

Origins
  of chemical engineering, see
      Chemical Engineering
      origins
Ostwald, W., 34, 234
Othmer, D.F., 214, 272-4

Paint industry, 128
Pennsylvania, University of, 6,
    127-37, 203, 353, 356
Petrochemical industry, vi, 25,
    293-352
Petroleum industry, 130, 236,
    264-5 (see also Naptha
    reforming)
Platforming, see Naptha reforming
Polymers
  blends, 405-18
  interpenetrating networks,
    405-18
Priestley, J., 243-57
Principles of Chemical Engineering, 7, 9, 11, 50, 187,
    199, 236, 353 (see also
    Walker, W.H., Lewis, W.K.,
    McAdams, W.H.)
Reactors (Chemical), 29, 54
  for ammonia synthesis, 29-54
  design, 380-1
  mathematical modelling, 353-87
  multiplicity and stability,
    389-403
  for reforming processes, 419-36
  stability and control, 377-9
Reforming, see Naptha reforming
Reverse osmosis, 451
Royal Institute of Chemistry, 20

SASOL process, 285-6, 291
Sludge treatment, 451
Soap industry, 19, 68, 129-30
        (see also Soda industry)
Society of Chemical Engineers
        (see also Davis, G.E.)
  first attempts to found, v, 20,
    26-7, 66, 353
Society of Chemical Industry (see
    also Davis, G.E.)
  founding, 20-1, 66, 235

Journal, v, 239
Chemical Engineering Group,
    23
Soda industry, 19, 30, 55-64,
    66-8
Solvay process, 19, 68
Standard Oil, 11, 419-36
Strathclyde, University of, 65,
    74, 91-4
Surrey, University of, 23, 65,
    87-9, 111-6
Swenson, M., 5-6, 165-175
Synthetic fertilizer industry,
    30

Technical University Karlsruhe,
    35

Ultrafiltration separations,
    449-58
Union Carbide, 11-14, 16,
    293-352
Unit operations, 1-18, 236
  origins, vi, 1-18, 51, 66,
    353
University College London, 23,
    65, 74-8, 96-103

Walker, W.H., 7-8, 11 (see also
    *Principles of Chemical
    Engineering*)
Wisconsin, University of, 6,
    159-94, 203, 205, 211,
    215-6, 358